YOGA ALIGNMENT
Principles and Practice

An anatomical guide to alignment, postural mechanics, and the prevention of yoga injuries

2nd edition of The Injury-Free Yoga Practice

Steven Weiss, DC, C-IAYT

AlignYoga 108 ™

All rights reserved. No part of this publication may be reproduced, stored in a retrieval system, or transmitted in any way or by any means electronic, mechanical, photocopying, recording or otherwise without the prior written permission of the copyright holder, AlignYoga108.

Yoga Alignment   Principles and Practice ©
An anatomical guide to alignment, postural mechanics, and the prevention of yoga injuries

ISBN-13: 978-0-9893272-1-3   Color Print Format

Published by: AlignYoga108   Steven Weiss, DC, C-IAYT   2019

Editing Staff:
Debra Gitterman
Esther Veltheim
Ronni Geist and GeistWriters

Design:
Ronni Geist, Geistwriters

Cover Design:
Carol Weiss
Danny Media

Illustrations:
Ben Schikowitz

Cover Photo:
Shutterstock_287266130

Photography:
Walter Fritz
Esther Veltheim
Carol Weiss
Steven Weiss
Laurie Troost

Photographic models:
Jaye Martin
Kristin Neisler
Catherine Barefoot
Karine Woodley
Steven Weiss
Kate Honig

For book and workshop information, please visit: **AlignYoga108.com**
Like us on FaceBook!

Table of Contents

Preface		1
Introduction		3
Chapter 1 **Yoga and Alignment**		5
Sthira Sukham Asanam		6
Hatha		6
Yoga Asana – an "alignment delivery system"		7
Alignment is not another style but the basis of every style		8
Successive Approximations		8
Chapter 2 **Alignment and the Body Systems**		9
The nervous system		10
Muscle fibers and myofascia		10
Body mechanics		10
Bone density		10
Wellbeing		11
General Adaptation Syndrome		11
Tissue repair and healing		11
Brain wave activity		11
The energetic body		12
Internal health conditions		12
You can't be neutral on a moving train."		12
Chapter 3 **Raja Principles of Yoga**		13
Yamas and Niyamas		14
Yamas	Restraints, rules, and control on behavior	14
Ahimsa	Doing no harm	14
Asteya	Non-stealing	15
Satya	Truth and wisdom	15
Brahmacharya	Avoiding unnecessary energy use	15
Aparigraha	Non-possessiveness	15
Niyamas	Basic life skills and practices	16
Shaucha	Cleanliness of body and mind	16
Santosha	Contentment	16
Tapas	Austerity and endurance	16
Svadhyaya	Spiritual study and awareness of life	16
Ishvarapranidhana	Surrendering to universal wisdom	16

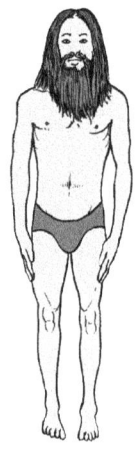

Chapter 4	**Foundations and Orientations of the Hips**	**17**
	Foundations provide freedom	17
	Hinges	17
	Foundations	18
	Open and closed hips	18
	"There can only be one… hip orientation"	19
	Examples of postures in each category of hip orientation	20
	Exceptions to the rule	20
	Linking poses into a flow	20
Chapter 5	**Principles of Alignment and Integration**	**21**
	Hatha's ultimate balancing act: movement vs. stability	22
	Interdependence	22
	Human architecture	22
	"First there is a mountain..."	23
	The "baby bear" intention	23
	What is the "just right" effort?	23
	Initiate movement from regions of least mobility	24
	Spinal Curves: Convex vs. concave	24
	The periphery of the body moves faster than at the core	25
	Taking a mobility inventory	26
	The hierarchy of the twist	26
	Hypermobility	27
	Bones approximate, muscles extend	27
	Para-physiological space	28
	Central axis of a joint	28
	Stretching tips	29
	The quality of Samasthiti	29
	Samasthiti – beyond balance	30
	Practice like a Bear or a Monkey?	30
	Moving with Samasthiti	30
	Identifying strengths and weaknesses	31
	Heyam Dukham Anagatam	32
	Alignment is not a "style" of yoga	32
	The most basic and essential principles of alignment	33
	Squaring the hips	33
	Summary of Chapter 5 principles	34

Chapter 6	**Form Follows Function**	**35**
	Form vs. Action	36
	Stretching basics	36
	Diaphragms	37
	Diaphragms and Bandhas align with central axis	38
	Spinal curves basics	38
	Whiplash	39
	Confusion about curve function	39
	Managing spinal curves in asana practice	40
	Paschimottanasana - Sitting Forward Fold	40
Chapter 7	**Anatomy and Physiology – Connective Tissue**	**41**
	Know your medium	41
	Components of connective tissue	42
	Ground substance	42
	Fibers of connective tissue – Collagen	42
	Fibers of connective tissue – Elastin	43
	Fibers of connective tissue – Reticular	43
	Fascia	43
	Myofascia	44
	Tendons	44
	Tendonitis	45
	Bursas	45
Chapter 8	**Anatomy and Physiology – Ligaments**	**47**
	Micro-pleating action of the ligaments	48
	Wrapping action	48
	How ligament action works	48
	Taking the heat - Friction	48
	A little goes a long way – elastin fiber and flexibility	49
	Two out of three is enough for ligament performance	49
Chapter 9	**Anatomy and Physiology – Muscle**	**51**
	Sacromeres and myocytes	52
	Stretching is forever	53
	Speed, time, and heat	53
	Strength, force, and efficiency	53
	Three types of muscle contraction	54
	The 10% rule	54
	Potential for muscle efficiency	55
	Muscles that unlock the power of another	55
	Overstretching	55
	Mixing yoga with athletics...maybe!	56
	Deep tendon and stretch reflexes	56
	Scar tissue	57
	Stretching and Habituation	58
	Fast and slow twitch fibers	58
	Muscles and aging	59
	The value of yoga for aging muscles	59
	Swaddling and hugging muscles to the bone	60
	Does stretching observe Brahmacharya and Ahimsa?	60

| Chapter 10 | **Anatomy and Physiology – Cartilage and Bone** | 61 |

    Cartilage   62
    Hyaline Cartilage   63
    Bone   63
    Bone Density   63
    Osteoporosis and Osteopenia   64
    Bone's absorptive qualities   64
    Bone re-modeling and the value of good posture   65

| Chapter 11 | **Align By Design** | 67 |

    Need GPS?   68
    The Alignment Grid   69
    Simple triple "S" alignment (skull, scapulae, sacrum)   69
    The floating ribs   70
    Design and function   70
    Scoop, Scoop, Draw in the Navel, Keeps the low back stable!   71
    Zip it!   71
    Beauty in its simplicity   71
    Lock and Load: Step-by-step stabilization for alignment   72

| Chapter 12 | **Anatomy of the Pelvis and Sacroiliac Joints** | 73 |

    The acetabulum   74
    The "sitting" bones   74
    The pelvis rocks!   75
    Pubis symphysis: the anterior joint of the pelvis   75
    Sacroiliac joints: the posterior joints of the pelvis   76
    Sacroiliac joints are shock absorbers   76
    Sacral function: an open and shut case   77
    Sacroiliac function   77
    Sacrum as a trap door   78
    Insult and injury to the sacroiliac joints   78
    Sacral pump   78
    Something else to chew on: the TMJ   79
    Equator of the body - the sacrum-coccyx juncture   79

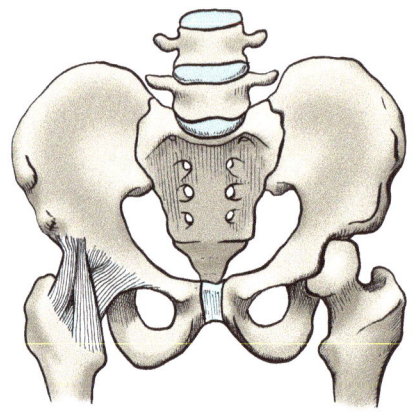

Chapter 13	**Pelvis and Sacroiliac Joints – Alignment Principles**	81
	Move from the Mula	82
	Pelvic integrative alignment	83
	Step one - Inward hip release	83
	Step two - Forward tailbone scoop	83
	Baddha Konasana and the bruised sacroiliac joint	83
	Causes of sacroiliac injuries	84
	Evaluating the function of the sacroiliac joints	84
	Marching in place	84
	Observing the sway	85
	Sacroiliac Therapeutics	86
	Therapeutic Supportive Bridge: Setu Bandha	86
	Asana that open fixated sacroiliac joints	87
	Gomukhasana – Cow face Pose	87
	Garudasana - Eagle Pose	87
	Twisting poses when the sacroiliac joints are unstable	87
	Asana and yoga therapy for stabilizing open sacroiliac joints	88
	Sacroiliac strap stabilization	88
	Baddha Konasana - Bound Angle	88
	Virasana and Supta Virasana with blanket or sacral block	89
Chapter 14	**Integrative Alignment of the Pelvis**	91
	Inward Hip Release - step one	92
	Effects of inward hip release	92
	Forward tailbone scoop - step two	93
	Effects of forward tailbone scoop	93
	Finding the sweet spot	94
	Putting on tight pants – how to engage pelvic integration	94
	The Human Pez® dispenser	95
	Pelvic positions	96
	Backward "Butt Walking"	96
Chapter 15	**The Lower Thorax**	97
	Method to engage lower rib cage integration	98
	Visualizing lower rib cage integration	98
	Scoop the breastbone, Scoop the tailbone, Draw in navel	98
	Urdhva Dhanurasana / Ustrasana	99
	Bhujangasana / Aldo Mukha Vrksasana	100
Chapter 16	**The Alignment of Sitting**	101

Chapter 17	**The Hip Joint**	**103**
	Are you *hip* to this?	103
	How far apart are the feet when spaced hip-width apart?	103
	A snug fit	103
	The swing of things	105
	Hips move in three axes of motion	105
	Mind the gap – deepen the hip crease	105
	Avoid the pinch	106
	Hips in action	106
	Prasarita Padottanasana	106
	Baddha Konasana	107
	Pinwheel, Agnistambhasana	107
	In an externally biased world, internal rotation is highly valued	108
	Virasana, Supta Virasana	108
Chapter 18	**Hip Extension**	**109**
	The hip's limited range of motion	110
	Experiencing the limits of hip extension	110
	Limited hip extension can injure the lumbar spine	111
	Muscles of hip extension	112
	Quadriceps stretch	112
	Iliopsoas muscle stretch	113
	Extenuating circumstances	113
	Bi-articular muscles	114
	Flexion before extension – the anterior pelvis	114
Chapter 19	**Alignment of the Legs**	**115**
	Benefits of leg alignment in yoga practice	116
	General alignment of the legs	116
	The Q-angle	116
	Compensation for a large Q-angle	117
	Specifics of leg alignment	117
	Shins forward-thighs back	118
	Shins in-thighs apart	118
	Lengthwise contraction	119
	Are you pulling my leg?	119
	Roman sandal strapping	119
	Various asana explorations and assists	120
	Reciprocal inhibition	122
	Lift your kneecaps!	122
	Co-activation	122
	Concentric vs. eccentric contraction	123
	Walking Well	124

Chapter 20	**The Hamstring Muscles**	**125**
	More ham, less string - muscle to tendon ratio	126
	The function of the hamstring muscles	126
	Basic anatomy of the hamstring muscles	126
	Semimembranosus	126
	Semitendinosus	126
	Biceps femoris, long and short head	127
	Why do tight hamstring muscles cause back pain?	127
	Tips and refinements for the hamstring muscles	128
	Extend thighs back, not straighten knees in these poses	128
	Uttanasana: a two-stage strategy	129
	Points to consider with yoga therapeutics	129
	Rehabilitative stretching of the hamstrings	130
	Points to consider with yoga therapeutics	131
	Hamstring pole stretch therapy	132
	"The poison is the cure"	132

Chapter 21	**Knee Alignment Principles**	**133**
	Muscles that flex the knees	134
	Extension	135
	Rotation	135
	Knee injury statistics	135
	Vulnerable knee positions	135
	Meniscus- the cartilage of the knee	136
	Roll and glide	136
	Ligaments of the knee – crosses and strips	136
	Hyperextension of the knees	137
	Into the fold	138
	Detailed instructions for knee alignment in asana	138
	The Squat	139
	Meniscus therapy	139
	Tibial torsion	140
	Avoid tibial torsion Keep knee as a simple hinge	140
	Oh my Goddess Pose!	141
	Eka Pada Rajakapotasana Pigeon Prep	141
	"X" and "O"	142
	Valgus/varus therapy	142
	The patella	143
	Baker's cyst	143
	"Don't it always seem to go…	144

Chapter 22	**The Ankle**	**145**
	The talus joint	146
	Inverted "T"	146
	Details of ankle alignment while standing	146
	No wrinkles!	147
	The powerful calf muscles	147
	Using accessory muscles to reduce foot sickling	147
	Motions of the ankle	148
	Ankle ligaments	148
	Sprain has sprung	148
	RICE, plus	149
	General guidelines for soft tissue healing response	149
	Ankle support	149
	Collapsed talus joint and foot pronation	150

Chapter 23	**The Feet**	**151**
	No wheelies, please!	152
	Muscles of the arches	153
	How the feet remotely control posture	153
	Additional details on foot mechanics	154
	"Don't stand so close to me"	154
	Eversion and inversion	155
	Plantar fasciitis	155
	Plantar fasciitis rehabilitation	156
	Bunions	156
	Twist walking	158
	The high arch	158
Chapter 24	**Anatomy of the Spine**	**159**
	Architecture of a vertebra	160
	Spinal Disc	161
	Rupture, herniation, bulge	161
	Shear madness	161
	Back pain - mechanical or chemical?	162
	The immune system's effect on disc injuries	163
	The Valsalva effect: Breathless in yogasana	163
	How many movements can the spine make?	164
	Independent facet movement	165
	Hypermobility, clicks and cracks	165
	Mobility – the curse or the cure?	166
	Changes in spinal curves throughout life	167
	The "Dead Zone"	167
	Scoliosis Dangerous curves ahead	168
	Scoliosis and mobility	169
	Essential yogic principles for scoliosis	169
	Scoliosis and yoga therapy	169
	Mapping out the spine	171
	Imbalanced spinal mobility	172
	Keep the lumbar curve intact	172

Chapter 25	**The Lumbar Spine**	**173**
	Motion of the lumbar spine	174
	The ilio-lumbar ligaments	174
	Happy baby, indeed!	174
	Stiff, or not stiff? That is the question!	175
	Spinal stenosis	175
	When to surrender	175
	Tip of the pelvis, wag of the spine	177
	Lower torso muscles	177
	Lower Crossed Syndrome	177
	Don't crush the egg!	178
	The Abdominal Obliques	178
	Keep the central axis aligned when twisting	179
	Abdominal six-pack	179
	The "black hole" of the belly	179
	Spondylolisthesis	180
	Supported Bridge for spondylolisthesis	180
	Sciatica	181
	Muscle-related back pain	182
	Move gently out of backbends	182
Chapter 26	**Yoga Butt**	**183**
	A flat lumbar curve can also initiate the yoga butt	184
	Lower Crossed Syndrome	184
	Springing into action	184
	Pain in the yoga butt	185
	The Righting reflex	185
	Steps to reduce yoga butt posture	186
Chapter 27	**The Psoas muscle**	**187**
	Iliopsoas	188
	The elusive psoas	189
	The dual nature of the psoas	189
	Psoas strength in asana	190
	Anterior pelvic tilt, inward hip release, and psoas stretching	191
	Hamstring muscles and the psoas	191
	A short, tight psoas muscle	191
	Psoas stretch increases sacroiliac joint mobility	192
	Asana for psoas stretching	192
	The psoas and the abdominal organs	193
	What is the sound of one hip snapping?	194
	Iliopsoas muscle stretch	194

Chapter 28	**The Thoracic Spine**	**195**
	Range of motion in the thoracic spine	196
	The rib cage	196
	The 12th thoracic vertebra	197
	Moving the thoracic spine	197
	Kyphosis - the rounded back	198
	Flat thoracic curve	199
	Chest expansion	199
	Rounding the upper back? Not on my watch!	199
	"What is round will roll!"	200
	See-saw rib cage	201
	Shoulder shrugging	201
	Thoracic spine extension with blocks	202
Chapter 29	**The Breath and the Bandhas**	**203**
	The diaphragm – primary muscle of respiration	204
	The diaphragm during respiration	204
	Diaphragm upon expiration	205
	Protruding lower ribs	205
	The abdominals	205
	Effective diaphragm engagement	206
	Paradoxical respiration	206
	Nose breathing	206
	Mouth breathing and stress	207
	The heart of the yogi	207
	Blood pressure and yoga	208
	Advanced yogic breathing techniques	209
	Basic three-part breathing	209
	Alternate nostril breathing	209
	Kumbhaka - breath retention	210
	The Bandhas	210
	Bandhas, from an anatomical point of view	211
	Ujjayi Pranayama - the victorious breath	211
	How loud is the Ujjayi call to victory?	212
	The Valsalva Effect and breath	212
	Back off the bandha!	212
Chapter 30	**Shoulder Anatomy**	**213**
	Shoulder mobility	214
	The three mechanical components of the shoulder	214
	The gleno-humeral joint	215
	Comparing the hip and shoulder "sockets"	215
	Biceps brachii tendon	216
	Joint clearance requires external rotation	217
	Shoulder dislocation	217
	The clavicle and its two joints	217
	Not a fashion statement!	218
	The acromio-clavicular joint	218
	The sterno-clavicular joint	218
	The scapulo-thoracic "joint"	219
	The scapula - fun facts	219
	Where go the palms, so go the shoulder blades	220

	Keep your angel wings folded in!	220
	Postural adaptations to shoulder limitations	221
	Upper-Crossed syndrome	221
	Serratus anterior	221
	Challenges of the serratus anterior	222
	Latissimus dorsi	222
	The rotator cuff	223
	Actions of the rotator cuff	223
	Rotator cuff injury	224
	Deep shoulder joint release	224
Chapter 31	**Integrative Alignment of the Shoulders**	**225**
	Shoulder ligament function	226
	Review: alignment principles for shoulders/whole body	226
	Preparation for shoulder alignment	226
	Thoracic region preparation for shoulder alignment	227
	Specific action steps for shoulder alignment	227
	Details: Action steps for shoulder integrative alignment	228
	Examples of asana to demonstrate shoulder alignment	230
	Avoid shrugging the shoulders	232
	Shoulder walking	232
	Simplified summary of shoulder integration and alignment	232
Chapter 32	**The Upper Extremities**	**233**
	Out on a limb	233
	Pre-requisites for upper extremity principles	234
	Move from the core, using the shortest levers	234
	Latissimus dorsi Use core upper back muscles to lift arms	234
	Bones draw to the midline, muscles extend out	235
	Triceps muscles rotate toward the midline	235
	Use the triceps muscle to extend the elbow	235
	Samasthiti - equal tension and balance	235
	The elbow	236
	Elbow flexion	237
	Elbow extension	237
	Hyperextension of the elbow	237
	The "eyes" of the elbows	237
	The forearm	238
	Anatomical terminology	238
	Upper extremity alignment	238
	Placing the hands	238
	Middle fingers in line with the seam of pants	239
	Elbows aligns with the side body rib cage	240
	Arm counter-rotation provides stability: Towel Twisting	240
	How wide apart are the arms in arm-balancing poses?	240
	Carrying Angle	240
	Carrying on with the carrying angle	240

Chapter 33	**The Wrists and Hands**	**241**
	Wrist and hand relationship	242
	Arch of heel of the palm	243
	Four-step hand placement for weight bearing	243
	Additional refinements in hand placement	244
	Keep the foundation stable	244
	Spider-fingers	244
	Flexion of the wrist	244
	Wrist extension - less than you think	245
	Wrist wrap	246
	Carpal Tunnel Syndrome	246
	Quick tips for reducing wrist strain	247
	"Popeye™ arms" therapy	248
	Sphinx Pose	248
Chapter 34	**The Head and Neck**	**249**
	A balancing act	250
	The cervical spine	250
	Atlas, the first cervical vertebra (C1)	250
	The axis (C2)	250
	The anterior compartment of the neck	251
	Can you see the collarbones?	251
	Foundation for the head and neck	251
	Head and neck posture	251
	Head games	251
	Managing the cervical curve	252
	Whiplash and cervical spine instability	253
	Hypermobility	253
	Is Headstand Pose safe?	253
	Alignment essential for weight bearing	254
	Where is the head placed for Sirsasana One?	254
	Considerations for the Headstand	255
	Musculature of the neck	256
	The hyoid bone	256
	Hyoid alignment - the smiling throat of Buddha	257
	Subtle actions for moving the neck	257
	How to align the head	258
	Therapeutic movements of the neck	258
	"Turtle neck" procedure	258
	Easy on the eyes	259
	Eye pillow for Savasana	259

Footnotes and References	**261**
Additional source material	**267**
Photographic Acknowledgments	**267**
Acknowledgements	**270**
About the author	**271**
About the illustrator	**271**

Preface

The term *yoga* derives from the Sanskrit root word *Yuj* and its interpreted meaning- to join, to control, to discipline. Classically, the harness that yokes together two oxen of an ox cart represents yoga. Both words - yoga and yoke - share the same Sanskrit root. With yoga, one ox symbolizes the body and the other, the mind. The yoke represents the binding of body and mind into a cooperative relationship that promotes an intimate awareness and consciousness of each other. As one acts, the other responds, providing constant feedback and communication that enables pure alignment in their behaviors. Spirit occupies the seat of the driver. Hopefully, Spirit is by nature compassionate and benevolent and can respect the forces it oversees. Through the yogic binding of body and mind, it is possible to achieve safety in action and wellbeing. Yoga enables integration and coordination of the body and mind. With that, Spirit can attain serenity and profound satisfaction; in other words, bliss.

Yoga practice follows a concept known as *Totalism*. This theory argues that to do anything, you must know everything. As with driving an automobile, you need to know more than merely steering the car before heading out on the road. You must learn everything about operating a car safely before you take your maiden journey. And, as it is with yoga, we can only learn a few principles at a time yet it requires implementing all of them to have a safe asana practice. It may take many months, often much longer, to master the many nuances of yoga alignment. This is why yoga is a practice. It requires patience and awareness and perhaps most of all, humility.

Anatomy is also filled with nuance. To understand anatomy, the subtleties of each structure must be unraveled and the relationship between each part identified and described in greatest detail. Anatomy, as presented in this book, will provide the underlying rationale for the principles of postural alignment. A more in depth study of anatomy could be provided for each topic touched upon in this book- but that would detract from its purpose. You are encouraged to explore beyond this text any and all topics that stir up your curiosity. Be assured that additional information will only enhance your yoga practice.

Throughout this book, there will be many cross-references and often a repetition of information. This is not a lapse or an oversight. It will occur when the alignment principles of a particular region cannot be fully understood without reviewing its relationship to other regions already discussed. Some readers may open the book to a specific topic and necessary information will need to be re-introduced as the prelude for adequately understanding alignment of that region and its linkage with the rest of the body.

In many ways, this book reads more like a novel than a how-to book. It is recommended that you start from the beginning and read to the end. For some readers, the sections on anatomy and physiology are challenging. If so, it is recommended that you give them a cursory read the first time and hopefully return to those chapters at a later point or as a reference to the material in the later chapters.

What this book is and is not

The principles of alignment presented in this book are not a ritualized set of instructions passed down from a specific yoga tradition. It is meant for all yoga styles and practices. Although most anatomical and mechanical concepts for yoga practice originated from the teachings of Sri B.K.S. Iyengar, the tenets presented in this book do not contradict the teachings of any school or tradition of yoga and can be applied to all asana practices, without exception. It will offer skill and insight into every posture and help determine if your specific actions are healing and invigorating - or causing injury.

Exceptions make the rule

There are many variations to the expression of human anatomy and not all exceptions can be covered in this one text. If you sense that a particular alignment instruction does not apply to your particular situation, allow your personal experiences to hold merit. Remember though, that although anatomy can vary, body mechanics seldom do. Most joints, ligaments, and muscles will function the same, regardless. Although the instructions in this book would unlikely be contraindicated, when in doubt, go slow and pay attention to what you already know about yourself!

My hope is that you will begin this journey with a beginner's mind. Allow the slate to be blank, free of attachment to what you expect to find or beliefs that you hope to reinforce. If you let the material in this book unfold story-like, you will get the most benefit from your effort. If the ideas that you read ring true and resonate with your own experiences, your yoga practice will powerfully transform.

Recognition of my teachers

Following yoga's top-down, oral teaching traditions, considerable information and terminology used in this book is compiled from countless teachers, workshops, and mentors. Much original thought does occupy the pages, as well. Please know that proper credit to all inspiring teachers will be shared, where possible and that acknowledgment is not an endorsement of any specific yoga tradition.

Introduction

With a mix of excitement and humility, I share in this book my personal and professional observations, compiled over forty years of practice, research, and discovery. Any contribution I may offer to yoga's growing knowledge base come, not only from my experiences as a yoga student, teacher and therapist but also from the unique perspective I bring as a doctor of chiropractic and an instructor in anatomy and alignment. My countless observations of the human form have brought awareness, not just to what makes us different, but also to the remarkable, underlying similarities we share. These commonalities reveal that, with only few exceptions, human bodies all follow a single, generic set of alignment principles. As the human body runs the gamut from individuals with rubber-like flexibility at one end to our stiffer brethren at the other, one thing remains consistent: alignment is central to the design of all body mechanics. Down to the cellular level, alignment organizes the very essence of what defines a living being. Alignment directly affects health and vitality. In physical rehabilitation, alignment is a primary, critical factor that determines either a favorable outcome or frustration.

Yoga asana practice holds a deserving place amongst bodywork and therapy. It is potentially a highly sophisticated and empowering method for creating structural alignment in the body. The outcome of yoga practice depends on one essential criterion: the successful application of alignment. Alignment is not an afterthought, nor exclusive to advanced postures. Alignment is essential at all levels of practice to ensure safety in action. The yoga poses themselves need not be overly advanced to be beneficial. In fact, basic postures are generally the more therapeutic since they are the easiest ones in which to find precise alignment. Precision is often what determines therapeutic success.

Yoga clearly occupies a main stage position on the cultural landscape. Its practice and its influence affect many people and their way of life. In prior decades when its popularity was emerging in Western culture, yoga practice was inseparable from its Vedic and Hindu roots. It had found an unlikely home in a rebellious counter-culture. And for some today, any departure from those "roots" is incredulous. The frenetic rush to nail down a definitive characterization of yoga will certainly continue.

With that said, abandoning the wisdom of yogic philosophy would be an injustice to all its practitioners. As will be discussed in Chapter 3, the foundation for an injury-free yoga practice rests upon its age-old principles, those that distinguish yoga from being just another exercise regimen.

Regardless of the belief system, how fast or slow one practices, the temperature of the room, or the name given to the poses, alignment is essential. Alignment crosses all philosophical and ideological lines. Any approach to yoga can result in either healing or injury. But regardless of the reason that a student practices yoga, getting injured would not be desired and hopefully, not tolerated.

Alignment does not happen by simply rolling out a mat or buying the latest clothes. Downloading a soulful chant may open the heart but not the hips. Practicing yoga alignment takes effort. It requires a willingness to apply alignment with dedicated precision at all times. Some practitioners often mistake alignment for merely an external appearance or a mechanical action. On the contrary, the holistic principles that guide asana alignment originate in yoga philosophy. The ancient yogic teachings offer guidance in not only how to live spiritually, but also how elegance emerges from asana practice through alignment.

This book introduces a concept referred to as *Integrative Alignment*. As its name implies, Integrative Alignment organizes the alignment of every region of the body in relation to the whole. It approaches the mechanical nature of the body as not something fixed or static, but organic and flowing, constantly negotiating vectors of external forces and internal tensions.

Yoga's greatest teachers

A pilgrimage to remote India will not help discover our greatest yoga teachers - the ones who guide our life and asana practice. Those teachers are right here, residing within ourselves. Pain and injury are our greatest teachers. They represent the *Sat guru tattwa*, our inner guru or guide that enables universal knowledge to flow freely to us. Rather than something to be cursed, pain and injury are an integral part of life experience and opens us to the potential of an empowered and enlightened life.

Injuries and their resultant pain inform alignment. When a posture is correctly aligned, pain lessens. If an injured area can be aligned and exercised without producing pain, it heals at an accelerated rate. Without alignment, healing may be impeded or unwanted compensations may develop.

Some yoga traditions view the body as a temple in which the spirit lives. Caring for this exquisite physical form is a tribute to spirit, the humble act of "sweeping the temple floors". The responsibility of all yoga teachers, both our inner teachers and the ones who stand at the front of the class, is to make asana practice a therapeutic and health-enhancing experience.

The yogic path toward greater awareness and higher consciousness must be free of all hazards that cause physical injury; misalignment is one hazard most common. This book serves as a guide for all on this spectacular, life-affirming journey called yoga.

In yogic philosophy, there is a subtle, exquisite point where energy merges with matter. The Sanskrit term *Sparsa* is used to describe this point where Prana, or life force, co-joins with the physical tissues of the body. Although Sparsa is conceptual, we can discover this deep alignment, not only in personal yoga practice but also in accompaniment to every human touch, therapeutic or otherwise. With each breath, our imagination can take us to that quantum point where this amalgamation, or Sparsa occurs.

1 Yoga and Alignment

From its origins in South Asia nearly 5,000 years ago, yoga has emerged and re-invented itself as the fastest growing practice for fitness and wellbeing in the world. [1,2] Between 2012 and 2017, the number of practicing yogis in the US has increased by 55%.[3] Yoga classes are offered alongside Pilates and kickboxing in practically every gym, school, and health club. Many of the principles of yoga, especially those regarding alignment, flexibility, and states of consciousness, are receiving high levels of acceptance and attracting research in science, medicine, and psychology.

The yoga studio has become a valuable center for personal growth and human potential. It is a place where students come together in a conscious, caring community. The ancient philosophies of yoga offer wisdom that can help us navigate the overwhelming demands of modern culture in which we face the dual challenge of a life of material excess and spiritual disconnectedness. These two predicaments contribute to a profound state of deep and unconscious exhaustion. Yoga's growing popularity is a testament to its ability to remedy the accelerated pace of life, which has become today's norm.

In mainstream media and popular culture, yoga is often categorized as a sport. This "takeover" has presented a serious challenge to many practitioners who yearn for a return to what they consider is yoga's roots. But if we look hard at its historical record, much of what occurs on the mat in today's popular yoga classes is actually a derivative of Scandinavian physical exercises, popularized in Europe in the mid 1800's. The British introduced these regimens to India for their military conscripts, young Indian boys being mustered into colonial service. As with Europe, these exercises became popular throughout India, finding a place in the Indian YMCA's "physical culture" system.

As a method for conditioning the body for long periods of sitting, this regimen was a perfect fit with asana practice. Great yogis of the day, such as T. Krishnamacharya and Sivananda Saraswati, quickly integrated these exercises into the traditional Hindu teachings of asana.[3] This history is a reminder that great ideas often arise from unlikely origins. They rarely arrive pre-packaged and fully actualized.

Of course, many practitioners want yoga to be magical and filled with ageless wisdom. Some expect it to be an all-inclusive, one-stop solution to the totality of life's problems. This desire is unrealistic. The good news, however, is that yoga reliably points a finger in the right direction.[4]

The gift that yoga brings forward and what makes it uniquely different from other physical exercise regimens is the intention with which it is practiced, Classical texts, such as the 2,000-year-old *Yoga Sutras of Patañjali* can elevate any physical endeavor from the popular "no-pain, no gain" sports mantra to the dignity and safety that a "no pain, no pain" approach offers.

Sthira Sukham Asanam   Take a steady, easy seat

The physical practice of yoga is called *asana*. It takes the student through a series of increasingly complex movements and forms. Asana comprises poses or postures and how they are sequenced is referred to as *vinyasa*. In Sanskrit, asana translates to sitting, or the seat. It refers to the position yogi's take in meditation and in other practices that require long periods of stillness. From a classical point of view, the purpose of these postures is to help the yogi develop the ability to sit effortlessly for the long periods of time necessary for the advanced practices of meditation and *Pranayama*.[5]

Few of Patañjali's yoga sutras apply directly to asana practice, supporting the likelihood that very few asana existed or were codified at the time of sutras' compilation. *Sthira Sukham Asanam*, which is a yoga sutra, suggests that sitting, yoga postures, and the transitions between them are to be steady (*Sthira*) and comfortable (*Sukham*).

Hatha

In today's most common usage, Yoga and *Hatha* have essentially the same meaning. Hatha is a system of yoga formulated nearly 500 years ago by Yogi Swatmarama in the text *Hatha Yoga Pradipika*. Most popular traditions of yoga practiced today have their origins in Hatha. From the Sanskrit terms *ha,* meaning sun and *tha,* meaning moon, Hatha strives unifies the energy of opposites, particularly those of body and mind.

In Sri B.K.S. Iyengar words, "Yoga practice is not only the preparation for spiritual awakening; awakening is inherent within the practice itself."[6] Iyengar regarded each individual posture as an "archetypal template." When precisely aligned within the constraints of these forms, like fitting into a cookie-cutter, a connection to something greater than the posture is achieved. As Iyengar expressed, "Yoga is equanimity. Equanimity is alignment and without alignment there is no equanimity."[7] Following his perspective, equanimity can be considered the combination of balance and integration.

Bringing the alignment component into the practice of yoga provides an essential tool for transforming asana from a simple floor exercise into a powerful exploration of mind-body awareness.

> The body does not attempt to mold itself into a determined form. Instead, the asana fits into the body's pre-determined, anatomical design

Yoga Asana – an "alignment delivery system"

In itself, there is nothing magical or divinely protective or healing about asana. The alchemy that many yogis experience as a result of their personal practice does not arise from something supernatural hidden in the form and shape of the poses themselves. Likewise, there is nothing inherently dangerous underlying any pose.

The determinant for the ultimate outcome of asana is precise alignment and whether or not it has been integrated into the postures. The magic that emanates from asana practice arises from the power of healing that alignment innately unleashes.

High achievement of human performance, such as being a concert pianist, ballet dancer, or proficient tennis player, requires constant attention to the smallest details and dedication to impeccable form. The effortless action that masters often demonstrate is the result of unrelenting discipline and hard-earned skills. In each field, mastery of alignment is fundamental to great performance.

A skillful yoga practice adheres to the sophisticated design of the human body and its underlying mechanics and alignment. Every yoga asana is an opportunity to put the body and its principles to the test. When successful, there is a clear sense of integration, elegance and refinement.

Yet yoga is not a panacea. Practicing asana can cause injury just as easily as it can heal. Orthopedists report increasing numbers of injuries directly attributable to the practice of yoga. The Consumer Product Safety Commission found that common injuries caused by yoga include repetitive strain and soft tissue trauma resulting from overstretching. Joint injuries in the spine and extremities are common. Its study also revealed increased health care costs and higher utilization of therapeutic services directly attributable to injuries from yoga practice.[8]

Why should so many strains, tears, and compressive injures result from yoga practice? If yoga asana declares itself as a spiritual and health-restoring practice, it seems contradictory for it to be potentially dangerous. Clearly, not all ways that yoga is practiced are equal.

Poorly aligned postures compromise body mechanics and ultimately cause yoga-related injuries. When yoga follows the principles of alignment and its actions are consistent with anatomy, the result is a practice that advances safely and is a constant source of therapeutic healing.

Alignment cannot be an afterthought. Most injuries occur while going into and getting out of a posture; when the foundations of the postures are not secure and stable. This makes practices with flowing transitions between asana more challenging for maintaining safety. From establishing the foundation to making subtle adjustments along the way, alignment principles lead the way.

> Asana practice is an opportunity to discover, establish, and maintain alignment as the body moves through increasingly complex movements and forms

Some yoga practitioners remain reluctant to focus on alignment during their practice. They view their time on the mat as a sacred experience, believing that preoccupation with alignment is antithetical to an intuitive and spiritual experience. They feel that alignment makes asana practice overly intellectual.

However, if a yogi were guided by their spirit alone or took a mind-*over*-body approach, asana practice would fail to follow yoga's most basic teachings.* Yoga requires *listening to* the body, not over powering it. The body communicates in great detail. Emotions are often considered a primal form of body language. Alignment is the clearest, most basic language of communication. Alignment creates a seamless and reciprocating interplay between body and mind. Without alignment, communication fails.

* From my own experience, my spirituality has heightened as my awareness of alignment has grown. Having more open hips has opened my whole self to a greater appreciation for life and a deeper sense of wellbeing. Correct alignment and freedom in my spine has opened my heart. My yoga practice is more sacred as my sacrum is better aligned.

Alignment is not another style but the basis of every style

There are many traditions and styles of yoga with many names and many ways to express each asana. Richard Rosen, a senior Iyengar yoga instructor from Oakland, California has researched the Vedic literature, particularly the 1934 text of Krishnamacharya, the *Yoga Makaranda,* where it is proposed that there are approximately 84 lakh of potential asana. A lakh is equal to one hundred thousand. This means that there are potentially 8.4 million asana that can be performed.

The outer appearance of asana is an expression of our inner anatomy, however, they are not always not always in sync with each other. Often, our inner mechanics must move in directions opposite an asana's final appearance. Initially, this may seem counter-intuitive but once the principles of body mechanics are understood, correctly engaging the inner actions becomes natural and the outer asana forms are expressed with greater skill, safety and ease. It is important to take the time to differentiate between form and action- outer appearance versus inner functional anatomy.

Successive Approximations

Anusara yoga teacher and educator, Betsey Downing, PhD, uses the term *successive approximation* to insightfully refer to alignment as an ongoing process that becomes more refined and exacting with each practice. It is not the final, position that a pose reaches that unleashes the value of yoga asana; instead, yoga's importance resides in the *intention* that is applied to reach that level.

Sri B.K.S. Iyengar was the most influential pioneer in yoga alignment and in developing yoga as a therapy. His life's work amassed an in-depth consortium of instructions that many teachers follow point-by-point. For some, application of these instructions had become exhaustive. Iyengar's concepts were further developed in the Anusara Yoga system, which codified alignment into a single, universal set of principles. This revealed the secret that the same alignment principles always apply in every asana. In this way, Anusara revolutionized yoga practice.[2]

2 Alignment and Body Systems

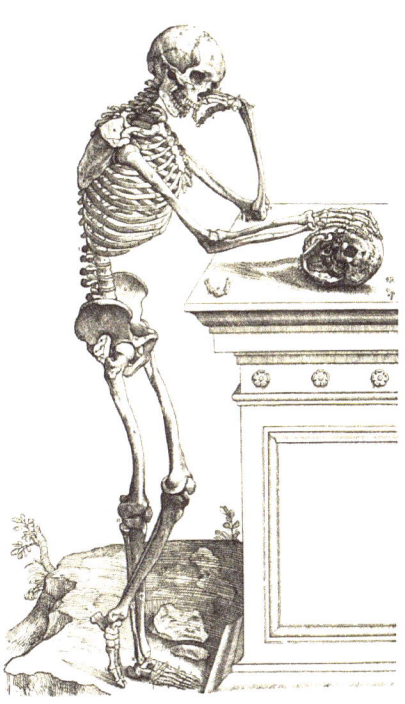

The body "loves" alignment. In addition to creating graceful appearances to our yoga poses, alignment provides considerable benefit to our health and wellbeing. Alignment generates a physiological re-boot and re-establishes its natural simplicity. When the body is aligned and fully integrated, our systems are more efficient and function more effectively. This chapter offers a modest review of the impact of alignment on some of the important systems of the body.

Asana practice is an excellent tool for mastering alignment. Asana takes the body through increasingly complex configurations while the yogi seeks discover the same fundamental principles in every form. Developing the resiliency to maintain alignment as the body shifts and adapts enables the various systems of the body to perform with greater vitality.

Any style of yoga that a student chooses to practice can potentially access asana's positive effect on health and rehabilitation. Of course, to obtain these benefits, skillfully aligned postures are required.

The nervous system

The nervous system has an influence on every one of the human body's estimated thirty-seven trillion cells. It facilitates communication from the brain throughout the body, from the muscles to the vital organs and to the smallest capillaries. All mechanical and organic functions rely on direct feedback from the nervous system. Our sense of balance and spatial awareness is nervous system controlled. A well-aligned spine is essential for optimal nerve function. When the spine is poorly aligned, the nerves exiting from between the vertebrae become physically compromised and their electro-chemical transmission hampered.

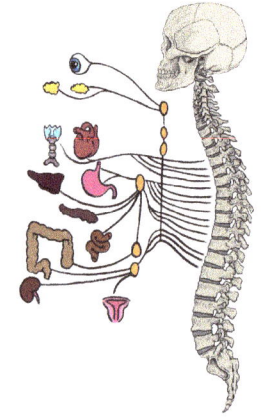

Muscle fibers and myofascia

When muscles contract, they produce forces across the joints they affect. Muscles are most efficient when their fibers align obliquely to the direction of that force. Good postural alignment allows the muscles to function most effectively and to develop for optimal performance. Myofascia, the collagen-rich sheathing that envelops every muscle fiber, is sensitive to structural tension. Poor alignment causes myofascia to torque, which impedes muscle's function and efficiency and increases its susceptibility to injuries of the tendons, ligaments, and joints. (Details in Chapter 9)

Body mechanics

A wide-ranging array of mechanical operations occurs simultaneously in the body. They depend upon proper alignment in order to be effective. Body mechanics and joint performance are most efficient when joints remain centered on their axes of rotation. This allows the greatest ease of motion and centers the body to best manage the forces of gravity.

In yoga, as in all exercise regimens, alignment and a stable foundation should be established before utilizing or building strength, flexibility, or agility. Attempting to develop muscle mass or skill sets upon a frame that is not aligned forces the body to operate from a position of stress. This will create unsuitable habits that are likely to produce injury.

Bone density

The body in motion is constantly negotiating the downward pressure of gravity with the upward forces of heel strike. At the same time, bone tissue is being stimulated by the contractions of muscles, hugging to its surfaces. Nerves located within the bones are delivering mild electrical stimulation to the many layers of bone. All of these forces directly affect bone density. Postural alignment distributes the impact of these various forces proportionately, helping to keep bone density uniform.

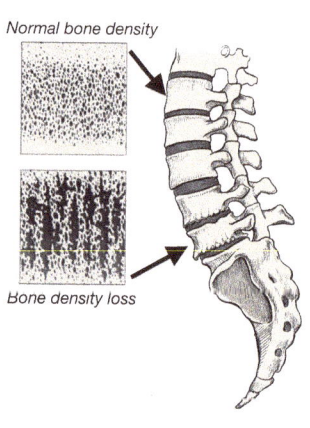

Normal bone density

Bone density loss

Wellbeing

Studies have shown that yoga practice can increase neurotransmitter levels. Endorphins that positively affect mood, improve self-confidence, and increase life satisfaction have been shown to increase with yoga.[1] Remarkably, these neuro-chemicals are present in higher concentrations at locations along the spine that correspond to the vortices (energy centers) associated with the chakra system.[2] Correct alignment increases the levels of neurotransmitters that produce positive states of being. In contrast, poor posture and improper alignment release stress-triggering neuro-chemicals into the body.

General Adaptation Syndrome

The work of Canadian Nobel Laureate Hans Selye, PhD., infers that poor alignment is one of many *stressors* that trigger a bodily response called the *general adaptation syndrome*. In the presence of uncompensated postural imbalances or unchecked muscle tension, stress hormones flood the body. This stress response increases demand on the vascular and hormonal systems. Chronic poor alignment can lead to adrenal exhaustion and eventually, a weakened, overtaxed immune system. This increases the potential for unchecked disease processes to develop.

The stress response of poor alignment is comparable to the reaction to falling backwards out of a chair, over and over again!

Yoga props are great tools that provide direct physical contact, which calms the stress response in the nervous system and attenuates (weakens) muscle contractive reflexes to allow deeper stretching. Supporting a suspended body part, such as placing a block under the forehead or a blanket under the buttocks, reduces the "hanging in mid-air" stress response on the muscles, the glands, and the nerves.

Tissue repair and healing

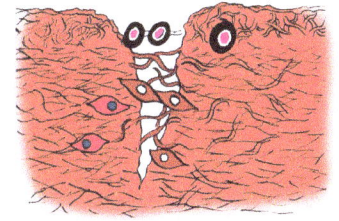

Alignment stimulates the repair of microscopic fibers and the formation of healthy, new cells throughout the body's connective tissue structures. Close contact between cells, which is enabled by aligning the tissues, is essential to the healing process in structures such as skin, muscle, bone, spinal discs and fascia. Effective alignment lessens scar tissue formation, reduces inflammation and prevents degenerative changes.

Brain wave activity

Yoga postures and the accompanying breath practices have been shown to reduce the high activity beta brainwaves and increase the alpha, theta, and delta waves. This are found when the nervous system is in a more relaxing and contemplative state. Practicing yoga has been well proven to reduce depression and anxiety and directly measured in brain wave activity. More details are provided in Chapter 29.

An overall calming of the nervous system occurs with alignment. When body tissues are not strained and are making good contact with other appropriate tissues, brain wave activity is more balanced. As an example, a deep calm is produced when muscles hug onto the bone, such as when wrapping and swaddling a baby induces relaxation, safety, and sleep.

The energetic body

The ancient Sanskrit texts that comprise *The Vedas*, or the Books of Knowledge, describe a system of 72,000 rivers of energy called *Nadis*. These channels run throughout and around the body, transmitting *Prana*, the energy force of life. Along the central core axis of the body, the popularly known *Chakras* are foci of Pranic energy.

Unimpeded flow of prana fosters health, vitality, and natural healing. Proper alignment, both of body structure and of each chakra with the others, allows Pranic energy to freely flow upward.

If the Nadi channels are blocked, pranic flow is restricted and *Granthis*, energetic knots, are formed. A granthi can develop in any of the Nadis or within the chakral channel. Alignment prevents Granthi formation. In seated postures, alignment runs through the central axis and keeps the Chakras vertically aligned.

Three main knots that can occur in the chakral channel are:

- Brahma Granthi - blocks instinctual actions; located at the floor of the pelvis, between the anus and genitals, the Muladhara chakra.
- Vishnu Granthi - blocks emotional/heart energy; located at the chest and the heart, the Anahata chakra.
- Rudra Granthi - blocks intellectual thinking; located at the middle forehead at the region of the "third eye", the *Ajna* chakra.

Internal health conditions

Studies using biofeedback, control groups, and peer scientific studies offer legitimate support for yoga practice reducing hypertension and stress.[3] Yoga has been shown to reduce cognitive decline following diabetes 2 and blood sugar stabilization.[4]

"You can't be neutral on a moving train!"   Howard Zinn

Postures are either in alignment or are misaligned. These are the only two options. An asana is aligned or moving into alignment; otherwise it is misaligned or moving in the direction of misalignment. There is no middle ground, resting point, or neutral position, nor is there anything in between. Balance and equilibrium are qualities that reside with alignment.

Identifying alignment is a practice unto itself. A yoga practice, regardless of the strength or flexibility of the student, cannot advance safely if there is misalignment built into the postures. This type of practice often results in injury. The healing and therapeutic value of yoga is available only when postures are aligned. The more precise the alignment, the more likely the desired rewards are received.

Alignment promotes healing - Misalignment causes injury

3 Raja Principles of Yoga

the tree of yoga

In the *Yoga Sutras of Patañjali,* yoga is portrayed as a tree consisting of eight branches. The *Astañga of Yoga* joins the words *astha,* meaning eight with *añga,* meaning limb. The tree of yoga represents *Raja Yoga,* the "royal union" of practices that challenge the body and mind and put the practitioner on a regimented path towards spiritual liberation. The yogi "climbs" the tree, the metaphorical framework of ethical, physical and spiritual values, developing mastery in the practices represented by each limb. The ultimate goal is ascension to the top and to reaching *Samadhi,* the post-meditative state of absorption and integration, putting everything together into complete wholeness.

Asana resides on the third limb from the bottom of the tree. It is the primary focus of this book. To practice asana, one must ascend the tree upon the two branches below. *Yama and Niyama*, these lower branches, serve as the foothold, necessary from where asana practice can truly begin.

Yoga teacher Richard Freeman, in his book *Mirror of Yoga* states, "The support that is built from the first two limbs, the yamas and the niyamas, is an interactive net of kindness and responsiveness to both oneself and to others."[1]

14 YOGA ALIGNMENT   PRINCIPLES AND PRACTICE

Trauma-yama is not a limb on the tree of yoga!

Yamas and Niyamas

What differentiates asana practice from other types of floor exercise programs is that asana is not merely a physical practice but is integrated into a broader system dedicated to spiritual awareness. The principles of Yama and Niyama are usually considered a series of ethical guidelines for how yogis behave and interact with people and the world around them. But these two lower-limb principles are the actual, practical foundation of the asana practice. Yama and Niyama are ever-present on the yoga mat and influence the physical postures themselves. Fully integrating the Yama and Niyama principles into asana ensures safety, dignity, and access to the therapeutic value of yoga.

Yamas   Restraints, rules, and controls on behavior

Taken from a Vedic story that chronicles the first death of a mortal, the first limb of the tree of yoga is called *Yama*. It refers to sins and punishments but can also be interpreted with positive characteristics. The Yamas, according to Patañjali, are comprised of five restraints on behavior. It is a code of conduct that details the "do not" rules for ethical interaction.

- Ahimsa   Non-violence, non-harming
- Asteya   Non-stealing, non-claiming
- Satya   Truth, pureness, and wisdom
- Brahmacharya   Avoiding unnecessary expenditure of energy
- Aparigraha   Non-possessiveness, non-greed, releasing from the grasp of the ego

Ahimsa   Doing no harm, non-violence

Perhaps the most fundamental principle of yoga is to do no harm, be non-violent, and avoid injury. It is easy to understand how this principle applies to the treatment of fellow sentient beings and affects choices of diet and lifestyle. What is less obvious is that it fully applies to asana practice. Practicing with ahimsa, not harming oneself, of course requires knowing what is actually harmful. Many yoga students who come from a background of sports and exercise bring with them the mindset of pushing beyond the safe limits of steadiness and comfort. It is not uncommon for new yoga students to force their poses and to ignore critical warning signals. Injuries are a common result of not following Ahimsa.

Alignment is an essential tool for the practice of Ahimsa (safety). Asana does not force joints beyond their current limitations nor does it stretch muscles beyond their capability. Practicing yoga without alignment is dangerous and a direct path towards injury. It is not Ahimsa. Without exception!

Practicing in accordance with the body's natural design is spirit in motion. An understanding and respect for anatomy, postural alignment and mechanics are expressions of spiritual pursuit.

Asteya — Non-stealing, non-claiming

This principle does not generally come to mind as applicable to asana practice, yet it triggers improper body mechanics and misalignment and becomes a subtle but common cause of injury. Asteya is violated when one part of the body overpowers or is simply used to perform an action intended for another. This is a pattern of stealing.

For example, a yogi with inadequate hip flexibility to correctly perform **Padmasana**, the Lotus Pose, might dangerously torque their knees and twist the ankles to claim the pose, essentially stealing the flexibility of the ankles to appear successful. A sign of a safely executed Lotus Pose is when the Achilles tendons are not wrinkled and ankles are not sickled. Protect the knees by pressing inferiorly through the inner heels.

Padmasana

Upward Bow, **Dhanurasana** is a hip extension pose. Because the hips anatomically only extend 10° with knees bent, the pose is often claimed by stealing the easily accessible hyperextension of the lower back.

Dhanurasana

Satya — Truth, pureness, and wisdom

Forced or poorly aligned asana have no truth or pureness, nor do they support the anatomical wisdom expressed in the body. Poor alignment ignores accepted truths and the misses the opportunity for healing. Satya requires students to shift from working hard to "working smart". Satya is attained and embraced when asana is performed with safety and efficiency. Ignoring alignment is not wisdom.

Brahmacharya — Avoiding unnecessary expenditure of energy

There are many ways to interpret this principle. The term literally means to occupy a seat in the chariot of Brahma, the Hindu god that represents absolute spirit and creation. Practicing Brahmacharya is to rise above the intrigues and machinations of life. It directs students to take a dispassionate view of the world and the trivial actions of the personal self.

Applied to asana practice, Brahmacharya is the intention to distribute our energy so it is not wasted or abused. Yoga practitioners learn to not over-effort and to remain efficient in every action and posture. Brahmacharya is expressed by the kinesiological process of *muscle efficiency*, which means getting the most power from the least amount of energy. Efficiency is referred to throughout this book and can be considered to be Brahmacharya in action. Alignment and integration are necessary to ensure efficiency and therefore, are direct expressions of Brahmacharya in asana practice.

Aparigraha — Non-possessiveness, non-greed, Releasing from the grasp of the ego

It is all too easy to unroll our yoga mats and immediately judge ourselves in relation to our classmates. Comparison triggers our competitive nature and drives us to go beyond levels of safety. For some, it may elicit an opposite reaction and produce sinking feelings of inadequacy and defeat. Although the ego can be powerfully motivating, it can also lead to actions that limit our potential or increase the likelihood of injury. Yoga invites us to engage our ego wisely in every situation and posture we take.

Niyamas — Life skills, practices, and disciplines

Niyama outline a set of life skills, observances and practices. The Niyamas establish a set of "to do" rules for conscious living.

- Shaucha — Cleanliness and purity of body and mind
- Santosha — Contentment in one's place and with one's achievements
- Tapas — Austerity, endurance, building up "heat"
- Svadhyaya — Spiritual study and awareness of life and brought into practice
- Ishvarapranidhana — Love, respect and surrendering to innate and universal wisdom

Shaucha — Cleanliness and purity of body and mind

Shaucha goes beyond cleansing practices and wearing clean clothing. It also refers to keeping the mat, props, and entire yoga space clean and orderly. Aligning mats in the classroom with each other and not blocking the view of the teacher from of other students also reflects this practice. Students who have spent years folding blankets in Iyengar classes know well this expression of Shaucha. As a Zen-like discipline, the mind can become free of distractions. Shaucha is ritually observed in the morning practice of *Kriya* yoga, a system of deep cleansing and hygiene.

Santosha — Contentment in one's place and one's achievements

Santosha is finding peace and contentment every day, in every practice, every posture, and in each moment. Santosha is cultivated when the student acknowledges gratitude for what they are able to accomplish and for the privilege of having a life that provides the opportunity to include yoga practice.

Tapas — Austerity, endurance, building up "heat"

The quality of Tapas is expressed by maintaining rigor, focus, passion, and intensity. Precise alignment is an expression of Tapas. Maintaining a hot (temperature) practice space is also a form of Tapas.

Svadhyaya — Spiritual study and awareness of life

Knowledge acquired from spiritual study is to be applied to asana practice. The act of implementing the Yama and Niyama principles in asana practice is the practice of Svadhyaya. Another form of Svadhyaya is becoming aware of the sensations the body emits during practice and being able to judge how to manage weakness and pain and gain perspective with power and pleasure.

Ishvarapranidhana — Love, respect, and surrender to Innate and universal wisdom

When a yogi surrenders and respects the endless challenges and the many opportunities that asana practice offers, Ishvarapranidhana is attained. The yogi cultivates patience with and acceptance of the concept of *practice*, which, by its nature, is *never complete*, perhaps infinite.

A further example of Ishvarapranidhana is establishing unyielding trust in the inner healing power of the body and the ability of asana to unleash it. Ignoring alignment is not a on the spiritual path.

4 Foundations and Orientations of the Hips

Foundations provide freedom

Birds use thermal pressures in the air as the foundation from which to soar. Yoga practice launches and liberates us from rigidity, not only in the structural body, but also in the mind and spirit. As with the human spirit, freedom flourishes best when it can maintain a sense of grounding as it elevates.

Many of the new, creative approaches to yoga have brought elements of dance and gymnastics to asana practice. With that, there has been a shift in focus from individual postures to the transitions and flow between them. Attention to stable foundations has never been more valuable. A stable foundation is not always evident in an asana's outward appearance. It becomes even more obscure as transitions between postures get more complicated and advanced. Foundations are easier to establish in basic, singular postures. Still, practitioners with advanced practices can effectively engage the necessary postural foundations and receive their great benefits. The art of Yoga is to advance one's practice while maintaining the groundedness and grace that makes it appear effortless.

Hinges

When initiating major movements, the body hinges at specific places:
- Forward flexion postures move hinge from the hips with the body moving like a jackknife
- Shoulders remain on the back body as the arms move from a stable horizontal axis
- Knees hinge anterior and posterior; the elbows can only hinge with flexion and extension

Foundations

The body's architecture contains multiple foundations, each individually established yet integrated with all others and with the body as a whole.

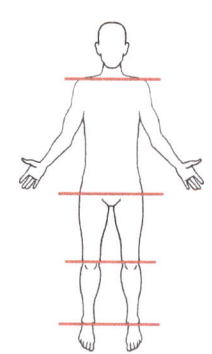

- The feet set the foundation for the legs
- The legs create the foundation for the pelvis.
- The pelvis forms the foundation for the spine
- Shoulders and upper back provide the foundation for the head and neck

Each yoga posture aligns one foundation with each other to establish a stable, integrated structure, most often starting from the ground and upward. If one becomes fatigued while holding a posture, it is best to come out of the pose before the foundations fail. This honors the body and helps avoid injury.

Open and Closed hips

The concept of Open and Closed hip foundations may be the most important and perhaps least understood alignment principle of asana practice. As with all alignment principles, it is the intention to set the correct foundation and not the pose's final appearance that matters!

Every yoga pose will adopt one of two possible hip orientations. All yoga postures, whether standing, sitting, or inverted, utilize either an *Open-hip* or a *Closed-hip* foundation. And importantly, the direction that the legs and feet open into follows the orientation of the hips.

The **Open-hip** position is performed as if the two hips can fit between two panes of glass, occupying only the two dimensions of horizontal and vertical. The primary direction that the body moves in these poses is lateral flexion (side-to-side). The legs open side-to-side along the *frontal plane,* forming a straight line from the front of the heel bone (or the ankles) to the center of the hip sockets. Open-hip poses usually set up along the long side of the mat. Not all yogis can perfectly open their hips and maintain these perimeters exactly but this is the clear intention in Open-hip postures.

Open Hip – Warrior Two

In Open-hip poses, the legs widen laterally to a distance between the feet where each ankle is vertically below the wrist of each outstretched arm. The rear foot aligns parallel with the back of the mat. It does not turned inward, which is often taught. Turning the rear foot inward will undesirably roll the rear hip and thigh anterior to the ankle. **Closed-hip** poses align front-to-back, or anterior-to-posterior along the *sagittal plane*. The hips, legs, and feet attempt to still face fully forward. This is challenging and generally not attainable, especially as the distance between the front and back legs lengthens. Intention is what makes alignment successful.

The front-to-back distance between the legs in Closed-hip postures is shorter than that for Open-hip postures, reducing the distance by 75-85%, or approximately one to two foot lengths shorter.

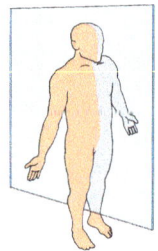
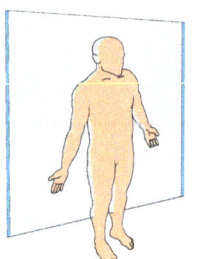

Sagittal plane   *Frontal (Coronal) plane*

FOUR: FOUNDATIONS AND ORIENTATIONS OF THE HIPS | 19

As will be presented throughout the book, the rear hip in Closed-hip poses most often internally rotates in order to properly release the hip ligaments and allow the rear leg to extend to its maximum degree. Internal rotation also assists the back foot to more effectively face forward and flatten to the mat.

In **Virabhadrasana One** and **Parsvottanasana**, the rear foot is unable, nor is expected to fully face forward. When attempting to square the rear hip to the front of the mat, relative external rotation in the rear hip joint occurs. This creates strain on its ligaments with possible injury to the hip and an unsafe torque on the knee. To release the ligaments, widen the inner hip joint, firmly internally rotate and draw the thigh back. Attempt to form a straight line along the outer (lateral) surface of the hip, knee and shin.

Closed-hip Virabhadrasana 1 (Warrior 1)   Closed-hip Parsvottanasana, (Pyramid Pose)

"There can only be one... hip orientation" [1]

Each yoga posture is assigned to only one of these two foundations. Yogis find themselves in trouble when, at the initiation of the posture, they do not establish its foundation based on the correct hip orientation.

Despite some yoga postures being linked together in an asana series or sharing similar names, they often do not have the same hip orientation. For example, Warrior One is a Closed-hip pose and Warrior Two is Open-hipped and require opposite foundations. Triangle is Open-hipped and Revolved Triangle is a Closed-hip posture. Transitioning from one posture to the other without re-adjusting the alignment of the feet, legs and hips can cause significant strain and trauma. This is a very common oversight that occurs in asana practices frequently and a cause of yoga-induced injury.

Begin each asana by setting the correct foundation with the feet and legs placed according to the hip orientation of the pose - Open or Closed. Often the hips do not have adequate flexibility to square fully forward and will benefit from internal rotation of the hip joints to increase flexibility. Other essential alignment steps to be included for hip flexibility are presented in Chapter 13.

Postures are not forced but remain within comfortable limits. In Warrior One, the hips rarely attain the full forward-facing orientation, especially as the front knee bends and approaches 90° of flexion. The more precise alignment is engaged, the more attainable is the posture.

Standing postures are powerful hip openers. For some yogi's aligning the legs in Open Hip poses is challenging. To alleviate, instead of forcing the front heel to align by bisecting the sole of the rear foot, the rear foot can set up more posterior so that the front heel aligns with the ball of the rear foot.

Examples of postures in each category of hip orientation

Open-hip postures

- Triangle — Utthita Trikonasana
- Warrior 2 — Virabhadrasana II
- Extended Side Angle — Utthita Parsvakonasana
- Half Moon — Ardha Chandrasana
- Wide-angle forward fold — Prasarita Padottanasana

Triangle Utthita Trikonasana

Closed-hip postures

- Revolved Triangle — Parivrtta Trikonasana
- Warrior 1 — Virabhadrasana I
- Pyramid Pose — Parsvottanasana
- Warrior 3 — Virabhadrasana III
- Lunge — Anjaneyasana

Revolved Triangle Parivrtta Trikonasana

Exceptions to the rule

Some yoga traditions teach a few of their poses with a foundation that is in between the two orientations. For example, the position of the hips in **Janu Sirsasana**, Knee Head Pose, may be taught with a foundation that is diagonal between the open and closed orientations. Other exceptions to the only-one hip orientation rule are very rare and would be unique only to a particular style of yoga.

Often when there is an exception, one direction of hip orientation is still predominant over the other. For example, Humble Warrior Pose usually is aligned on an Open hip foundation, although it is not always obvious.

Janu Sirsasana

Linking poses into a flow

Most asana practices move through a flow or sequencing of yoga postures, often called a vinyasa. **Surya Namaskar,** the Sun Salutation, is the most well known vinyasa. Virtually any group of postures can be linked together; however, it is necessary to properly establish the correct hip orientations for each posture and adhere to all other principles of alignment. The more a vinyasa shifts from Open to Closed hip foundations, the more challenging it becomes to set clear and definite foundations for every individual pose. The foundations are never to be abandoned, regardless of the speed of the flow. Less experienced students become disorientated by rapid changes in hip position and unable to negotiate the abrupt changes in leg and foot positions that are required. They are often caught with an improper or lack of foundation and likely to become injured. If a yoga teacher chooses to create complex asana sequences, it is a primary responsibility to guide students skillfully and precisely to safely establish their foundations. It is best to design vinyasas in sets of postures that use the same hip orientation. The closer the outer form matches the inner function, the safer and more therapeutic are the asana.

> Alignment is the intention and action engaged, not whether an external goal or outer appearance is reached

5 Principles of Alignment and Integration

What is the best way to bring alignment to your asana practice? Can reviewing pictures of yoga poses in books, blogs, or journals be enough? Can instructions that were once heard in a class or workshop be applied? Are they credible? What will honor your personal skill level and physical condition? For a system of alignment to be appropriate, it must be based on reliable principles. It should consistently increase the skill and refinement of asana practice. It must enable students to become stronger and more flexible and most importantly, remain injury-free. If the way alignment is applied is supported by human anatomy, it will consistently provide all these essential criteria.

Alignment integrates every part of the body together within a harmonious relationship that supports precision and cohesiveness that can be applied to all asana. This is called *Integrative Alignment*. It can readily transform a beginner's practice into one that demonstrates the grace and refinement of that of an advanced practitioner. Although it may at first seem daunting, the alignment principles and how to engage them is not overly complex. The journey is best taken slowly and step-by-step.

A common confusion regarding alignment is the question; "if anatomy has so much variation, how can one set of alignment principle be applied to everyone?"

It is important to know that human anatomy has so many variations to its form that it could be said that abnormality is the norm. It is also important to know that alignment is not based as much on anatomy as it is on functionality. The mechanics of muscle and joint physiology, which are referred to as the "actions" are what determines alignment. Despite a wide variety in forms, shapes, and sizes, human bodies all follow the same "owner's manual". Modifications may be necessary to accommodate some unique, anatomical variations but these adaptations rarely contradict basic body mechanics. Alignment does not change from person-to-person or pose-to-pose. Instead, it offers a reliable set of generic principles that apply to all students and importantly, every tradition and style of yoga. Integrative alignment follows the inner actions and mechanics of the human body and not compromised by the often-conflicting outer form and appearance that an asana may take.

Hatha's ultimate balancing act: Mobility vs. Stability

Yoga postures negotiate a perpetual balancing act between movement and stability. Movement requires flexibility and stability depends upon strength. Specific mechanical actions will enable either movement or stability. Integrative alignment principles guide how to correctly engage the body and prevent injury from unintended actions. The skill of yoga practice is in determining what is required from each region of the body - either movement or stability - and how to employ the correct actions.

Mobility   *Stability*

Interdependence

Each region of the body moves independently, following its own unique rules for alignment. At the same time, all regions maintain an integrated relationship with all other regions called *interdependence*.

Interdependence allows the body to be dynamic, fluid, and responsive to the ever-changing positions and stresses it must negotiate. Essentially, the body mechanics is a sophisticated system of organic cogs and wheels, pulleys and levers that influence not only large joints and muscles but also structures down to a cellular level. Relationships between body regions are not static or fixed, nor are they forced to rigidly perform a pre-determined set of actions.

Some structural examples of interdependence are obvious. Movement of the shoulder contributes to the function of the hand and wrist. Those who keyboard for many hours a day may have learned that lack of shoulder engagement results in arm fatigue and eventual tendonitis in the forearms and wrists.

Interdependence can also cause injury and may need oversight. In Dancer's Pose, **Natarajasana**, the rear hip extends. The tendency of the lower back to interdependently extend needs to be controlled.

All yoga postures demonstrate interdependence in their mechanics and all are linked to their common, originating form - **Tadasana**, the Mountain Pose.

Human architecture

The human frame follows an underlying design. This blueprint informs all asana, enabling the yoga student to establish stable foundations that provide support and load bearing. This universal blueprint is the go-to guide for asana alignment. Following the blueprint enables the body to experience long-term resiliency and longevity while resisting injury and breakdown.

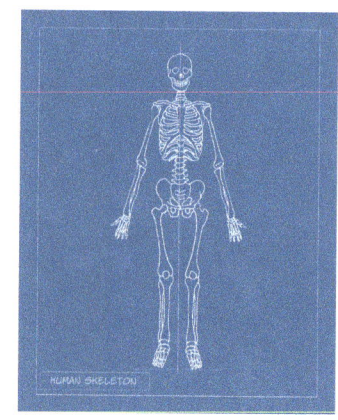

Basic postures most closely follow the body's blueprint. The simpler the posture, the easier alignment can be established without the additional challenges of balance, strength, or acrobatics. Since it is alignment that ultimately produces successful healing, yoga therapy primarily employs basic yoga postures.

If you cannot read the blueprint, the structure cannot be built safely!

"First there is a mountain, then there is no mountain, then there is." Donovan

Mountain Pose or **Tadasana** follows the basic blueprint that outlines the principles of alignment used in all other asana. **Tadasana** expresses the body and its completely integrated design in its simplest form.[1] It can be argued that there is only one asana in all of yoga and that is Mountain Pose. All other postures are simply variations of it. **Tadasana**'s basic alignment in some bodywork systems refers to it as the *neutral* posture.

As asana becomes more complex as a practice advances, the blueprint of Mountain Pose is obscured. This makes the practice of yoga one of discovery – to find **Tadasana** alignment in every posture and position.

Tadasana

The "baby bear" intention

In Astronomy the concept called the Goldilocks[3] Zone distinguishes regions in the universe that meet the criteria for life to exist on another planet or moon. They present the "just right" conditions. Yoga practice has its own baby bear, "just right" standard. Yoga students constantly face the predicament of discovering how deeply to move into an asana in order for it to be challenging and yet still remain safe. Yoga teacher Suzie Hurley of Tacoma Park, MD tells her students, "It's not how far you go; it's how you go far!"

Kofi Busia is a master yoga teacher, Sanskrit scholar and one of B.K.S. Iyengar's first Western students. His Santa Cruz, CA, yoga classes are well known for long-held poses. While challenged with a ten-minute headstand, the class may be prodded by Busia with these stirring questions, "If you have already come out of the pose, why have you come out? But, if you are still in the pose, why are you still in it?" This perplexing inquiry fills the minds of the students who are hoping to find the peaceful middle ground between over-efforting and giving up. As in the children's story *The Three Bears*, asana's journey begs the questions, *"What is too much? What is too little? What is just right?"* The art of yoga practice resides in mastering the "just right" degree of effort.

What is the "just right" effort?

Yoga postures are safest and are most therapeutic when they do not exceed the ability of the yoga practitioner to maintain a steady and stable foundation. The yogi must retain alignment in the most challenged regions of their body and keep their overall posture fully integrated. Asana practice that does not satisfy these basic guidelines becomes, unfortunately, a trauma-inducing experience. In Chapter 9, a physiological point of view is shared where the degree of effort should not exceed 10% beyond a base or stable level; this number based upon the properties of connective and its primary protein, collagen.

Initiate movement from regions of least mobility

An important principle in asana practice is to begin each asana by aligning and moving from the parts of the body that are the most challenging to engage. Postures require a coordination of movements that occur in multiple regions of the body, yet some regions move more easily than others. As a rule, engage the least mobile regions first before engaging and perhaps exploiting the regions that are more mobile. As an example, integrate the core structures before engaging the extremities.

Examples of coordination movement from least mobile regions:

- The upper thoracic spine is significantly less mobile than the shoulders. When moving the arms, initiate movement of the shoulder girdle first. Extend the less mobile upper spine first and let hand movement be the final action. This action is sometimes expressed as "leading from the heart".
- The hamstring muscles both flex the knee and extend the hip, the latter being more limited. Once the knee is flexed, hip extension can become reduced by as much as 50%. In the transition from Three-Legged Dog to Wild Dog, first rotate and stack the hips vertically, then bend the knee.

Two major considerations when attempting to move first from the least mobile regions of the body:

1. The curves of the spine
    a. The deeper the curve, the less mobile; therefore move from the deeper curves first
    b. Before moving a curve, first lengthen that curve to decompress the spine
2. Initiate movement from the core
    a. Move the regions of the body that are closer to the midline (core) first. Then, move the peripheral regions after the core has had its chance to fully participate. This prevents faster or freer moving body parts from exploiting slower ones (e.g. chest before arms)

Curves of the spine: Convex vs. concave

Convex- outward curve   Concave- inward curve

Geometric principles dictate that vertebrae on the convex side of a curve move faster and farther apart compared to the vertebrae on the concave side. In keeping with the principle of initiating movement from least mobile areas first, initiate spinal movement by first lengthening the side being bent toward, the concavity. Often in a lateral bend it is incorrectly practiced to lengthen the side becoming the top, or convexity. However, the convex side is already lengthened and initiating the side bend from this movement compresses the spine on its concave side. This can result in nerve or spinal disc damage. When using yoga as therapy for scoliosis (lateral curvature), this principle is invaluable.

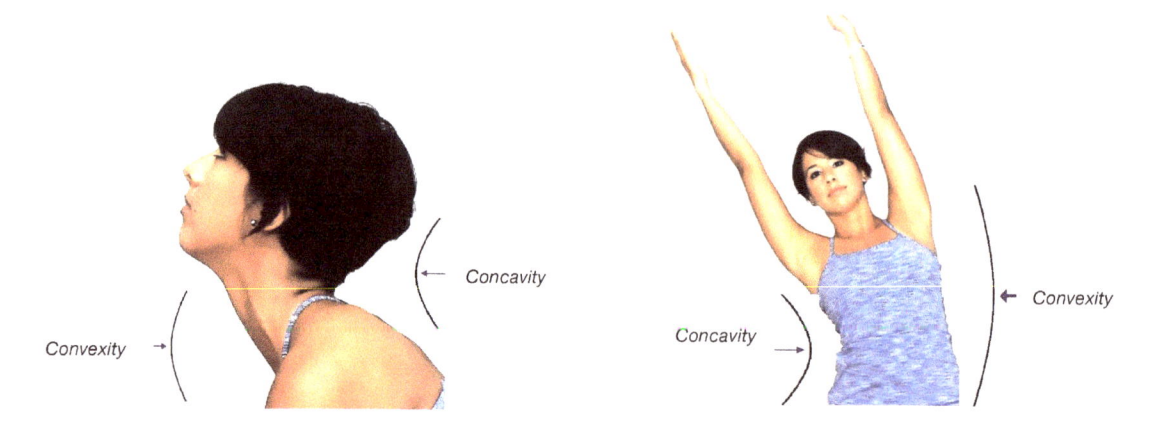

> When moving into a pose, first lengthen the concave side of the spine
> to prevent nerve and disc compression

Based upon the anatomical design of the spine, extension of the thoracic region is significantly less than in the lumbar region. For this reason, "backbend" postures should initiate from the upper thoracic spine and attain as much mobility as is possible before the lower back participates. Also, the neck is especially mobile in extension and should be the last spinal region to engage.

By habit and often by teachers' instruction, the head, neck, and especially the eyes misguidedly initiate backbend postures. The most important region to be mobilized, the thoracic spine, is often overlooked or neglected.

The periphery of the body moves faster than at the core [5]

The regions closest to the midline of the body have less mobility and move more slowly than the regions farther from the core. When initiating a yoga posture, start movement from the midline of the body. Continue to engage the rest of the body, following an outward progression. If the easier, mobile peripheral regions are exploited, injury is likely to occur. Injury is usually not immediate but results from poor mechanical habits and patterns practiced over long periods of time. To prevent injuries, focus on core engagement and build the posture from there.

In **Utthita Trikonasana,** Triangle Pose, enthusiastic students often raise their arms overhead almost immediately before the chest has rotated upward and the shoulders have stacked vertically. Triangle pose initiates by lengthening the lower side body while the chest flattens and rotates. The shoulders align vertically, one over the other. Finally, the upper arm and hand are extended straight above, staying in line with the center of the chest.

The arm should never move further and faster beyond the position of the chest or the front of the shoulder. Should the arm move posterior to the shoulder, the head of the humerus is forced anterior in the shoulder joint, which is a common misalignment and cause of shoulder injury.

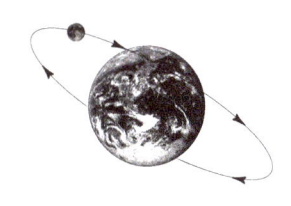

In **Urdhva Hastasana**, or Mountain Pose with Arms Overhead, first lengthen the side bodies and thoracic spine; draw the shoulders onto the back before the easier action of lifting the arms.

Parsvakonasana, Extended Side Angle, requires pure lateral flexion (side bending) of the torso. Often, students first forward flex the spine to the inside of the front knee to bring the torso toward the front bent leg. Once the easier-to-engage forward flexion has begun, it is more difficult to move the posture into lateral flexion. Lateral flexion is necessary to safely stack the shoulders vertically, revolve the chest upward, and keep the arms in line with the shoulders.

Taking a mobility inventory

When setting up for a yoga posture, the first order is to *establish a stable foundation*. Standing poses usually form their foundation with the feet. In sitting poses, it is the pelvis.

Once the foundation is stable, moving into the pose asana initiates from the least mobile regions of the body first. This is a step-by-step process. An assessment, or inventory is made as to which regions must move before the others, based upon which regions have the least amount of flexibility. Joint mobility and muscle flexibility are evaluated and "ranked" from the most limited to the most freely moving. Each region moves as fully as possible without creating strain or compromising alignment anywhere in the body. It continues from lesser to greater mobility until finally, the most freely mobile region gets to participate in the pose, as well.

At first glance, this assessment approach may seem remedial and tedious. It is, however, comparable to Yoga Nidra, a self-guided meditation that brings focus and awareness to the subtleties of the body. Personal assessment and step-by-step procedures are used with any endeavor worth mastering. The seasoned airline pilot runs through a checklist before each flight. World-class concert pianists prepare with basic scales before each performance. Over time, the taxing, mental efforts fall away and the process becomes almost unconscious.

The hierarchy of the twist

Seated twist is an excellent pose for exploring a step-by-step initiation of movement that begins with the least mobile region.

It is important to be aware that twisting postures occur exclusively in the spine and not the shoulders or the pelvis. In a typical, non-injured spine, the pelvis or shoulders form the stable foundation of the pose. The spine lengthens in its entirety to create joint space and increase ease of motion. For most yogis, the least mobile region involved in the twist will be the upper thoracic spine and movement begins from there. The lower thoracic spine is usually the next least mobile, followed by the mid-lumbar spine, then the mid-cervical region, and finally the upper cervical vertebrae. The eyes, being the most mobile body parts, are the last movement, looking toward the direction of the twist.

Unfortunately, it is common for teachers to lead students into revolved postures by instructing to first look into the direction of twist. Unless the goal is a deliberate neuromuscular exploration as used in the Feldenkrais Method® or something similar, it is not appropriate or safe body mechanics.

Revolved poses can reveal underlying imbalances in spinal mobility. Poor posture, misalignment, or flawed habits in body mechanics are often at the cause. Following the hierarchy of movement for the twist as described above can significantly increase mobility and ensure safe action.

Immobility may be the result of an anatomical variation, an abnormality in spinal design, or the result of injury. These less mobile, hypomobile regions, rather than being forced to move early in a pose, need to instead be stabilized. Attempts to move these regions often produce pain. It is important to become aware of any variations in body structure and to understand the nature and extent of all injuries. This body awareness is a required skill of yoga that brings the Raja principles of Yama to Asana into play.

Hypermobility

When a joint can easily move beyond the ranges of its design, it is *hypermobile*. If a joint moves less than its capability, it is *hypomobile*. When joint range of motion is "just right", its motion is considered to be within normal limits.

Hypermobility does not conform to standard body mechanics. Although its origin is usually genetic, hypermobility can result from injury or repetitive movements that strain tissues and lead to instability.

Hypermobile joints move farther than what is safe. They overstretch and weaken the supportive structures of a joint, particularly the ligaments. Other regions of the body that have a strong interdependent relationship with a hypermobile joint may be forced to compensate and modify their actions. These compensations sometimes create strain and injury to a seemingly unrelated and otherwise, unexpected region.

The illustration of a loose door hinge is similar to what happens to the joints of the body. Excessive, aberrant movement and misalignment will eventually damage the surrounding supportive structures.

To be effective and not a continued source of aggravation, hypermobile joints need to be identified and their instability controlled. Hypermobile joints should always be kept aligned, especially when force is applied during muscle contraction or deep stretching. Contracting (squeezing) the small muscles that surround the hypermobile joints can reduce joint play and provide important stability. When ranking the hierarchy of movement, hypermobile joints are engaged last, whenever possible.

Hypermobile joints often inflame and will eventually suffer degenerative damage. After the degenerative changes occur, hypermobility reduces and the joint ultimately becomes hypomobile. The ironic result of long-standing hypermobility is the protective mechanism of the body called *osteoarthritis*. Yoga asana practice can be an excellent therapy for arthritis as long as alignment is precise. Modifications to the poses and practice are often necessary to accommodate for the severity of the condition, such as reducing any repetitive or aggressive movements of the damaged joints.

As a general rule, initiate movement from the least mobile regions of the body. However, it will take greater effort to prevent hypermobility from dominating asana. Learning how to limit the movement of hypermobile joints is perhaps more challenging than learning how to move the less mobile ones. With hypermobile joints, stabilization is the primary focus in every posture.

Bones approximate, muscles extend [6]

"Reach, stretch, lengthen, extend!" These are popular prompts that yoga teachers use to encourage students to deepen their postures. On the surface, these instructions seem straightforward and make sense. Upon closer inspection, however, they are often contrary to body mechanics.

If a student reaches and extends their arms and legs outward, the joints separate and weight-bearing responsibilities shift off of the bones and joints to the soft tissues that surround them. Separating the joints strains tendons that are designed to anchor muscle to bone. It can tear or sprain the ligaments that regulate joint mobility. Neither of these tissues can be safely overstretched. Once overstretched, ligaments are unable to provide stabilization, a situation that quickly becomes permanent. Repetitive overextending and overstretching sets the course toward osteoarthritis and joint degeneration.

The skeleton is analogous to the wooden framework of a house; both structures designed for support and stability. As unsound it is to support a house by its walls and sheetrock, the body's soft tissues - ligaments, tendons, muscles, and fascia - are not designed to provide the amount of weight bearing needed to support joints of the human body.

Consistent with this anatomical principle, B.K.S. Iyengar instructed his students to energetically draw bones toward the midline of the body. Once the posture is securely supported on the skeleton, the muscular tension that was used to draw the bones inward is then softened and released outward, allowing the muscles to extend out toward the periphery.

The most efficient and safest way to either stretch or contract muscles is to engage from the muscle bellies and not from the tendons and ligaments. Muscles provide flexibility or power. Bones provide stability. The tendons transfer flexibility or power between muscle and bone. The ligaments assist the joints in whichever is needed - flexibility or stability. Correct engagement of all the structures is essential to avoid injury. (See Chapter 8 for more details on ligaments)

Drawing the bones together at their joints and toward the core of the body is effective rehabilitation for inflamed tendons (tendonitis). Hugging the muscles onto their bones (in whatever way seems possible) reduces strain on both the muscles and the tendons. The drawing and hugging actions increase muscle efficiency and flexibility and safeguards during resistance training or heavy lifting activities.

Para-physiological space

The natural-occurring gap between the bones of synovial joints called the *para-physiological space*. It allows for joint lubrication and reduces friction when a joint is in motion. This micro-space provides the skeleton with pressurized fluid cushioning that helps support body weight and reduces compression.

Confusion may arise with this concept and the previous instruction to draw bones toward their joints. Drawing the bones inward stabilizes the joint and reduces strain on the tendons and ligament tissue. However, excessive, forceful drawing together of the bones, especially when attempting to mobilize the joint, can produce unwanted joint compression. For example: when holding an egg, the fingers clasp around it to keep it secure but not so tight as to crush it.

Maintaining the micro-space in the shoulders and the hips is especially important to avoid damage to the cartilage collars or gaskets that surround the joints known as the *labrum*. Space for healthy joint function is naturally present but yogis should avoid overexerting during the drawing in phase. Once the mobility phase begins, muscle contraction around the joints can soften and the muscles can begin to stretch, extending from the center of their fleshy bellies and outward toward the periphery.

Central axis of a joint

Joints are designed to function most safely and with full range of motion when centered and aligned along their central axis. When centered, they are most stable and best able to handle the demands of weight bearing. The central axis can range in size from pinpoint size, as in the hips, to between 1-2 inches, as with the shoulder joints.

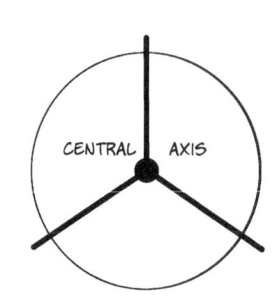

Stretching tips

- Shorter muscles have more strength and are more efficient than longer muscles. As an example, the masseter muscle of the jaw is one of the shortest and most powerful muscles in the body. When a muscle contracts and shortens, it has less flexibility.
- An injured, aged, or weakened muscle shortens to gain more power. In the process, it becomes tighter and less flexible. From a survival point of view, it is worth the tradeoff. See Chapter 9 for extensive details on stretching and muscle physiology.
- Muscles that attach to the concavity of a curve will be shorter, stronger, and tighter. Muscles along the convexity of a curve will be longer and weaker. This is observed with the lateral curvature condition know as *scoliosis*. Chapter 24 will discuss this concept further.
- For maximal flexibility, initiate a stretch from the thick, fleshy center of the muscle, referred to as the *muscle belly*. Continue the stretch by extending outward from the muscle belly toward the joints and the tendon attachments.
- Likewise, when engaging a muscle for strength, contract the muscle belly first. The muscle belly will pull on the tendon to reduce the angle of the joint. If the tendon initiates contraction, the force can tear the tendon at its boney attachment, especially if the contraction is quick or overly forceful.
- Stretching (lengthening) a muscle while it is contracting is referred to as *eccentric contraction*. At first it might seem counter-intuitive to open a joint angle while contracting the muscles. It is very safe and therapeutic and used in many forms of rehabilitation for muscle injury.

> Move from the core - Teach from the core

The quality of Samasthiti

Equal standing, Balanced stillness

From the Sanskrit:
- Sama: same, upright, straight
- Sthiti: stillness, steadiness, standing

In the Astañga Yoga tradition, *Samasthiti* is used interchangeably with the term **Tadasana**, or Mountain pose. Another way to view Samasthiti is as a characteristic or quality expressed in a posture. Samasthiti is achieved when all tension in the musculature is equalized and balanced. Samasthiti generates a sense of equilibrium. Equal length, equal tension, balanced weight, and postural centering are established along every body surface and in every direction that can be engaged. Every muscle and tissue is balanced with equal tone in every direction, right down to the outer edges of each nail bed.

The quality of Samasthiti is a valued goal in every asana. It is especially important when yoga asana is used to therapeutically address scoliosis.

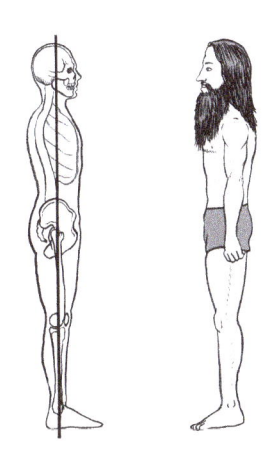

Alignment through the Coronal Central Axis

Samasthiti – beyond balance

Although we like to imagine human structure as balanced, its asymmetry and disproportions are numerous. Right or left-handedness alone creates imbalance throughout the body. Applying Samasthiti wherever these imbalances can be found will greatly improve the safety and performance of asana.

A few examples:

- The muscles of the chest and anterior shoulders are approximately 30% stronger than the muscles of the upper back and posterior of the shoulders. The shoulders are also almost 30% more flexible in movements toward the front of the body as compared to movements toward the back. Lifestyle activities that are predominantly forward facing reinforce this imbalance. To create Samasthiti, extend (flatten) the thoracic spine when moving the shoulders to access greater strength from the upper back muscles more fully.

- The calf muscles are 30-40% stronger than the thin musculature of the front shin. Pointing the toes increases calf strength. Pressing inferior through the heels strengthens the front shin muscles. Practice Samasthiti by engaging "neutral feet", which presses through the inner heels. Approach all postures in this way whether they are weight bearing or non-weight bearing.

- A knowledgeable weightlifter does not do sets of chest presses for the chest's pectoral muscles without balancing his routine with Rows and Pull-downs for back muscles. This too, is Samasthiti.

Practice like a Bear or Monkey?

In humans, the chest and anterior shoulder girdle exhibits 30% more strength and flexibility than the upper back and shoulders,. This is consistent with our front-of-body lifestyle and making our actions "bear hug-like". In contrast, a monkey swinging through the treetops keeps its shoulders on its back body and utilizes the strength of its upper back musculature, particularly its powerful the latissimus dorsi while the rhomboids are only marginally developed. To find Samasthiti, consider practicing in a more monkey-like than bear-like fashion by engaging the upper back more fully.

Moving with Samasthiti

Samasthiti engaged while moving attempts to move from every direction in as balanced and integrated a fashion, as possible. Movement is careful, controlled, and precise, performed as if carrying a full pot of boiling water safely across a room, not wanting to tilt or wobble. Imagine the focus needed to control balance when setting up into **Vrksasana**, The Tree Pose.

For example: when first lifting the arm, the shoulder is drawn back. The outer part of the shoulder moves farther and faster than the inner part. This can actually force the humeral head to shift forward and externally rotate. To move the shoulder correctly, draw the inner head of the arm bone back from its inner armpit surface with equal effort to what is applied to the outer surface. This expression of Samasthiti keeps the shoulder ligaments lax to enable safer movement and greater range of motion.

Examples where Samasthiti is not applied in an overall sense:

- A yogini kicks up into handstand, using her more agile and stronger leg but never practices using her more challenging side
- A flexible yogi exploits his flexibility immediately as the pose initiates before he establishes a stable asana foundation
- A yogi can easily transition from pose to pose in a vinyasa but is unable to hold postures for long periods. Will he modify his practice to develop the strength to hold poses longer or only practice fast-flowing styles that let him skip over the required strength?
- A flexible yogini can open her hips in forward flexion and extension as in **Hanumanasana** (Full Closed Hip Split), but cannot externally rotate the hips to stack the knee over the ankle as in **Agnistambhasana** (Fire Logs Pose). Will she address her limitations and practice hip rotation asana or will she avoid them and continue to only develop her flexion/extension postures?

Hanumanasana

Agnistambhasana

Identifying strengths and weaknesses

Yoga can be transparent, stripping away our facades and revealing our strengths and weaknesses.

"Always put your best foot forward!" is a familiar exclamation and a proclamation rarely questioned. It seems natural to take advantage of personal strengths to be successful. This way of thinking, however, should be reconsidered when stepping onto the mat. In fact, this initial misstep may account for many injuries occurring in both yoga and sports. It may be the reason an athlete fails to reach their potential. From a yogic point of view, dependence on our personal strengths does not encourage long-term balance and integration. Instead, it exploits our existing skills and achievements and does not allow our less developed capabilities to emerge and improve.

Yoga is best practiced by humbly identifying the nature of our weaknesses, limitations, and challenges. If we initiate our poses from the more recalcitrant areas and enable them to develop unhindered, our overall yoga practice will improve greatly. Samasthiti can be re-established in the body.

If we can let go of the urge to reach for the highest bar at any cost, identify our limitations without judgment and focus solely on their resolution, our personal journey becomes one of grace.

As with a traveling train, it is not the fastest car but the slowest that determines the quality and pace of the trip. Increasing the power or efficiency of the engine cannot overcome the inherent restrictions caused by a rear car that has a rusted axle or broken bearings.

Heyam Dukham Anagatam

"The pains that are yet to come can be, and are to be, prevented"

This Sanskrit phrase is one of the *Yoga Sutras by Patañjali*, codified almost 1700 years ago. This axiom advises that asana practice should not be the cause trauma; instead, it is protective and preventative for injuries. This edict and the guiding principles of Yama and Niyama (described in Chapter 3) set the ground rules for the safe and integrated practice of yoga.

Alignment is not a "style" of yoga

The term "alignment-based" is often used to describe a style of yoga distinct from other approaches. However, the use of this term is misguided. Alignment is not a unique style or belongs to a specific tradition. It is fundamental to all approaches to asana practice. It does not matter how fast asana transition or how contorted they may get. All asana, in every tradition require an aligned foundation and a precisely integrated body. The alignment principles and any specific language used to instruct them are not reserved for a particular type of practice but apply to every form a yoga posture takes.

Students have many reasons for practicing yoga. Some wish to improve their flexibility. Others know yoga's value in increasing core strength, gracefulness, and stamina. Breath capacity and control may be another valuable goal. Increasing blood flow can improve organ function and body detoxification. Yoga is invaluable for building vitality and longevity. All of these are worthy pursuits.

Yoga asana is also an opportunity to spend an hour or two each day administering a skillful form of physical therapy for existing injuries or to prevent potential ones. To be effective, serious study and dedication to the application of alignment in asana is required.

Certainly, students should choose any style of practice that they are drawn toward. In theory, any pose in any tradition can be performed correctly. The responsibility of studentship is not only to learn and develop the proper skills for an auspicious practice but also to discern whether the instructions presented by a teacher are correct and promote alignment and safety. It needs to be clear to every student that the focus of asana practice is completely on the intention of applying correct principles and not on what the final appearance of the asana may look like.

The most basic and essential principles of alignment

If limited restraints of time allowed one to teach or apply only the most basic principles of alignment and yet receive the greatest benefit, these basic principles are clearly the most critical alignment directives. They offer immediate refinement and safety to asana and will make the beginner and most basic practices look skillful and refined. All principles in this list are discussed in detail throughout the book.

1. The chest is on the front body and the shoulders on the back. The chest expands forward and broadens while the shoulders remain on the back. The lateral borders of the shoulder blades, found at the outer back of the armpits, glide inward to bring the shoulder blades toward the spine.

2. The shoulders align with each other. Hips align with each other. Either shoulder or hip does not lift higher or drift forward or back in relation to the other.

3. The arms align with the shoulders. Legs align with the hips. The arms attach to the shoulder blades on the back body. They stay aligned with the shoulder blades and do not drift forward inline with the chest. The legs align with the hip sockets that are located posterior on the pelvis. The legs draw back to align with the hip sockets and not forward toward the pubic bones.

4. Twists occur in the spine, not the shoulders or hips. Twists do not come from or exploit the highly flexible shoulders. Twists do not disrupt the hips that attempt to remain aligned.

5. Each pose has only one possible hip foundation: Open or Closed. The hips and feet face either square to the front of the mat or facing the side. All foundations of poses align with one or the other and not in between.

6. Maintain a straight and aligned Central Axis. A straight line is created through the body's midline, both front-to-back, the sagittal plane, and side-to-side, the coronal plane. The Bandhas and the diaphragms of the body align through the central axis.

Important caveats:

- When moving into or out of an asana, the body will often shift out of alignment. This corrects in the final position. Example: when binding the arms, the shoulders might initially roll forward of the chest.

- The final appearance of a posture does not always display perfect alignment. If the intention and attempt to create the posture are in the correct directions to move and align the asana, it will be safe and gracefully refined.

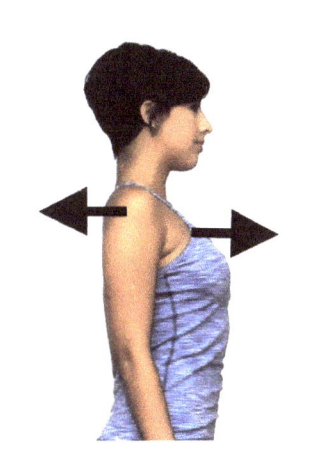

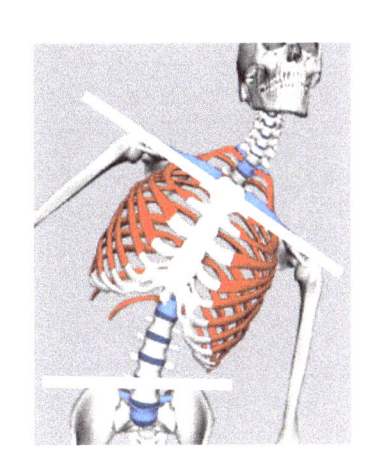

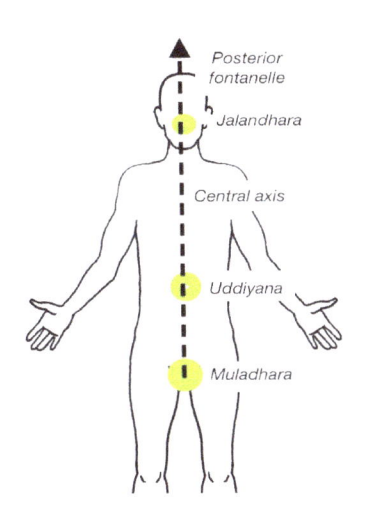

Squaring the hips

This term is used often when describing alignment in a posture, especially when the hips face forward for a closed-hip foundation. Squaring the hips occurs in three dimensions, which translates to the hips positioned in the same horizontal plane with neither being anterior or posterior to the other.

Summary of Chapter 5 principles of alignment and integration

- Integrative Alignment: establishes position of each part of the body for ideal function. Every part integrates into a coordinated, unified, singularly functioning whole
- Movement /flexibility vs. stability /strength are in constant interplay
- Interdependence: Each body region aligns independently but remains integrated with other parts
- **Tadasana** (Mountain Pose) contains of all alignment principles needed for all other postures
- The art of practice is to determine how far to push a pose and how much to try harder
- Initiate movement from regions of least mobility. Start with the least mobile regions and continue until the most mobile participate
- Twist occur in the spine; not the shoulders or hips
- Hypermobility: A joint that moves beyond its designed range of motion; easily subject to injury.
- Convex vs. concave: straight curves are more mobile but are less stable and strong. Concave, compressed curves are less flexible but muscles are more efficient and stronger. Curves that are too deep can cause compression to spinal discs and nerves. Lengthen concavity before moving it
- Periphery of the body moves faster than the core
- Move from the core to integrate movement
- Bones approximate, muscles extend
- The most efficient and safest way to stretch (or to contract for strength) is to initiate from the muscle bellies, not from the tendons and ligaments
- Para-physiological space is the natural micro-space in synovial joints; allows for joint lubrication and mobility to occur without compression
- Joints function most safely and fully when they centered and aligned along their central axis
- Shorter muscles have more strength and more efficient than longer muscles, but less flexible
- Aged muscles shorten to increase their overall efficiency
- Stretch from the thick, fleshy center (belly) of the muscle. Continue stretch by extending outward from the muscle belly toward its tendon attachments
- Eccentric contraction, stretching while contracting, is a valuable therapeutic tool
- The quality of Samasthiti: Equal length, tension and balance of weight are established along every body surface. Find this quality in every posture
- Strengthen and engage greater the upper back muscles to find Samasthiti
- Calf muscles are 30% stronger than musculature of front shin. Press through heels to strengthen front shin muscles
- A swinging monkey requires stronger upper back muscles than the bear where chest musculature has great power and strength; practice like a monkey to balance upper body
- Control the movement of a joint from every possible direction in a balanced and integrated fashion
- Heyam Dukham Anagatam: Yoga can prevent pain and injury.
- Alignment is not a "style" of yoga but fundamental to every style and tradition
- Square hips horizontally and vertically.

6 Form Follows Function

The fields of biological science, architecture, and technology uphold the 19th Century adage, "form follows function"[1]. Things *look* the way they do because of *what* they do. A primary driver of how form takes shape is nature's impulse to create efficiency. Our human form is the outcome of innate adaptive forces that strive to make the body function as efficiently and effectively, as possible.

There are countless forms of asana, all having the potential to demonstrate efficiency alignment and good postural mechanics. Poor alignment is a distortion of the body's natural design. Poor design is intuitively observed by others and signals dysfunction and compromised health. Vitality and vigor are also intuitive and instinctively visible. This is perhaps a tool for species survival. Good posture and alignment are clear and obvious expressions of health, energy, and a welcoming spirit.

Some yoga poses have acquired names that represent a form that is not typical to our human design: think Fish or Camel. It is adherence to human design and our anatomical requirements, not an asana's name that takes priority over any attempt to imitate a ritualized demand that a name may suggest.

Finely tuned by evolution, the structures of our body function in specific, predetermined ways. Some tissues are designed to stretch while others provide support. Asana practice should not force the body to operate contrary to its design. Otherwise injury is inevitable.

Other than the muscle fibers, the body's structural tissues are comprised primarily of a class of organic material called *connective tissue*. Connective tissue takes many forms; each designed to accomplish a specific set of functions. Bone, for example, is formed from connective tissue. Bone must be rigid and firm in order to provide a stable framework upon which the human body is suspended. Since this rigid frame must also be mobile, joints between the bones are necessary. To allow the joints to function, another specialized form of connective tissue, ligaments, had to develop. Details of the anatomy and physiology of the various connective tissues will be reviewed in the four chapters following this one.

Form vs. Action

The proverb, "it's what inside that counts" is appropriate to yoga practice. It is not the outer form or appearance of a posture that makes it refined or safe. Instead, asana's value is achieved by correctly applying the body's inner actions, its anatomy and physiology.

At first, it might be assumed that the outer forms and inner actions of poses are the same. For many, the very idea that there is a difference may never have been considered. The outer forms of asana are usually stylized poses passed down through a yogic tradition or sometimes just a teacher's personal design. The inner actions of a pose, however, do not waver based on style or choice. Inner actions are well-defined alignment principles that are always consistent with body mechanics.

Clearly, it is possible for most forms of asana to be performed safely. However, the farther the outer form deviates from the inner actions, the greater is the risk for injury. The closer form and action match, the more therapeutic asana remain. With that reasoning, yoga therapy is best delivered by using mostly basic asana forms.

This awareness creates a fundamental responsibility to not cause injury. Yoga teachers must shift the focus of their instructions from simply describing the outer form of postures to one that teaches from an action-based, mechanical point of view. Should yoga practice explore poses that expand or deviate beyond their required inner actions, the instructions and cueing must be precise and exacting.

This book provides guidance to both students and teachers and the tools necessary to align postures accurately. Much of the information utilizes new research from exercise physiology and biomechanics that was not developed when yoga first surged into the world of health, wellness and fitness. Although early Vedic principles contain much brilliance, the physical aspects of yoga asana are relatively new.

Stretching basics

Stretching is often thought to be the function of muscle tissue alone. Other structural tissues also play an important role. Understanding the capabilities of these different components and how to access them allows flexibility to develop carefully and effectively and without injury.

The following chart illustrates that muscle tissue stretches easiest. Muscle tissue is surrounded by myofascia, which forms a collagen-rich sleeve of connective tissue that envelops and binds every strand of muscle from the smallest fibrils to the largest fibers. Myofascia can safely stretch to only 10% of its resting length. This is considerably less than muscle tissue's ability to stretch to nearly 200%. *Myofascia is the limiting factor in stretching.* Injuries caused by overstretching that are often assumed to be the result of muscle trauma are more likely injury to the myofascia.[2]

Safe Percentages for Tissue Stretching

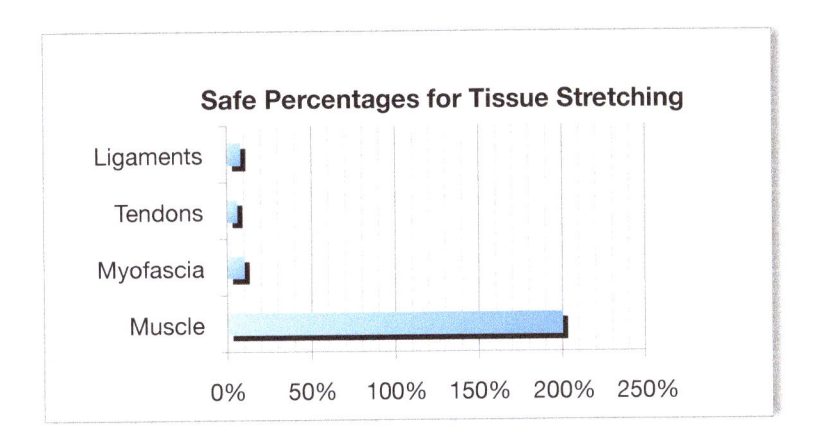

Tendons anchor muscles to bones. Comprised mostly of collagen, tendons are thick and inflexible cables that transfer the power of muscle contractions across the joints. As with myofascia, tendons are not designed to stretch. They only stretch 4% at their boney attachments and 8% where they join with muscle. In yoga practice, overstretching tendons is a common injury, referred to as tendonitis.

The design of ligaments is that they do not stretch. Ligaments provide stability to joints by binding bone to bone. They keep joints aligned and prevent aberrant movement. Ligaments must provide this stability yet also allow joints to experience their full range of motion. Additional discussion on ligament tissue physiology is found in Chapter 8.

Diaphragms

A structure is considered to be a diaphragm if, like a trampoline, it is suspended and does not maintain complete contact with the underlying structures. Diaphragms are composed of a combination of muscle and connective tissue. Cavities (spaces) are often present above or below them. Most well known is the thoracic diaphragm. It is the primary muscle of respiration and separates the two thoracic and abdominal cavities. Other tissues that are diaphragm-like are the plantar fascia, perineum, diaphragm, soft palate, the fascia of the palm of hands, as well as the eardrums.

When the body is aligned and integrated, these diaphragm-like structures do not bear significant weight and can contract with minimal tension. In yoga practice, the diaphragms correspond with the bandhas, a system of energy locks that build up Prana. It is believed that Pranic energy transmits through the diaphragms, vibrating like the auditory membrane of a stereo speaker.

The palms of the hands and the soles of the feet, with their diaphragm-like qualities, are considered to be sites at which Prana enters and exits the body. More details regarding the bandhas are presented in Chapters 14 and 29.

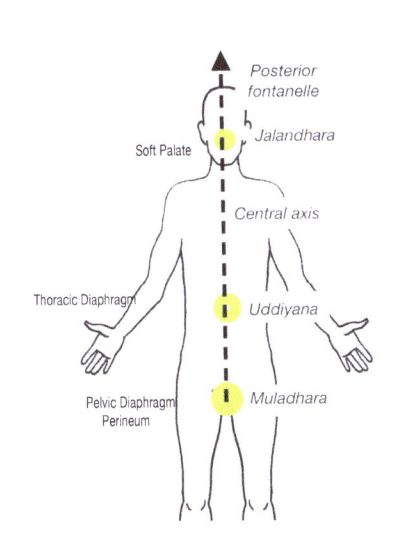

Diaphragms and Bandhas align with central axis

Along with their corresponding diaphragms, three major bandhas are located along the body's central axis. When the diaphragms muscularly engaged, their associated bandhas are also activated. As the diaphragms align and lift, the bandhas energetically lift, continuing energy upward that eventually exits through the posterior fontanelle, located at the superior/posterior part of the skull.

- Pelvic floor (perineum)   Mula Bandha
- Thoracic diaphragm   Uddiyana bandha
- Soft palate (roof of mouth) Jalandhara bandha

Spinal curves basics

Mountain Pose is considered to be the body's *neutral* position and default posture. The amount of curve that the spine forms in the neutral position results from both genetics and attempts by the body to adapt to the stresses of life, habits and injuries. Spinal curves are not static. They constantly shift to adapt to changing movements and positions of the body. Changes in the depth of the curves directly affect the balance between mobility and stability in the spine.

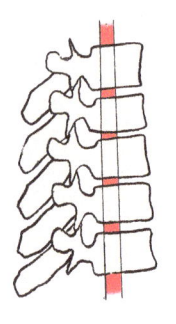

Flat curve

Greater mobility - less stable - less supportive

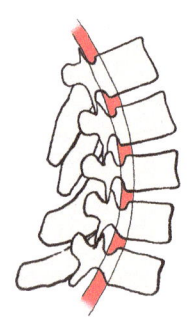

Deep curve

Less mobile - more stable – more compressed

Some people naturally have straighter spines that have flatter curves. A straight spine increases flexibility but is unstable and often hypermobile. To compensate for the instability, an opposing curve elsewhere in the spine may deepen. A common example is the lumbar curve deepening to offset the effect of a flat, upper thoracic curve. In backbend postures, this compensation can unwittingly cause lumbar disc compression injury.

Some students who have a flat lumbar curve believe that they may have a stiff and inflexible spine. However, when the lumbar spine is flat, it is actually hypermobile. This confusion can be settled when, in the classic Cat-Cow sequence, the students are able to easily modify their lumbar curves from flat to deep. The stiffness that they experience is not caused by lack of spinal mobility but the result of muscle weakness and tension that typically occurs when a curve is flat.

Whiplash

A forceful injury that whips the head forward and back can cause the cervical curve to flatten or reverse. A flat or reversed curve is unstable and hypermobile. Severe muscle spasms that accompany whiplash injuries are the body's attempt to brace the neck, which is incapable of supporting the head without help. Hypermobility in the neck increases the risk of debilitating damage to the cervical spinal discs and nerves.

Confusion about curve function

When a fishing pole bends, its distant end comes in closer to reach. At first thought, the pole seems to be more flexible when rounded. What actually happens, however, is that as the pole bends, it becomes stiffer and more stable with a build up of *potential energy*, enabling it to better resist the pull of a fish. In contrast, when the pole is straight, it is flexible, unstable, and wobbly as it releases its *kinetic energy*.

In understanding spinal mechanics, similar confusion arises about curves. In a forward bend posture, a rounded, upper back might allow students to reach forward more readily. However, rounding the upper back limits thoracic flexibility rather than increasing it. When the lower back begins to round, the lumbar curve straightens and supplies mobility- but at the cost of the ligaments being overstretched and unstable. If the lumbar spine is rounded on a regular basis, its vertebrae become hypermobile and its supportive muscles stretch, weaken and spasm. The process causes the muscles, spinal curves, and alignment throughout the rest of the spine to compensate. Over time, the lumbar spine becomes immobile as structural degeneration sets in, attempting to heal the damaged area.

Aging changes the spinal curves. After many years of life, some people appear to shrink. This indicates that the curves are deepening. It is often accompanied by the collapse and fracture of vertebrae. Deepening its curvature is the body's attempt to stabilize a weak and deteriorating vertebral structure. It is the trade-off made to protect the delicate nerves the spine protects.

Habitually rounding the upper spine forward weakens the muscles of the upper back. Strengthening the upper back muscle groups along with and extension postures are vital for rehab of the conditions of an aged spine.

There are some types of arthritis, such as ankylosing spondylitis, and other age-related conditions that cause the spine to become rigid and immobile. These may cause the spine to flatten instead of becoming deeply curved.

For these and most other types of structural change, yoga is a valuable rehabilitation aid. Aged students benefit from all alignment efforts and undulating poses, even if they seem frustratingly ineffective. Cat-Cow and **Astangasana** (Knees-Chest-Chin Pose) are two examples of poses that gently support the spinal curves.

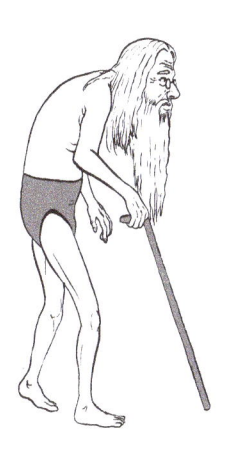

Details on the spinal aging process are available in Chapter 24 and 25.

Managing spinal curves in asana practice

Most human activities take place on the front of the body. The predicament is that overtime this can create structural imbalances. One common result of a front-body lifestyle is the rounding forward of the upper spine and shoulders. Correctly aligned yoga postures discourage spinal rounding. Instead, they lengthen the spine and draw the shoulders onto the back body.

The first directive when moving the body, once a stable foundation is set, is to lengthen the spine. The longitudinal stretch of the spine straightens the curves and increases the space between the joints, creating greater potential for mobility. Lengthening the spine continues during movement. Once the desired position of a pose has been reached, lengthening ceases and the spinal curves deepen to return to their neutral position. In postures that involve weight bearing, the curves might continue to deepen beyond neutral.

If the spine is going to carry a significant load in a weight-bearing position, the depths of the curves are increased to provide spinal stability. If the spinal curves do not deepen but remain lengthened and flat during heavy lifting, injury can easily occur. More details on spinal mechanics and safety in movement are presented in Chapter 24, 25, and 28.

When the thoracic spine rounds, its mobility becomes considerably limited. This position rolls the shoulders forward, forcing them to also round. This limits the shoulder's ranges of motion, increasing the likelihood of injury. Some students mistakenly assume that they have limited shoulder mobility when the actual issue originates in the upper thoracic spine.

When the shoulders and upper back round forward, the chest collapses and breathing become impaired. The foundation for the head and neck that is created by the shoulder girdle is lost and the bandhas and diaphragms are unable to align along the central axis of the body.

Paschimottanasana   Sitting Forward Fold

In all forward folding asana, the primary action is hip joint flexion. It is not shoulders and upper back flexion. Initially in Forward Fold, the entire torso maintains the alignment of Mountain Pose. The spine actively lengthens. The lower back forms a curve the size of a small egg. As hip flexion begins, the upper back lengthens and lifts as the thoracic curve flattens. The chest presses forward while the shoulders remain on the back. After the hips have fully flexed, the torso releases toward the thighs with this movement leading from the chest. Once reaching the maximum of forward movement that is possible with a flat upper back and a curved lumbar spine, the pose can release into its final form by allowing all muscular tension in the back to surrender and allow the curves to round. With arthritis, it is best to skip the final position of the pose and maintain some degree of curvature.

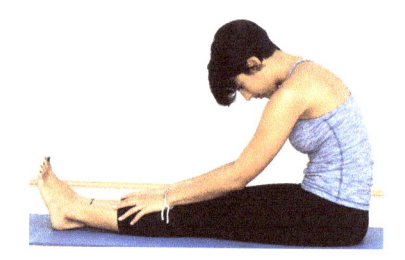

Incorrect rounding of the upper back and shoulders

Limited flexibility of the hamstring muscles is a common challenge in this pose. Tight hamstrings prevent flexion of the pelvis, which interferes with full expression of the posture. It is best to limit the degree of fold or to bend the knees as not to force the pose beyond the constraints of the hamstring muscles.

Correct position for the upper back and shoulders

7 Anatomy and Physiology Connective Tissue

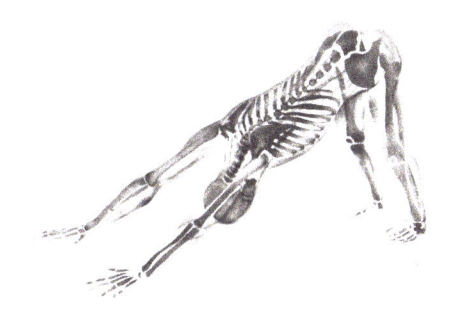

Know your medium

Every artisan or designer involved in creative processes compiles extensive knowledge of the materials and instruments used in their work. Material tolerances and compatibilities must be well understood to change the formless into expressions of organic beauty. This is true in the art of yoga. How well yoga students understand anatomy and physiology directly impacts on their ability to create a masterful asana practice. Knowledge brings wisdom and the ability to recognize the intelligence of the body.

The medium most impacted by asana practice is connective tissue. Connective tissue is the type of material used to construct most of the mechanical and structural components of the body.

We begin life formless as an aggregate of embryonic connective tissue. The structures of our body emerge from a thick, primordial soup, much like a sci-fi creature appearing out of a dark, slimy swamp. Throughout our development and entire life, connective tissue surrounds and sticks to every structure of the body. It covers every blood vessel and organ, bone and muscle, forming a contiguous, head-to-toe cocoon. Connective tissue interweaves muscle fibers with bone so completely that a severe muscle contraction often fractures the bone before it pulls the tendon free.

Ligaments and tendons, cartilage and bone, myofascia and fascia, and skin are all connective tissue. Muscle is a distinct, separate class of tissue and not considered connective tissue although connective tissue components are completely woven into its general structure. [1]

Components of connective tissue

Connective tissue is composed of three primary elements: cells, fibers, and ground substance. The percentage of each of these elements in the tissue determines what type it is and how it functions. Minerals infiltrate into most connective tissue, creating distinct qualities and changing the tissue into unique structures, such as cartilage or bone.

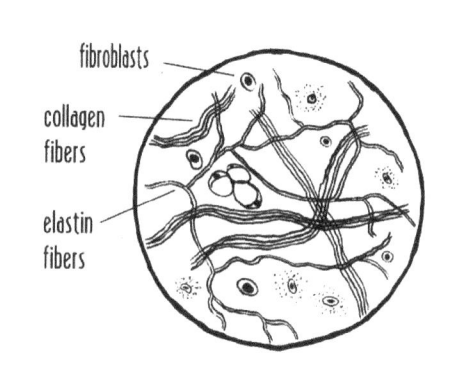

Ground substance
- Water
- Sugars
- Proteins

Cells
- Fibroblasts
- Immune cells
- Fat cells

Fibers
- Collagen fibers [2]
    - Type 1   Tendons
    - Type 2   Cartilage
    - Type 3   Bone
    - Type 4   Muscle lamina/ cell membrane
- Elastin fibers
- Reticular fibers

Ground substance

Ground substance is a watery, glue-like gel that makes up the base structure of all connective tissue. It acts as a diffuse, sticky web that supports the other connective tissue components. It is a viscous suspension made up of mostly water and protein-sugar compounds called proteoglycans and glycosaminoglycans. Ground substance is 85% water in infants and reduces to 70% in adults. It continues its water content loss with aging. Connective tissue has a limited blood supply and depends upon the watery, viscous nature of ground substance to serve as a primitive circulatory and nervous system. Ground substance also serves as a nutrient transport system and promotes electrochemical communication throughout the tissues of the body. [3]

The myofascial sheathing that surrounds muscle fibers is comprised of connective tissue. Ground substance abundantly coats the inner surfaces of myofascia to provide lubrication for moving muscle. The greater the amount of ground substance in connective tissue, the more flexible it becomes.

Fibers of connective tissue

Collagen fibers

Collagen is the most abundant protein in the body. It exists in very high concentrations in connective tissue. Collagen protein fibers provide great tensile strength but are resistant to stretching. Collagen is made up of the amino acids proline and glycine. The proteins string together to form long chains that wrap into a triple helix shape. Collagen fibers are rough, with small, unfettered strands. Hook-like projections reach out to surrounding fibers and cross-link with them to inhibit motion. This design specifically allows them to resist stretching. This design contrasts with muscle fiber's cross-linking mechanism that is designed with smoother strains that propel the fibers to promote motion, not limit it.

To be powerful, cable-like, and tensile (resistant to stretching), the average collagen fiber can stretch only 10% beyond its resting length before it ruptures or tears. Collagen also becomes damaged if repetitively compressed, crimped, or bent. Once stretched, changes in collagen length are permanent. Wrinkling of our skin is an example of permanent collagen stretching.

Many collagen-based tissues require some flexibility. To accomplish this, tissues such as ligaments strategically wrap around joints and *micro-pleat,* a process of folding and unfolding at predesigned locations. These design features allow a "pseudo-stretching" that does not overpower the collagen proteins. More details on ligament physiology are presented in Chapter 8.

There are four major types of collagen although at least sixteen are known to exist. Few can stretch beyond 10% of their resting length. One obscure type of collagen, however, is significantly more elastic than the others. If a person inherits a high percentage of this type of collagen, they will have more flexibility than the norm and can perhaps exhibit contortionist-type abilities.[4]

Elastin fibers

Elastin is a recoiling, spring-like protein that comprises elastic fibers. Whereas collagen proteins enable connective tissue to be rigid and inflexible, elastin permits connective tissue structures to be elastic. Elastin, or elasticin (post-puberty) is found in structures such as skin, lymph and blood vessels, the heart, lungs, intestines, tendons, and ligaments.

Elastin fibers can elongate to 1.5 times their length and return to their original size and shape and up to 200% before rupturing. Elastin fibers are smooth with a double helix configuration. Muscle is formed mostly from elastin fibers, making it distinctively different from its connective tissue neighbors.[5]

Reticular fibers

Reticular fibers are thin, delicately woven strands that form the soft meshwork that supports lymph tissue and bone marrow. Instead of being classified separately, they are sometimes categorized along with collagen Type 3 fibers.

Fascia

Derived from the Latin term for *band* or *bandage,* fascia is a tough membrane made of densely packed collagen fibers. It is the underlying framework and support system for the body. Fascia sheaths encase and delineate the borders around every structure within the body, from the smallest vessels to the larger organs and bones. Fascia's enveloping design allows it to transmit and distribute nerve impulses and circulate vital nutrients to local tissues.

Fascia

Collagen fibers in fascia generally organize in a longitudinal direction, running parallel to the structures they cover. In this layout, they can effectively provide structural support to the tissues they surround. The collagen composition of fascia limits its ability to stretch beyond 10% of its resting length.

In some locations, fascia runs horizontal or perpendicular to surrounding structures, such as in the thoracic diaphragm, the pelvic floor, the soft palate, the palms and the soles of the feet. In these locations, fascia is a more independent structure and can function with a trampoline-like springiness.

Still, fascia has too much flexibility to provide significant weight bearing. It provides only a modest degree of support.[6] As an example: plantar fascia supports the foot, allowing it to vault and forms its arches. If the bones of the foot spread apart or if its intrinsic musculature of the foot weakens, the fascia cannot support the required amount of weight bearing. The fascia overstretches and the arches collapse. When fascia is forced to provide an excessive amount of structural support, it can become swollen, inflamed, or torn; what is called *plantar fasciitis*.

Myofascia

As a subgroup of fascia, *myofascia* is specifically associated with muscle tissue. It comprises 30% of the overall mass of muscle. The smallest muscle filaments, called *fasciculi*, to the largest bundles are enveloped in myofascia. Myofascia binds muscle fibers together. It spreads the impulses of muscle contraction and reduces friction that develops between muscle fibers.

In the center of the muscle, the muscle belly, myofascial fibers are loose and randomly arranged. At the ends of the muscle, they get progressively denser, more tightly packed, and more parallel in their alignment. At the tapered ends of the muscles, the multi-layered sheaths of myofascia conjoin into a seamless tendon.

When an entire muscle stretches, the limit to its length is controlled by the high collagen content in its layers of myofascia. Research shows that collagen protein creates more than 40% of the overall resistance to stretching.[7]

In weight lifting and other resistance training programs, when muscle fibers enlarge, they become compressed against the myofascial sheaths and their flexibility reduces. To rectify the "muscle bound" phenomenon, a dedicated stretching program is a necessary tool for the athlete.

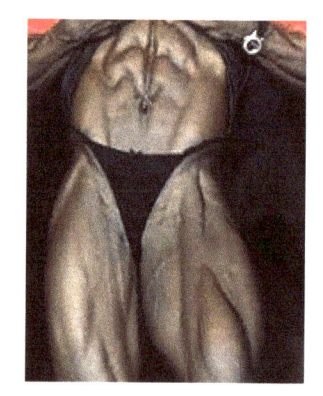

Tendons

As muscle and myofascia taper toward the bone to where they attach, the individual myofascial sheaths that cover each muscle fiber bind together to form a common cord, or tendon. Tendons are thick, fibrous and cable-like. The high percentage of collagen fibers in tendons provides great tensile strength, or resistance to stretching. Their role is to anchor muscles to bones and to transfer the power of contraction across the joint. The muscle contracts and pulls on the tendon that hoists the bone.

Tendons are not designed to contract with the muscle. The limited elasticity of tendons enables them to efficiently convert muscular energy into mechanical action on the joint. Tendons stretch in a limited range from 8% where the tendon originates from the muscle to only 4% where it attaches to the bone.

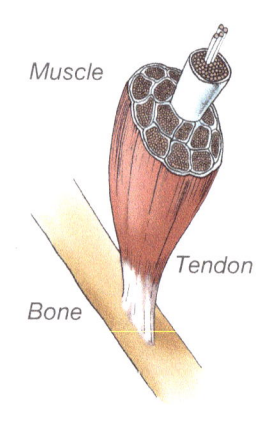

Tendonitis

Tendons are generally not designed to stretch beyond their limited tolerance. A small numbers of muscle fibers, however, migrate from their bellies into the tendon's tapered ends, providing enough contractile tissue to give the tendons the marginal ability to stretch and contract. Overuse can easily result in tendon strain. If overstretched, the myofascial sleeves covering the tendon become torn and inflamed, a common condition known as *tendonitis*.

Normally, the muscle fibers that taper into the tendon sheathing move freely, similar to a plastic straw encased loosely inside a paper wrapping. If tendon sheathings that are inflamed or stretched, they can tear like paper wrapping on a straw after it has become wet and matted.

Prolonged or repetitive inflammation or extensive tearing of a tendon will cause it to become permanently overstretched and loose. Besides pain being produce with use, the tendon's muscle loses efficiency and is less able to transfer power across the joint. Even with a minor tendonitis, the healed tendon never fully returns to pre-injury effectiveness. Baseball pitchers often develop tendon laxity in their throwing arm, motivating some athletes to receive a surgical procedure that shortens the tendon to restore its tensile strength.

A yoga therapy for tendonitis is use a strap and splint across an injured tendon. This simulates a closer-in muscle attachment and reduces the mechanical pull and tension on the inflamed or torn tendon.

Cryotherapy, the use of ice, reduces inflammation. It is the first step in protecting the tendon from damage. Because micro tears recur during rehabilitation, ice therapy can continue well beyond the first few days.

Bursas

There can be many muscle tendons crossing a joint in close relation to each other, exerting powerful forces in multiple directions. Bursas form out of an out-pocketing of myofascia and filled with a serum-like fluid. Bursas act as spacers between the individual tendons to reduce the friction that can build up between the bones and tendons. There are approximately one hundred and fifty bursas throughout the body in high-friction, mechanically stressed locations. Bursas can become inflamed or damaged by excessive strain and joint misalignment. Bursitis is an extremely painful condition. Movement of the involved joint exacerbates the pain.

Bursitis responds to ice therapy, which reduces inflammation. Joint alignment is also very important. When precisely aligned, the joint's musculature is balanced across the strained area, allowing the irritated bursa to calm and eventually recover.

The principles of *integrative alignment* presented in this book are excellent tools for the yoga student suffering from bursitis. Often, introducing slight adjustments to the alignment of an inflamed joint will reduce pain and improve mobility. As with all yoga therapy, paying careful attention to pain and the actions that reduce it provides the best guidance.

8 Anatomy and Physiology Ligaments

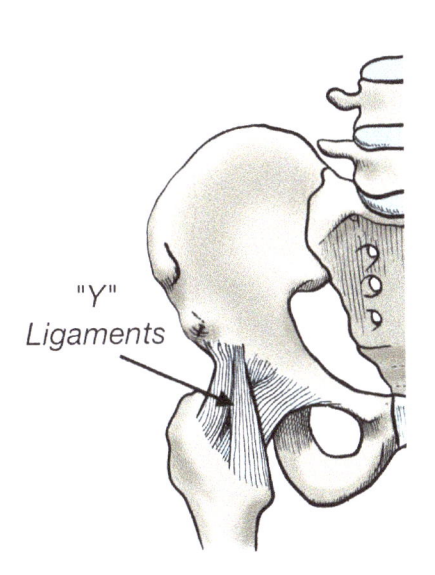

"Y" Ligaments

Compared with the animated images that often portray the skeleton, our actual, bare-to-the-bone frame would appear as a random pile of bones. When re-assembled, the bones fit together into joints that enable the skeleton to transition between motion and stillness. To coordinate both flexibility and stability, a complex ligament system has evolved that allows joints to rapidly adjust to these changes.

The structural ligaments of the body are thick fibrous straps that bridge from bone-to-bone across the joints. They are positioned to maintain ideal joint alignment and allow full and safe ranges of motion. Made up mostly of collagen fibers, the average ligament can stretch to approximately 8% of its resting length. As described in Chapter 7, collagen provides ligaments with excellent resistance to stretching but it is quite unforgiving when crimped or improperly bent.

Micro-pleating action of the ligaments

To provide either flexibility or stability without damaging its non-elastic, collagen-rich tissue, a network of extremely small folds, or *micro-pleat*s is built into each ligament. In a design similar to an accordion-type window blind, the micro-pleats enable ligaments to shorten or lengthen safely by folding or unfolding at pre-determined locations.

Pleated window shade

Wrapping action

Along with micro-pleating, ligaments also wrap around the joints. They wind and unwind in concurrence with the micro-pleating action. When the ligaments wrap, they torque around the joint and become taut. When the ligaments unwrap, they become loose and lax.

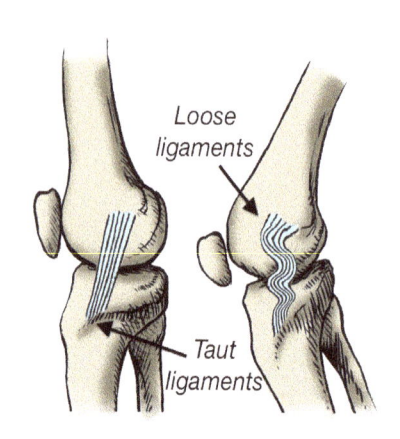

How ligament action works

When ligaments pleat (micro-fold) and unwrap, they loosen, unbinding the joint to allow for movement. Ligaments loosen when a joint moves in any of these three directions: *internal rotation, flexion, and adduction*. The more that each of these individual directions can be engaged, the freer the joint will be for flexibility and movement.

When ligaments un-pleat (unfold), they are stretched and wrapped around a joint. They tighten and become taut, which produces stability. Un-pleating and wrapping occurs when joints move in any combination of *external rotation, extension, and abduction*.

In many situations, the movements that engage the ligaments are actual, physical, concentric actions. Other times, they are isometric or eccentric and sometimes only an "energetic" intention.

This chart outlines the directions of joint movement that either loosen or tighten ligaments. It follows the function of ligaments and illustrates how our anatomy directly influences yoga practice. These concepts may be confusing at first so it is invaluable to review them often!

Ligaments Loosen
- Flexion
- Internal rotation
- Adduction

Ligaments Tighten
- Extension
- External rotation
- Abduction

Taking the heat – Friction

Friction is created by the collagen fibers of ligaments as they pleat and un-pleat. This produces a sizable amount of heat. Heat assists the ligaments in joint mobilization but only in low amounts. Heat is more beneficial for muscle tissue and blood circulation. The greater the collagen fiber content in a ligament, the more heat that is produced. Elastin fibers, on the other hand, produce considerably less heat than collagen fibers. For this reason, most ligaments contain more elastin than do the tendons, which can better utilize the extra heat for muscle function. Heat produces inflammation and ultimately can also damage the cartilaginous surfaces of joints. It is however, beneficial for muscles and tendons. Although the collagen-to-elastin ratio between ligaments and tendons is marginal, it provides a critical difference between joint and muscle physiology.

A little goes a long way – elastin fibers and flexibility

Small increases in the number of elastin fibers in ligaments produces a significant increase in flexibility. The proportion of collagen to elastin in a ligament is consistent with the mechanical demands of the joint it supports. Examples of ligaments with high elasticity are the cervical sections of the ligamenta flavum and ligamenta nuchae, located along the spine in the neck and upper back. The high elastin content of these ligaments increases their ability to stretch to as much as 25% of their resting length, markedly beyond the 8% stretching capacity ligaments typically display. In some four-legged animals, such as cows and dogs, these two ligaments are large and well developed in order to support the weight of their heads in a suspended position while their elastin content enables the needed wide ranges of motion.

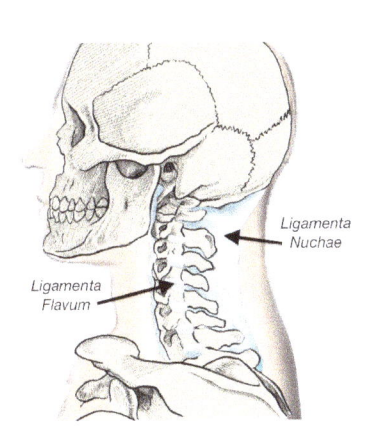

In contrast, the collateral ligaments of the knee have minimal elasticity. Their mobility relies almost exclusively on the mechanics of wrapping and micro pleating. If the knees habitually hyperextend, their ligaments become overstretched and weak. Overstretching any ligament results in permanent laxity and instability. Chronic swollen knees from injury will stretch and permanently weaken the ligaments.

Genetic predisposition also determines the ratio of elastin to collagen in ligaments. Slight differences can significantly affect flexibility. To some observers, a relationship may exist between heredity and the flexibility although no assumed affiliation can be made between race and ligament anatomy.

Two out of three is enough for ligament performance

In the full expression of **Virabhadrasana Two** (Warrior Two), the front leg flexes to 90° angles at the hip, knee, and ankle. Many students find this aspect of Warrior Two challenging, causing the muscles of the thigh to burn and the leg to fatigue. With the joints in flexion, the ligaments of the front hip and knee loosen. This makes the leg less stable and forces the quadriceps muscles to overwork. Since front leg flexion is built into the form of the pose, it cannot be changed. However, the student can engage abduction and external rotation to enable the ligaments to tighten and the joints to stabilize.

To tighten the ligaments in the front leg, align the outer buttock with the outer thigh. Track the kneecap over the small toe side of the foot. These positions produce abduction in all of the joints. External rotation can be engaged isometrically, as well. Assist both actions by contracting the musculature that runs along the outer aspects of the joints.

Virabhadrasana Two

In **Baddha Konasana**, the Bound Angle Pose, the outer appearance of the asana is contrary to how the ligaments of the hip and knee function. In the pose's final form, the hips open fully and the knees release toward the floor. These are the directions of external rotation and abduction, both causing the ligaments to tighten and restrict joint mobility.

Flexion, which loosens the hips, is fortunately in the design of the pose. To add to flexion, draw each femur bone up into its socket and deepen the anterior hip crease. Internally rotate and widen the hips from the groin. Lift the gluteal folds of the buttocks and draw them posterior and oblique. Adduction, which loosens the ligaments, can energetically be engaged by drawing the inner kneecaps toward the midline. At first, these actions may seem counter-intuitive. With practice, moving in sync with ligament mechanics will feel natural and the only way to perform hip opening postures.

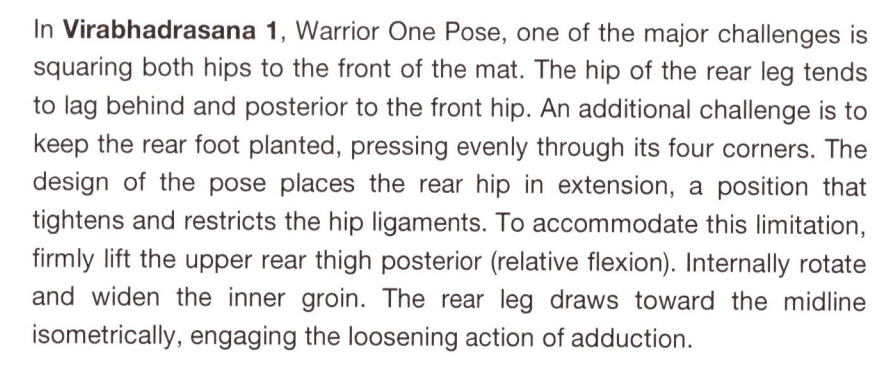

Baddha Konasana

In **Virabhadrasana 1**, Warrior One Pose, one of the major challenges is squaring both hips to the front of the mat. The hip of the rear leg tends to lag behind and posterior to the front hip. An additional challenge is to keep the rear foot planted, pressing evenly through its four corners. The design of the pose places the rear hip in extension, a position that tightens and restricts the hip ligaments. To accommodate this limitation, firmly lift the upper rear thigh posterior (relative flexion). Internally rotate and widen the inner groin. The rear leg draws toward the midline isometrically, engaging the loosening action of adduction.

To loosen the front hip ligaments and to best square the hips forward, initially internally rotate and deepen the front hip. Once fully in the pose and stability is require, the front hip and knee externally rotate and abduct as is performed in Warrior 2 and in a mostly isometric fashion.

Virabhadrasana One

Vrksasana, the Tree Pose, is a popular balancing posture. The weight-bearing leg must be solid and stable and maintain correct ligament at the hip, knee and ankle. To provide ligament stability for the straight, otherwise immobile leg, isometrically contract the muscles around the joints in the directions of external rotation and abduction. Squeeze the outer hip and knee to engage the additional, supportive musculature.

Initially, lift the front bent knee forward into flexion. The inner knee rolls internally toward the midline and the femur head draws into its socket. These are all three actions that loosen the ligaments that enable hip flexion. Once the lifted leg is placed against the inner thigh of the standing leg, stabilize the front hip by engaging abduction and external rotation. Squeeze the outer aspects of the joints in the same fashion used for the standing leg.

Vrksasana

9 Anatomy and Physiology Muscle

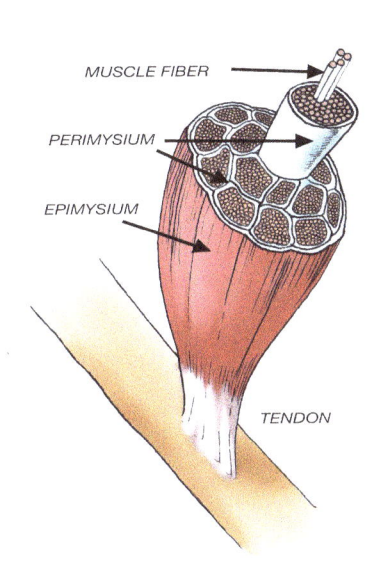

MUSCLE FIBER
PERIMYSIUM
EPIMYSIUM
TENDON

Three types of muscle tissue are identified in animals - smooth, skeletal, and cardiac. Smooth muscles are found within internal organs and large blood vessels. Their contraction and relaxation governs the diameter of blood vessels and propels food along the gastrointestinal track. Cardiac muscle is similar in function and design to smooth muscle but unique to heart tissue. Although it is not scientifically confirmed, some yogis have been known to be able to control these types of muscle. As essential as they are, these two muscle types are not directly engaged when structurally aligning yoga asana.

The muscle type that is fundamental to asana alignment is skeletal muscle. Skeletal muscle is engaged constantly in asana, fluctuating between stretching for flexibility and contraction for strength. Although muscle is not a form of connective tissue, it cannot function without its intimate connective tissue compatriot, *myofascia*.

The anatomy of skeletal muscle tissue resembles that of a thick telephone cable. Its internal structure consists of tiny strands or filaments of muscle tissue, each enveloped by sheaths of myofascia to become muscle fibers. Every fiber is banded together into small bundles. The small bundles form into increasingly larger bundles, eventually becoming the muscle itself.

The tiny filaments of muscle tissue are called fasciculi. The massive and stronger muscles have large numbers of fibers bundled together. Smaller muscles, such as those of the hand that provide delicate and fine motor movements contain fewer fibers per bundle.

Myofascia encapsulates each fiber and binds together the bundles of fibers, creating an elaborate, interwoven relationship between muscle tissue and myofascia. Myofascia that surrounds the outermost muscle fibers is called *epimysium*. The bundles closer to the bone are covered by *perimysium*.[1]

Sarcomeres and Myocytes

Derived from the Greek word for "fleshy part", *sarcomeres* are the contractile units of muscle tissue. They are protein-based and found in the cytoplasm of muscle cells (*myocytes*). Muscle cells are long and fibrous with a tubular, fusiform shape. Sarcomeres can be seen microscopically as alternating dark and light bands. Muscle cells consist of two parts: *myoplasm*, the chains of sacromeres; and the *sarcoplasm* that makes up the remainder of the muscle cell's gel-like cytoplasm.

The contractile portion of a sarcomere is composed of myofilaments made from two proteins. The protein *myosin* forms thick, dark "A" bands that lay end-to-end between two "Z" lines. *Actin*, the other protein, makes up the thinner myofilaments that form the "I" bands. Muscle contraction occurs as the proteins overlap. Small hook-shaped strands that overhang the myosin protein latch onto the actin proteins and pull to shorten the width of the cell. When muscles stretch, the strands unlatch and release the overlapping protein myofilaments, allowing the sacromeres to lengthen.

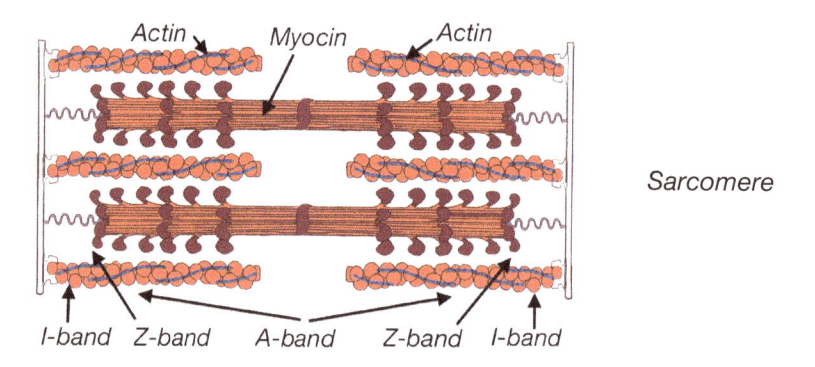

Sarcomere

Muscle fibers average from 3-30 cm in length to 10-1000 μm in diameter. Individual muscle fibers rarely extend the entire length of a muscle. Instead, they collectively form an overlapping, almost haphazard network of fibers that span the bones. Fibers usually align themselves obliquely to the line of force that a muscle exerts on a joint. This orientation provides the greatest strength and efficiency.

Regardless of size, every fiber and sarcomere function the same. When a muscle stretches, some of its fibers lengthen while others remain at rest and "go along for the ride". The final length achieved by muscle stretching depends on the number of fibers that participate. The more fibers engaged, the greater the length. The principle is the same for muscle contraction. Overall strength depends on the number of fibers that contract, what is called recruitment.[2] It is speculated that recruitment of fibers increases when an individual's focus and intention increases. This can be observed with Olympic weightlifters; their intense focus most likely enables a greater number of fibers to engage and their performance to excel.

Stretching is forever

A muscle fiber reaches its maximal stretch when all of its sarcomeres are fully elongated. The myocyte's gel-like sarcoplasm expands into the additional space created by the stretch. Muscle cell expansion is called *elastic* or *plastic elongation* and, in theory, can repeat indefinitely. If a stretch is held repetitively for a prolonged period of time, a permanent change occurs in the length of the muscle cells.[3]

Although muscle fibers have a near unlimited capacity to lengthen, it is the myofascial sheathing interwoven between every muscle cell that is the actual determinant. Since myofascia can only safely stretch to 10% of its resting length, muscle tissue cannot exceed that limitation without traumatizing the myofascia.

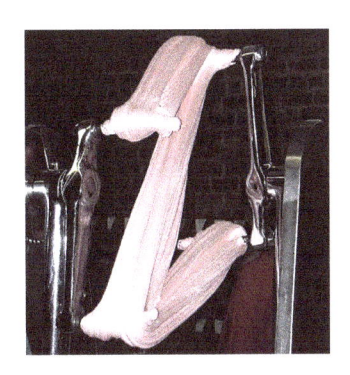

Plastic elongation is like stretching taffy: slowly, repetitively, warm

Speed, time, and heat

Increasing the length of time that a stretch is held allows the muscle tissue and myofascia to extend in the direction of the stretch and the muscle cells' sarcoplasm to expand into the space that is created. When a muscle is rapidly stretched and released, its length and shape remain unchanged. Bouncing or sporadically releasing a muscle while it is being stretched also triggers neurological reflexes that cause the muscle fibers to quickly contract, blocking any possibility of any long-term, permanent elongation.

Warming up muscles before a yoga class or any deep stretching produces beneficial amounts of heat, which is created by the muscles' collagen-rich, myofascial sheathings and the pleating/unpleating of collagen fibers in the muscle tendons.

The amount of time necessary for plastic elongation to be effective varies amongst individuals. Most students will reach maximum benefit and level off at about five minutes of constant stretch. Long-held yoga postures, such as those practiced in *Yin Yoga*, can produce permanent muscle elongation. Five minutes of holding a seated forward fold posture and repeated regularly can increase the length of the hamstring muscles.

A common strategy in the Iyengar tradition is to hold poses for shorter periods, thirty seconds to one minute, and repeated, usually three times in a row. This approach has been successful in deepening poses and muscle stretching. Which approach is best is part of a yogis' personal exploration of yoga.

Permanent tissue changes come with a caveat. Plastic elongation can decrease the overall power that a muscle can elicit. Also, since muscle elongation follows the lines of force applied, prolonged stretching practices should only be performed while engaging precise alignment. Overstretching in unintended or misaligned directions may weaken the tensile strength of the tendons and ligaments. Poorly aligned muscle stretching can also aggravate already injured muscle tissue.

Strength, force, and efficiency

Strength is considered the amount of force that a muscle can exert, regardless of the energetic cost. Efficiency takes it further; it measures the relative trade off between muscle force and the amount of energy needed to create it. The practice of Brahmacharya is essentially getting the most strength from the least amount of energy. Yoga favors efficiency over absolute strength. Think Prius over Hummer!

Percentage of Connective Tissue Stretch

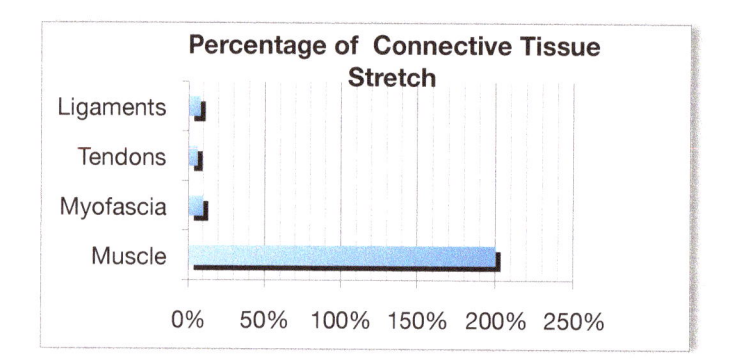

(bar chart showing: Ligaments ~5%, Tendons ~5%, Myofascia ~5%, Muscle ~200%)

Three types of muscle contraction

Muscle contractions act upon joints to allow flexion - bringing bones closer together and reducing the angle of the joint; or extension - widening bones apart and increasing the joint angle. Either action can be engaged with any of the three types of contraction.

- Concentric Contraction
    - Muscle shortens opposite direction of resistance
        - Example: contraction pulls weight closer
- Eccentric Contraction
    - Muscle lengthens in same direction as resistance
        - Example: weight moves away while contracting
- Isometric Contraction
    - Muscle contracts but remains fixed in length and position
        - Example: no movement of weight

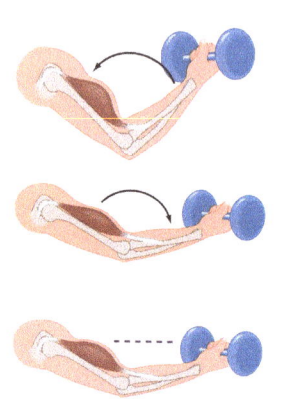

Concentric contraction is the most common and type easiest to understand. The muscle shortens as it contracts. Eccentric contraction, a muscle lengthening as it contracts, is less obvious. As an example: the iliopsoas muscle concentrically flexes the thigh to the torso. In Tree pose, the iliopsoas on the side of the standing leg will instead eccentrically contract to stabilize the hip to resist flexion.

Eccentric contraction while also attempting to stretch a muscle is sometimes referred to as *resistance stretching*. It is an invaluable rehabilitative tool for strengthening injured, weak, or ruptured muscles and used in many types of therapy. When a muscle is lengthened without any contraction, it is called *passive stretching*. This action typically occurs in restorative postures, Yin Yoga, in yoga therapy, and with some assists.

Chapter 19 will explore muscle contraction and its application in yoga alignment. Topics such as reciprocal inhibition, antagonist vs. synergists, and co-activation will be presented.

The 10% rule

Among exercise professionals and trainers, it is a commonly observed lesson that physical activities such as running, stretching, and weightlifting can result in injury when the increase in demand exceeds 10% at any one time beyond current base levels for that activity. This rule corresponds with the physiology of myofascia, for which its maximum limit of tissue expansion without injury is 10%. It should not be confused with the 20% muscle efficiency concept discussed in the following section.

Potential for muscle efficiency

Muscle contraction is most efficient when it is stretched to 20% of its resting length. Similar to the lengthening a rubber band, stretching increases a muscle's kinetic energy. Beyond the 20% "sweet spot, efficiency decreases as a result of the proteins of the sarcomere separating beyond their capacity to sufficiently overlap and provide maximum contraction.[5]

Muscles that unlock the power of another

Supraspinatus

Shorter muscles are generally more efficient than longer ones. As an example, the masseter muscle of the jaw is one of the shortest and also most efficient muscles in the body.

When beginning to move a joint from its at rest, neutral position, many larger muscles are not efficient enough to initiate the movement by themselves. Instead, they rely on short, highly efficient muscles that act as "keys" to unlock the power of the large muscles. As an example, the supraspinatus, a rotator cuff muscle, has a key-type relationship with the deltoid muscle in initiating shoulder abduction. This relationship is easily observed in someone with the common rotator cuff injury of a torn supraspinatus and is unable to abduct their arm from the side of the body. Other examples include the relationship between the anconeus muscle and triceps brachii in elbow extension; also the popliteus muscle being the key to unlock the power of the hamstring muscles in flexion and rotation of the knee. In each case, the smaller muscle initiates the movement of the larger one.

Muscles will often shorten in length and decrease in mass with aging or illness. A shortened fiber is a way to improve efficiency and compensate for a general loss of strength. As muscles shorten, flexibility reduces. This phenomenon can be observed in the shortened gait of our long-retired population, pejoratively referred to as the "senior waddle".

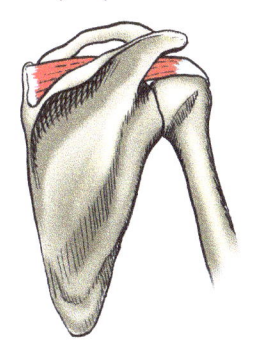

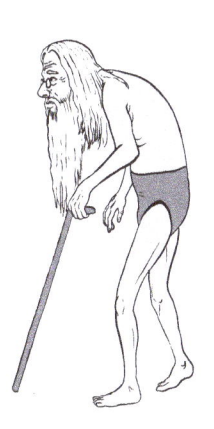

Overstretching

Stretching is not the simple act of reaching the arms overhead. A degree of skill is required in order to avoid overstretching. As presented earlier, overstretching can tear the myofascial sheathing that surrounds muscle tissue. Collagen-rich tendons and ligaments can become permanently unstable after prolonged stretching that repetitively exceeds their anatomical limits.

Habitual overstretching, especially engaged during the years of active body development, may result in hypermobile, unstable joints. This risk calls for consideration when children participate in dance and gymnastic activities. At any age or condition of health, the most skillful way to avoid overstretching is to implement the integrative alignment principles as presented in this book. The instruction to *draw bones closer together into their joints and extend muscles out from their thick bellies* is perhaps the best preventative for overstretching of tendons and ligaments (See Chapter 5 for more details).

> Draw bones together and extend muscles apart to prevent overstretching

Mixing yoga with athletics...maybe!

If muscles being short and compact enhance efficiency, the question arises whether yoga, or stretching in general, is appropriate for athletes. That specific question is being explored in competitive sports training and whether stretching causes injury. Some coaches in track and field are discouraging athletes from engaging in the static stretching; long-held stretches as are popular in yoga. They purport that static stretching reduces speed and power by lengthening muscles. Shorter muscles are more powerful, creating a shorter "piston" distance that transfers more energy across the joints.[6] These are valid physiological considerations and extensive stretching before sports activity may hinder performance.[7]

Although this may seem to warn against competitive athletes practicing yoga, there are other factors to consider. Increases in flexibility, balance, and agility obtained through yoga practice can overshadow the downside of a small loss of contractile power. Increased body awareness that yoga provides the athlete may prevent traumatic sports injuries, which are more likely the real challenges for a long successful career and ultimate performance.

Scientifically studying the effects of yoga is challenging. Current research provides no definitive answers. Typical methods of assessment, which might include controlling the minutes of practice or testing an isolated set of yoga postures, would be inadequate for well-designed experimentation.[8] Countless variables challenge a scientific study of yoga. Were the postures performed with correct alignment? What level of intensity did the subjects bring to their practices? Was yoga approached as another exercise discipline or were there other intentions? What are the physiological effects of mind-body interconnectiveness and other "spiritual" elements on the performance of asana?

Personal note: As a life-long athlete, I have found a working balance between my yoga practice and sports that allow them to be complementary. Practicing yoga for twice the time as running allows my muscles to continue to develop flexibility. A ninety-minute yoga class balances my forty-five minute run. It is a reasonable way to keep both activities safe and consistent. Yoga is best practiced at times separate from other high-demand physical activities. Running and other sports naturally cause small tears in the muscle fibers. Practicing yoga or doing extensive stretching is not recommended immediately after sports activity. And, since yoga practice itself can cause muscle micro-tears, a full practice is not recommended immediately before intense sport activity either.

Deep tendon and stretch reflex

Muscles and tendons are often subject to abrupt or accelerated stretching. At other times, they must instantaneously contract as when unexpectedly catching a heavy object. Both types of sudden change place the muscles and their tendons at risk for injury. To protect against injury, embedded in muscle and tendon tissue are special sensory nerve cells called *proprioceptors*. Proprioception is an intricate system that operates like a trip wire to trigger the spinal reflexes.

Muscle activity is triggered when sensory information is sent to the brain to determine an appropriate response. Reflexes are designed to skip the lengthy pathway to the brain and instead loop directly to the spine and back out to the body. Reflexes enable the high-speed response often critical for safety.

Located in the bellies of muscles are sensory neurons called *muscle spindle cells* that prevent overstretching. They record changes in muscle length and the velocity of a stretch, sending signals to the spine to relax a muscle before it can over-contract or overload and become damaged. The more abrupt the change in muscle length or load, the stronger is the response. This is called the stretch or *myostatic reflex*. Besides this protective feature, muscle spindle cells serve to maintain the muscle's resting state and tone.

In tendons, spinal reflexes are under the control of *Golgi tendon organs*, or *GTOs*. They are designed to sense subtle changes in muscle tension and signal muscles to neither overstretch or forcefully contract beyond safe limits. Golgi tendon organs are interwoven into every ten to twenty strands of the collagenous end-fibers of muscle tendons. When triggered by a forceful or abrupt tendon contraction, the GTOs send sensory nerve signals from the tendon directly to the spine, creating a reflex that rapidly *inhibits* muscle contraction and reduces its intensity. During locomotion, the GTOs can also excite muscle contraction rather than be an inhibitor.

In a physical examination, a health care provider uses a small rubber-tipped hammer to gently strike a patient's knee or elbow. This strike is given to the tendon and tests the reflexive firing of the stretch receptors that produce the *deep tendon reflexes*. It evaluates the path of the spinal nerves from the spinal cord to the muscle it controls. The impact of the hammer on the muscle tendon stimulates the stretch reflex to fire, which triggers a contractile response from the muscle. These innate reflexes have had a useful place in human evolution, offering protection from injuries resulting from abrupt, forceful physical stresses that regularly impact the body. The reflexes provide protection in asana practice where improperly stretching, bouncing, or changing position commonly occurs. Excessive movements may cause muscles, myofascia, tendons, or ligaments to overstretch if not being protected by the reflexes.

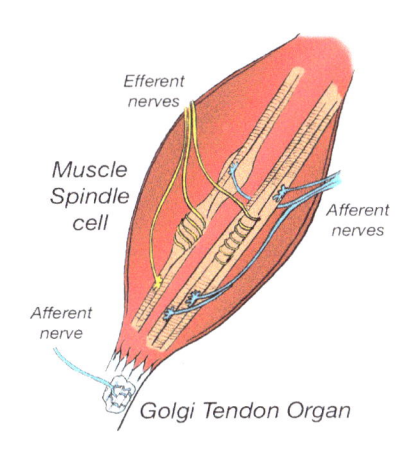

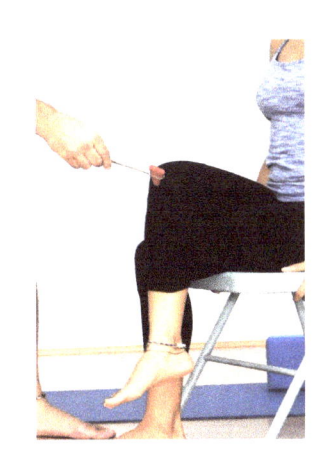

Scar tissue

Scar tissue can form when connective tissue fibers grow into an injured area. Fibers arrange mostly in a disorganized and random fashion. Scar formation is easily visible in skin tissue but also forms within other injured, connective tissue structures. Although scars are sometimes inevitable, their development can be minimized. Slow, long held stretching without bouncing encourages the fibers to align properly. Precise alignment through injured muscle better organizes the tissue and less scar tissue is produced. A mild amount of heat during healing is also beneficial in reducing scar formation.

Stretching and Habituation

When held for a long period of time, stretching can significantly change the length of collagen-rich soft tissues. Prolonged, slow stretching trains the muscle spindle cells not to react, becoming attenuated to being fired. Eventually, the alarm signals being sent to the nervous system diminish or shut off. This suppression of neuromuscular reflexes is called *habituation*. Yin and Restorative yoga practices utilize habituation effectively by taking advantage of low-velocity stretching utilized in those approaches.

To best access habituation, a muscle is stretched slowly but is stopped *before* the point there would be a contractive response to tension or pain. The position is held still until the feeling of reaction has subsided; then stretching can proceed and possibly go deeper. Incrementally, the stretched muscle safely reaches a deeper, more flexible state [9]. Although habituation can increase flexibility and have a positive rehabilitative value, its effectiveness can also be used improperly. If poorly aligned while being stretched, habituation can cause the muscle tissue to lengthen in inappropriate directions, resulting in muscular imbalance or joint instability.

Some sports and weight-training experts have disagreement regarding the effect of habituation on the Golgi tendon organs. They posit that the repetitive nature of weight lifting makes habituation conflicting for strength building. To that end, they discourage yoga and all deep stretching for their athletes. "It is theorized that overuse (of muscles) using forced repetitions on very heavy weight may teach muscles to prematurely fail. Strength training involves a neurological adaptation to motor development, contraction efficiency, as well as a morphological adaptation. Repeated use of forced repetitions on very heavy weight may prematurely activate the Golgi tendon organs." [10]

Other trainers and sport professionals disagree and contend that the Golgi tendon organs are "...nowhere near as powerful an inhibitor of muscle activation as many in the fitness industry believe." Crago et al.[1] This controversy may be of high importance to those involved in serious bodybuilding pursuits where every advantage that can be received from training is explored. It clearly poses little concern, however, for the yoga student striving to increase their general flexibility and overall strength.

Fast and slow twitch fibers

We are born with nearly equal numbers of fast-twitch and slow-twitch fibers in our muscle tissue.

Fast-twitch fibers (type-two) are large and provide speed, power and strength. They contain fewer capillaries and their cells contain fewer mitochondria, the cell's powerhouses. Because of this, fast-twitch fibers easily fatigue. Fast-twitch muscle fibers are analogous to the "white meat" of fowl.

Slow-twitch fibers (type-one) are smaller than fast-twitch fibers but have greater endurance. They can better initiate muscle contraction and able to perform well beyond the point where fast-twitch fibers fatigue. Slow-twitch muscle fibers correspond to the "dark meat" of fowl muscle tissue. Their cells are dense with mitochondria and they have a high blood flow, making them darker in color. Slow-twitch fibers receive more oxygen, produce less waste, and are more efficient. For these reasons, slow-twitch fibers are most suitable to older age.

Muscles and aging

Fiber by fiber, muscle strength and endurance does not significantly decrease with age. Once activated, the cell proteins in the sarcomeres overlap to the same degree in advanced age as they have during the time of one's peak vitality.

Muscles, however, demonstrate clear changes with age. Older muscles require a longer recovery time between episodes of exertion and tend to re-fire more slowly. This process is known as *contraction fatigue*.

The overall mass of muscle decreases with aging, a condition referred to as *sarcopenia*. More muscle fibers are lost in the lower extremities than the upper body. Between the ages of twenty and eighty, the average person experiences a 25% decrease in the overall number of muscle fibers. Fast-twitch fibers are lost in greater numbers than are the slow-twitch type. Developing a larger mass of muscle in the early years of life helps preserve the fast-twitch fibers.

The primary fibers lost with aging are motor units - fibers that directly communicate with the nervous system. Muscle loss from aging almost exclusively occurs in the fast-twitch fibers, whereas their slow-twitch counterparts undergo little change. By attrition, muscle loss increases the relative percentage of slow-twitch to fast-twitch fibers. It has also been theorized that in aged muscles, slow-twitch motor neurons may actually replace or rescue and repair the lost fast-twitch motor units.[12]

The value of yoga for aging muscles

Yoga offers countless benefits for aging. Yoga improves muscle pliability and increases strength. It also stimulates muscle's ability to self-repair, and with that, increases muscle fiber longevity.[13]

Building muscle mass is called *muscle hypertrophy*. Yoga, as with most physical activities, stimulates hypertrophy. Arm balances, standing poses, and single-leg balancing postures are recommended to be included in a daily practice to significantly increase the mass and strength of muscle. Since asana practice provides these benefits, the earlier in life a yogi builds muscle, the greater the number of fibers are at their disposal in later years. A larger muscle mass raises the baseline of muscle density before the long-term effects of aging and the loss of muscle mass inevitably set in.

To additionally offset aging, yoga helps maintain posture, flexibility, and balance. Balancing poses develop core strength and challenge agility. They keep the nervous system stimulated and muscles adaptable and responsive. Yoga is invaluable to the elderly student in its ability to increase blood circulation and improve arterial health. A gentle, flowing vinyasa series can increase cardiovascular capacity and bring greater blood flow to the muscles. The heart benefits greatly from yoga's nasal breathing techniques. Nose breathing is shown to be healthier for the heart muscle than the more explosive type of breathing typically experienced during intensive sport activities. More details on this topic are presented in Chapter 25.

In practical terms, extra muscles fibers ensure the lifting and enjoyment of more soothing cups of tea!

Swaddling and hugging muscles onto the bones

A concept shared in yoga workshops is to firmly hug muscles to the bones with which they attach. This action relieves muscles from strain, aligns individual muscle fibers, and enables efficiency. A deep sense of calm is produced when muscle contact is made with associated tissues. Greater health and healing is enabled with greater contact. This concept of hugging muscles onto the bone is apparent when wrapping and swaddling a baby and the almost immediate calm it produces.

Contact between parts of the body or with the ground can calm the nervous system. Placing a hand on the chest or belly slows and balances the breath. Using yoga props is another valuable way to attain this connection. Placing a blanket under a suspended hip in **Janu Sirsana** is an example. The tight, suspended hamstring muscles releases tension and stretches more effectively **Paschimottanasana** if a blanket supports the back of the thigh. If the head does not touch the floor in **Upavistha Konasana**, it can be supported on a yoga block or a folded blanket to enhance the calming effects to the nervous system that this pose can produce.

Props are great tools to provide contact that calm the nervous system and the muscle reflexes to allow deeper stretching. Supporting a suspended body part, such as placing a block under the forehead or a blanket under the buttocks reduces "hanging in mid-air" stress reactions on the muscles and nervous system.

Does stretching observe Brahmacharya and Ahimsa?

Stretching is fundamental to yoga practice and practitioners take for granted that stretching is always in full congruence with the basic principles of yoga. However, if we consider that muscle efficiency is an expression of Brahmacharya, then stretching a muscle beyond its point of efficiency, or 20% in length, may become a violation of this important Yama.

Likewise, practicing with the goal of reaching muscle efficiency quickly often surpasses tha 10% rule of not tearing the myofascia. This too, is a Yama violation that does not observe Ahimsa, not doing harm.

These considerations are the essence of what determine whether physical a activity is the practice of yoga or just exercise. All activities can become yoga but yoga asana can also become simply exercise if our awareness is not brought to the mat.

> It's not how far you go – but how you go far!

10 Anatomy and Physiology
Cartilage and Bone

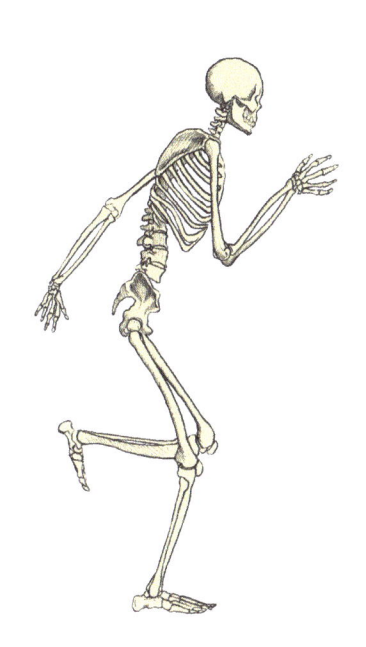

This chapter reviews one of the most easily identified parts of the human body, the skeleton. Bones and cartilage are classified as connective tissue. They form the framework on which muscles attach and internal organs are suspended. The skeleton also protects internal organs from external injury. The infant skeleton consists of over 270 soft bones that are made up of mostly cartilage. Cartilage converts to bone through a process of calcification and fusion that continues into our early twenties, leaving a tally of 206 bones in the adult skeleton. Bones are a basic component of the mechanical apparatus of the body. Yoga postures place great emphasis on the skeleton and make no bones about that!

Cartilage

Various types of cartilage can be found throughout the body. Flexible cartilage is the major constituent of the ears, nose and bronchial tubes. Fibro-cartilage is a fibrous, firm cartilage that comprises the outer rings of the intervertebral discs and menisci of the knees. Hyaline cartilage, a firm, dense, and pearly blue tissue, envelops the joint (articular) surfaces of bone.

Cartilage is produced by *chondroblasts*. They are specialized cells that release chemicals called protoglycans that infiltrate the collagen fibers and transform them into cartilage tissue.

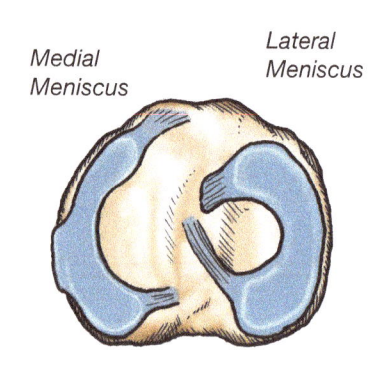

Although not desirable for the adult skeleton, cartilage-to-bone conversion is essential for bone growth and maturation in children. Blood vessels grow and expand into a child's cartilage-rich skeletal system and slowly orchestrates its transition into bone.[1] As an infant develops muscle tone, cartilage-to-bone conversion is stimulated by the electro-mechanical energy produced by muscle contractions. As an example, when a baby first begins to stand, its lower back muscles contract and stimulate the eight different growth centers on each lumbar vertebra to mold and form into bone.

There is no significant blood supply to adult cartilage. It is essential for cartilage to be free of minerals. If blood were present, it would deposit elemental minerals and the mineral salts of calcium, silicon and boron. Cartilage exposed to these minerals readily converts into bone; something undesirable on joint surfaces where healthy cartilage is necessary for mobility. The wisdom within the body's design keeps blood separated from cartilage after bone formation is complete in adults.

In fully formed bone that is without blood supply, cartilage imbibes nutrients directly from the synovial fluid. Not only does joint fluid lubricate joint surfaces, the short, squeeze-and-release compressions created by joint movement provide a pumping action that cleanses cartilage. It allows vital nutrients to enter and waste products to exit. This lavage method of supplying cartilage with nutrients is slow and inefficient. It does, however, prevent cartilage from coming into direct contact with blood. Because adult cartilage has this rudimentary and inefficient circulatory system, the healing process of cartilage is substandard. Damaged cartilage rarely heals without consequences such as scar tissue.

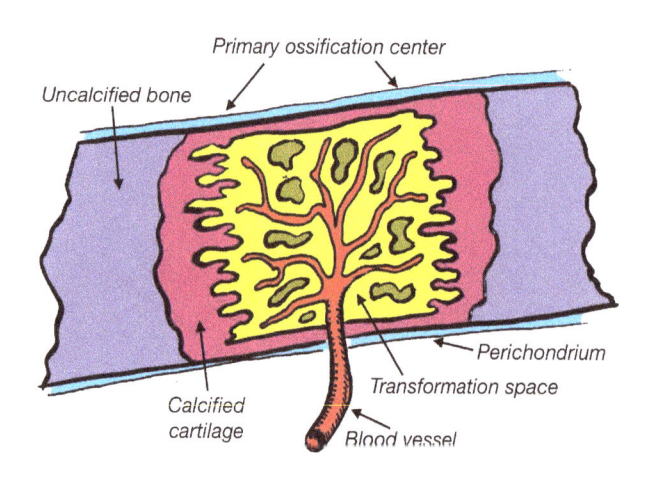

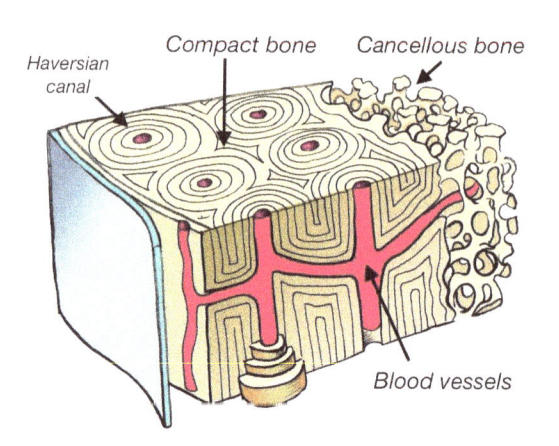

Blood supply in forming bone tissue

Mature bone and blood supply

Hyaline Cartilage

Hyaline cartilage encapsulates the ends of long bones. Its blue-colored, smooth contours reduce friction between joint surfaces. Hyaline cartilage corresponds to the gristle found on the ends of bones eaten by our carnivorous cohorts.[2]

Hyaline cartilage plays an important role in keeping the joint space free of blood contact. It provides a barrier between the rich blood supply located in the *periosteum*, the outer layer of bone, and the synovial membranes on the inner linings of the joint. In a joint injury, blood vessels dilate and cause excessive swelling in the tissues and spaces. The trauma may cause blood to enter the joint and make contact with the hyaline cartilage. This can cause calcification of the joint cartilage, arthritic spurring, and the deformation of joint surfaces. Swelling is a method of immobilizing an injured joint, splinting and protecting it from greater injury. All too often, the impatient yoga student will force their swollen joints to move before they have properly healed, causing the hyaline cartilage to shear and tear.

Bone

Mature bone is a living tissue. It is as strong as cast iron yet flexible and lightweight. Bone has tensile strength (able to resist being stretched) but buckles when excessively compressed. Its construction is of collagen fibers infiltrated by an ideal balance of organic proteins and inorganic minerals that form a strong, flexible matrix. There are two types of bone, compact and spongy (cancellous). Compact bone forms the shafts of long bones and the outer surfaces of all of the bones of the skeletal system. Spongy bone is located in the heads of long bones and in irregular bones, such as vertebrae and the plates of the skull. A series of canals course through a bone's many layers, carrying blood vessels and nerves. Blood supplies minerals and other essential nutrients to the bone. Nerves enable sensation and provide electrical stimulation to the bone, which helps maintain their density.

Bone density

Calcium and other inorganic mineral ions are constantly circulating throughout the body. The calcium located in the femur bone one week might be found in the stomach two weeks later manufacturing its digestive acids. As described previously in relation to cartilage, minerals such as calcium can easily penetrate collagenous soft tissue and stimulate bone formation, at times in undesired places. Bone spurs that accompany arthritic joints are an example of this process.

The primary stimulus for bone density is mechanical force on the bone itself. Muscle contraction, gravity and heel strike forces all deliver mechanical signals to bone to maintain density. In normal physiology, bone is constantly being broken down, rebuilt, and remolded to accommodate the changing tensions and stresses to which bone is subjected.

Nerves constantly measure the mechanical stresses that pass through bone. The autonomic nervous system coordinates the action between the *osteoblasts*, cells that build bone, and the bone destroying cells, the *osteoclasts*. Bone cells in stable bone tissue are called *osteocytes*.

Bone becomes markedly thickened in response to muscular contraction. In athletes and workers with physical occupations that promote muscular imbalances, bone density can become uneven. If an injury or disease causes significant weakness in the muscles contiguous with a bone, osteoclast activity will dissolve bone tissue on the weakened side, while the osteoblasts will build bone on the stronger side.

Yoga practice can improve bone density. The Anusara yoga instruction to *"hug muscles to the bone"* creates significant mechanical tension on bone. For students with advanced bone loss, an entire yoga practice may consist of the basic sitting and standing postures while holding isometric contractions that hug the muscles to the bone. Precise alignment and *Samasthiti*, the principle of balanced tension, are essential tools for practitioners needing to increase bone density.

In cases of scoliosis, the spine abnormally curves and the musculature becomes imbalanced. This can alter the mechanical stresses on bone and modify its density. Alignment and Samasthiti are essential tools to reduce long-term damage to the spine with scoliosis and other types of curvature.

Resistance training (weight lifting) is one of the best methods for delivering mechanical tension to the bone and increasing its density. Studies have measured mineral mass to be highest in the femurs of weightlifters while considerably less in swimmers. Prolonged periods without mechanical stimulation such as during bed rest can result in as much as one percent loss of bone per week.[3]

Osteoporosis and Osteopenia

Osteopenia is a serious, and yet practically epidemic bone condition in first world cultures. It is characterized by the deterioration or destruction of the protein-matrix of bone. As a result, minerals are without a place to embed. Calcium supplements offer little benefit to bone tissue when its matrix is lost. When mineral concentration is also measurably reduced, the condition is called *osteoporosis*. Often, these bone-loss conditions result in a structural collapse in the form of compression fractures.

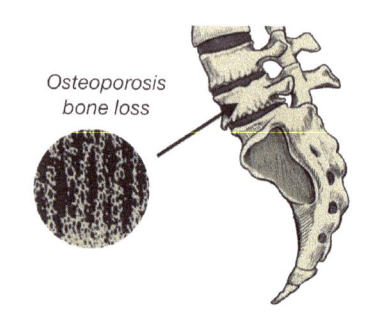

Osteoporosis bone loss

Hormonal changes, particularly in the reduction of estrogen and adrenal hormone levels are a common cause of bone loss. Hormones regulate the levels of minerals in the bloodstream and their absorption into all tissues of the body. This is one reason why bone loss frequently occurs with aging.

Bone's absorptive qualities

As described, bone's collagen protein easily absorbs trace minerals. This metabolic process is there to ensure that bone receives the micronutrients needed to maintain its density.

Substances besides those that build healthy bone structure may also become absorbed unwittingly into bone tissue. As we know from dental treatments, the teeth can readily absorb fluoride minerals. After treatments, the levels of fluoride in bone tissue have also been shown to increase. Although fluoride makes teeth and bones stronger, it causes bone tissue to be more brittle.[4] Heavy metals and toxic chemicals, from nicotine to DDT, get absorbed and stored in the collagen proteins of the bone matrix. When bone loss occurs, as is commonly seen with aging, a release of these sequestered metals and toxins can send them back into the blood stream, potentially triggering serious illness.

Avoiding heavy metal exposure, such as to lead and mercury, is an obvious preventative for bone toxicity. A diet high in organic fruits and vegetables with ample whole grains helps remove toxic substances that are already present. Starting a healthy, high-fiber dietary regimen early in life can protect the body from toxic substance storage in bone.

> ### Technically Speaking
>
> Bone matrix is made up of bone fibrils formed from a suspension of *hydroxyapatite*, a complex form of phosphate and calcium- Ca5(PO4)3OH. The matrix is then infused with micro-crystals mineralized with calcium, boron, silicon and other trace minerals. Bone is ceramic by nature, providing excellent tensile resistance to stretching. It can withstand moderate compression and shearing stresses. The bone fibrils that form the bone matrix align in relation to the forces placed upon them by gravity, weight, and muscular tension.

Bone re-modeling and the value of good posture

Mechanical forces and neuro-electrical stimulation not only affect bone density but also can also modify and remold bone's actual shape. Laboratory experiments have demonstrated the ease with which the shape of bone can be manipulated. In clearly a non-yogic experiment, rabbits had the bones of their legs placed in movable casts that were slowly twisted during a six to eight week period. Bones were forced to remodel from straight to curved. The procedure was then reversed and the bones returned to being straight.[5] An adaptation of this principle is used in orthopedics for treating fractured bones. Mechanical and electrical devices have helped heal bone fractures utilizing the same physiological properties demonstrated in the lab tests on those unfortunate rabbits.

Yoga and yoga therapy influence bone re-modeling. The forces of gravity and muscle contraction that are constantly delivered constant in asana practice can re-shape bone. If the forces are correctly distributed and integrated through alignment, bone can re-model into a stronger, more desirable form. Although no scientific studies have directly tested or confirmed yoga's specific ability to re-model bone, most experienced yoga teachers and students can attest to yoga's effectiveness in this regard. Anecdotes abound of yoga practitioners who have applied alignment principles to help straighten bowed legs and attaining positive results with a dedicated, many-years practice.

Yoga provides an excellent opportunity to maintain bone density and alignment along with the overall organic health of our human frame. Long-time practitioners have first-hand experience that confirms the value of yoga and its ability to assist in bone health and repair.

11 Align By Design

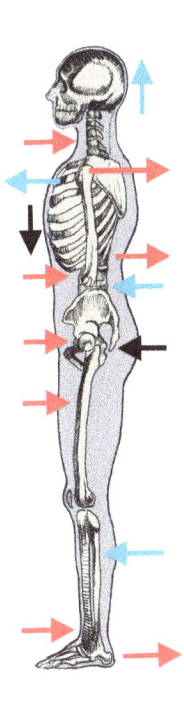

The outer appearance of the human form varies widely. Vast differences are apparent in our sizes and shapes, hues and colors. Body types range from the sleek and flexible savanna dweller of the African continent to those that have emerged from large-boned European peasant stock. Individual differences can also be found in our internal anatomy. The presence or absence of certain muscles, nerves and blood vessels fluctuates at high levels in the human population. The location and profile of ligaments and tendons, at times, seem haphazard and makeshift. For those determined to challenge the notion of one universal system for alignment, our anatomical differences may be central to their point of view.

And yet, the genetic make-up between all humans is virtually identical. Less than one tenth of one percent of human DNA varies between the most diverse members of our species. Mechanically, our joints and muscles and even our posture virtually function the same with all individuals, despite the many anatomical variations. Alignment is founded upon body mechanics and physiology. The subtle anatomical differences are relatively inconsequential to how our body best moves and functions.

Regardless of body type, anatomical variation or history of injury, a yoga practice is appropriate for all. Despite differences in physical capabilities between individuals, there is only one "owner's manual" for everyone. All yoga students follow the same *universal blueprint* that decodes the basic design of the body and how best to establish alignment.[1]

The good news is that it is not necessary to learn a new set of alignment instructions for each asana. Principles of alignment are not based on the poses; they are based on the body. Yoga postures are expressions of the body. The body operates the same way even as postures become more varied and complex. Of course, adjustments and modifications may be necessary to account for subtle anatomical differences or to accommodate for injuries. Adjustments support the body as it attempts to find its way into alignment. Modifications to poses will affect the outer appearance of a pose from pone person to another but do not modify or nullify the underlying alignment principles. The closer a student adheres to alignment, the better protected they are from potential injury.

Technically Speaking

One anatomical variation of the body occurs at the vertebral joints called *facets*. Typically, the joints hinge along the same plane. A small percentage of people have facets that face in opposing directions. This anomaly is referred to as *asymmetrical facets*. Radiological imaging is necessary to detect the presence of asymmetrical facets.

When affected vertebrae move, an undesired twist of the segments may occur. Should a yogi have this design anomaly, knowing its location and configuration is valuable. He can learn to adjust his practice to reduce where there is limitation or excessive twist or torque in one direction relative to another. Implementing the principles of alignment and Samasthiti can minimize aberrant movements and limit the damage that this anatomical anomaly can otherwise cause to the spine.

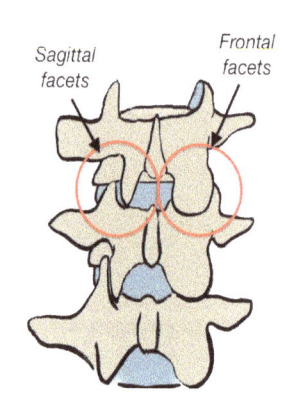

Sagittal facets *Frontal facets*

Need GPS?

Do you know what specific directions bring a posture into its correct alignment? Ironically, it is easier to be out of alignment, especially when fatigued. Alignment is like a navigation system where a vessel veers off course constantly is brought back on-track and in line.
Every part of the body, large and small, is designed to move in a specific direction to come into alignment. Alignment may require either an actual physical movement or may only entail an isometric muscular contraction with an energetic intention. Movements that bring the body precisely into alignment have positive, therapeutic qualities. Movements that shift the body further from ideal alignment become dangerous and set up the body for injury.

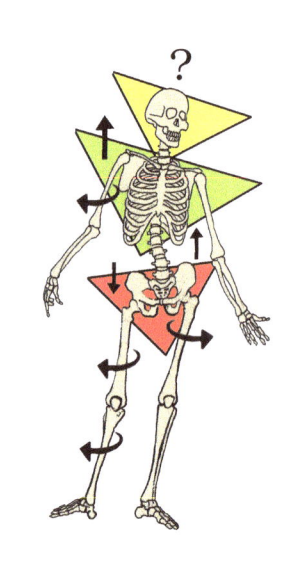

> The principles of alignment are applied to the body, not to the asana
> Asana is the expression of alignment, even when the outer form varies

The Alignment Grid [3]

The most fundamental posture in yoga is **Tadasana**, the Mountain Pose. As implied by its name, a mountain embodies the characteristics of groundedness. Mountain Pose is solid, yet fluid and organic. *Samasthiti,* the quality of balanced tension, is fully expressed in Tadasana.

The components of alignment used for all other asana are present in Tadasana. Although Tadasana is perhaps the first posture a novice student will learn, it requires practice and skill in order to master all of its subtle actions. The precision applied to Tadasana often reflects the intention that the student brings to their entire practice.

The directions that bring every asana into alignment follow the actions depicted in the Alignment Grid.

- Heels draw back
- Ankle creases draw back
- Shins move forward
- From the side, the hip socket (greater trochanter) is vertical over ankle
- Thighs draw back
- Hip creases and navel draw back
- Low back at L4 - L5 level moves forward to form egg-sized curve
- Coccyx scoops forward
- Lower rib cage draws back
- Side body rib cage lengthens evenly on both sides
- Chest extends forward
- Inner and outer armpits equally draw the head of arm bones back
- Tip of breastbone (xiphoid) drops down and scoops toward navel
- Throat draws back (mid level of neck)
- Roof of mouth is horizontal, in line with the center of the ear canal
- The base of skull lifts vertically, meeting the line from the ear canal

Every movement is coordinated and integrated with all other movements. When one region moves into alignment, it does not offset or disrupt any other previously aligned region; instead each action further contributes to the overall integrated posture. Yoga teacher Betsey Downing, PhD. describes moving into alignment in asana practice as *successive approximations*. It is not the appearance of a pose but the intention to move the body correctly that refines and advances yoga practice and makes it safe and therapeutic.

Simple triple "S" alignment (skull, scapulae, sacrum)

The back of the skull, shoulder blades (scapulae) and sacrum align on a vertical plane.[2] This triple "S" organization aligns the body's central axis, bandhas, and diaphragms and configures the spinal curves to provide their greatest strength and weight bearing.

To practice the Triple "S", sit or stand with the back against a wall. Press the sacrum to the wall, then the shoulders. If the back of the skull does not touch, do not force it. Instead, bring the throat and inner armpit ribs back to naturally bring the skull closer. If the lumbar curve is greater than egg-sized, reduce the degree of curve by drawing the lower rib cage posterior. If the student has a naturally larger buttocks and a deeper lumbar curve, a larger curve may be appropriate.

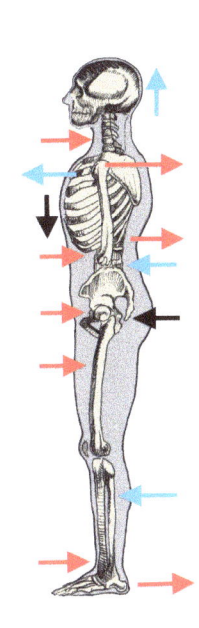

Correct S alignment 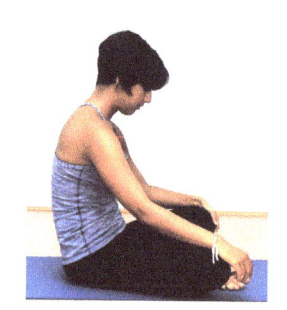 *Incorrect posture*  *Ardha Matsyendrasana*

Sitting closer to the pubic bone tilts the pelvis forward and helps to form the lumbar curve. Without the lumbar curve, the rest of spinal alignment is difficult to achieve. **Ardha Matsyendrasana**, or Seated Twist, aligns according to the Tadasana alignment grid. Frequently but incorrectly, the shoulders roll forward of the chest and the upper back rounds. Pressing an elbow forcefully into the knee or rounding the shoulders when attempting to bind the hands can misalign the pose. Twists always occur in the spine. The shoulders remain as square with each other, as possible.

The floating ribs

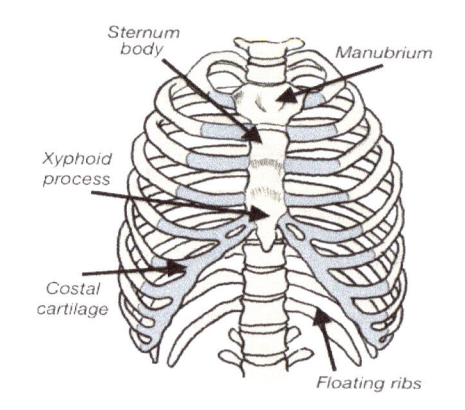

The lowest two sets of ribs are called *floating* ribs because they do not attach to the breastbone or to any other anterior structure. They only attach to the skeleton at the spine's transverse processes of the 11th and 12th vertebrae. The floating ribs cover and protect the kidneys from trauma.

The floating ribs align in three directions:
- Draw posterior
- Widen across the back
- Lift up the back

B.K.S. Iyengar instructed students to broaden the kidneys across the back body. Anusara yoga addresses this action with the principle of the *Kidney Loop*.[4] Aligning the lower ribs can be difficult to both isolate and engage, especially for more flexible students and those who have deep lumbar curves. Full details on lower rib cage alignment are presented in Chapter 15.

Design and function

The sacrum is centered between the two hipbones. The breastbone is located centrally in the anterior rib cage. From the side view, both the sacrum and the sternum have a similar curve and appearance. Each also has a small tail at its tip - the coccyx on the sacrum and the xiphoid process at the bottom of the sternum. Although both structures are not specifically related, the both function similarly by performing a scooping, inward action that engages their respective bandhas and diaphragms.

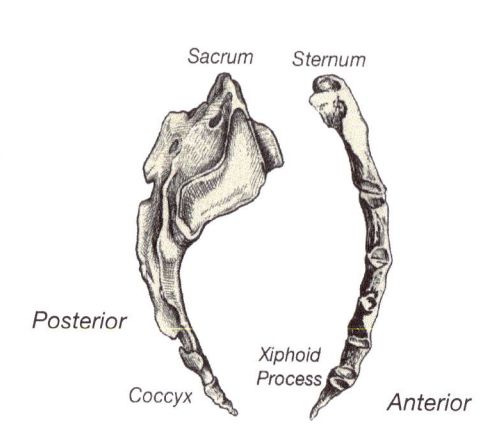

Scoop, scoop, draw in the navel – keeps the low back stable!

A major challenge in asana practice is the integrative alignment between the legs, pelvis, and lower spine and torso. This method, as described below, is a user-friendly way to establish this alignment that is essential for safe function of the pelvis and health of the lumbar discs and spine:

- Scoop the tailbone inward (Mula Bandha): stabilizes the sacroiliac joints
- Scoop the breastbone inward (Uddiyana Bandha): stabilizes the lumbar-thoracic region, including the spine's primary swivel point at T-12
- Draw in the navel (Manipura chakra): stabilizes the lumbar spine

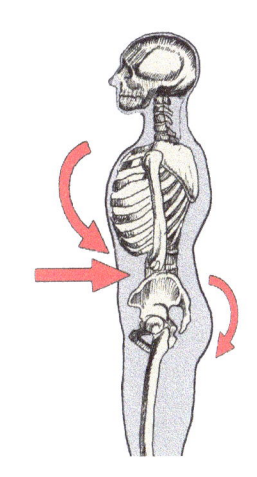

> Anusara yoga teacher Jaye Martin describes the quality of the action of *drawing in the navel* as imagining that a ripe strawberry is placed in the belly button and it is drawn inward about one half inch. The action is gentle so as not to crush it - but firm enough to not drop it!

Zip it!

Another useful visualization for integrating the lower torso and pelvis is offered by senior Iyengar yoga teacher Joan White. White suggests that the student imagine zipping up a pair of pants. This has the same effect as a *forward tailbone scoop*. Borrowing White's metaphor, the student can also zip down from the xiphoid process to the navel as visualization for scooping the tip of the breastbone.

Tight jeans: the action is similar to pulling up a tight pair of jeans and zippering them up. The thighs first draw back when putting on the jeans; then the pants are zipped up, which is the action of scooping of the tailbone forward. These details are discussed in Chapter 13.

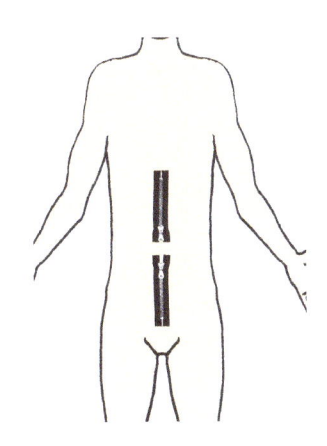

Beauty in its simplicity

As this book emphasizes, it is unnecessary for yoga students to learn a complex list of alignment instructions to apply to each and every individual asana. Every yoga pose follows the same set of directions. Once comfortable with applying the Alignment Grid and other supporting principles, the initial mental efforts being applied ease up and become almost unconscious.

New styles and hybrids of yoga are emerging rapidly. Some of these new yoga incarnations are brilliantly performed while others lack the intention of alignment altogether. Students will remain safe with these new approaches by remembering to reference alignment to the body and not to the poses or flows. Yoga is the practice of retaining alignment in every variation in form and posture.

Lock and Load: Step-by-step stabilization for asana alignment

Stability in asana begins by setting the foundation for each pose. This is a critical first step and precise alignment is essential. Typically, there are multiple foundations in an asana. Foundations are set, step-by-step from the ground upward with each one locking into place, subsequently after the other. This locking and loading method in setting each foundation is analogous to the closure of canal locks to enable each region of the body to be stable, secure and balanced.

In many cases, foundations overlap with the bandhas and diaphragms of the body. As you learn to engage those elements, the foundations become more stable.

Each step of the Alignment Grid is also engaged as a foundation and set in this lock and load fashion; once again usually setting each alignment from the ground and upward.

Examples of setting the foundation:

- In standing poses such as Warrior 2, aligning through the four corners of each foot creates a firm foundation. The front knee uses this foundation and bends toward 90° and externally rotates, even if isometrically. The torso lifts from the front thigh to vertical, relying on the lower foundations to be strong and stable in the pose. Each side body rib cage establishes equal length and tension in relation to each other (Samasthiti). The chest presses forward and the shoulders align on the back, becoming square to the side of the mat. From this foundation, the head and neck align vertically and turn toward the middle finger of the front hand.
- In Downward Facing Dog, precise placement of the hands allows the arms to be stable, the forearms to roll inward, and the humeral heads to externally rotate. From this stable foundation, the shoulder blades hug firmly to the posterior ribcage.
- Also in Downward Dog, the shins press forward, which causes the knees to micro-bend. The thighs use the resistance of the shins pressing forward as the foundation to draw back and align without the forcing the knees going into hyperextension.
- In seated postures, sitting closer to the pubic bone tilts the pelvis forward and shifts the spine's foundation forward. This is usually essential in forming the lumbar curve. Without the lumbar curve established, the rest of spinal alignment and correct posture is difficult to achieve.

12 Anatomy of the Pelvis and Sacroiliac Joints

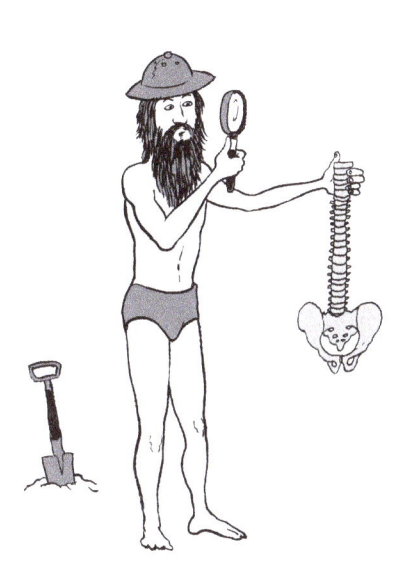

Pelvic bones are an inspiring find to archeologists. The smallest remnants provide a treasure trove of information about the life and behavior of their owners and of human ancestry, in general. A prehistoric pelvis can reveal untold details that may unravel the evolutionary journey toward human development. Did this ancestor live mostly in trees or on the land? Was it bi-pedal and what was its gait like? What was its sex? What degree of uprightness did this species achieve? The pelvis is not only a fascinating part of the human frame to study but plays a central role in body mechanics.

The pelvis is a bowl-like structure constructed of three multi-sectional bones: the two outer hipbones (os coxae) and the centrally located sacrum. The pelvis forms a solid ring that protects the organs of the lower abdomen. The sacrum is positioned as an inverted triangle with its wide base facing upward and provides the foundation for the vertebral spine. The hip sockets form at the lower, outer portion of the pelvis.

The acetabulum

Each hipbone is formed from three anatomically distinct bones: the ischium, ilium, and pubis. All three bones converge at the center of the hip joint where they form a deep, cup-shaped hip socket called the *acetabulum*. By bring all three bones together at the center of the socket, an even distribution of the forces of heel strike and gravity can reverberate evenly through the pelvis.

For most individuals, the acetabulum completes its formation into one solid bone between the ages of twenty and twenty-five.[1] Some women complete bone formation sooner if their menstruation began earlier.[2] Beginning a yoga practice at an age before the hip sockets completely form may be beneficial for increasing lifetime hip flexibility by being able to mold a broader-shaped acetabulum.

Adults that start a yoga practice after their bones are completely formed should not be discouraged. As we have seen, bones retain a subtle ability to remodel throughout a lifetime. Of course, as bones and joints age, remodeling tends to be degenerative as a result of daily wear and tear. Bones, however, can potentially remodel into healthier, more desirable shapes. A well-aligned yoga practice can direct the many forces on bone to build bone density and mass and help maintain good, functional shape. There are reasons to be optimistic!

Encapsulating the acetabulum is the labrum. It is formed from a strip of articular cartilage, a gasket that collars around the head of the femur bone where it hugs snuggly into the joint. The femur head is centered in the joint to achieve stability and support for weight bearing while minimizing the amount of joint play or wobble. This snug arrangement in the hip socket has a "vacuum effect" that suctions the head of the femur into the hip socket. In comparison, the shoulder joint, which also has a labrum, has one to two inches of joint play along its central axis that allows significantly greater ranges of motion. Greater shoulder joint mobility comes at the compromise of less stability than found in the hips.

The "sitting" bones

The German word *Sitz* translates into seat. It references the curved, lower surfaces of the pelvis that are used for sitting. Anatomically, the sitz bones refer to the *ischial tuberosities*.

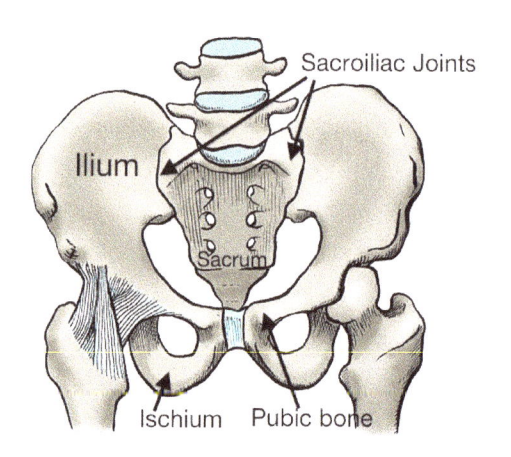

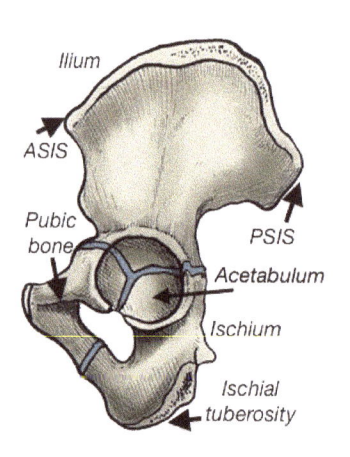

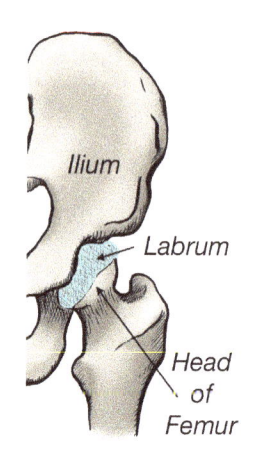

The pelvis rocks!

The curved-bottom ischial tuberosities function similarly to rocking chair rails. By rocking forward on the ischial tuberosities, the sacral base and pelvis tilt forward. This deepens the lumbar curve and shifts the body's central axis and center of gravity forward.

Conversely, rocking back on the ischial tuberosities brings the pelvic floor and sacral base more horizontal, bringing the central axis posterior and causing the lumbar curve to flatten. A common postural response to rocking back on the ischial tuberosities is rounding of the upper back and shoulders.

Good alignment requires finding the "sweet spot" where rocking on the ischial tuberosities brings the body's central axis vertical to gravity and the curve of the lumbar spine is egg-sized.

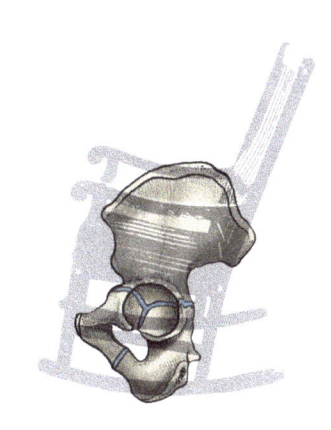

Tight hamstring muscles or limited hip external rotation can significantly limit forward pelvic rocking, especially when flexing the pelvis forward if attempting a straight-legged forward fold. Tight hamstrings interfere with the formation of a healthy lumbar curve. Limited hip external rotation flattens the pelvis when most forward movements are attempted. Students with these challenges should sit on a support or blanket to tilt the pelvis into a more anterior position and enable it to rock more freely. Sitting closer toward the pubic bones allows the student greater control over pelvic rocking and the ability to control the lumbar curve.[3]

Pubis symphysis: the anterior joint of the pelvis

The two pubic bones comprise the anterior ring of the pelvis. They join together at the anterior midline of the body by a hard, fibrocartilage disc called the *symphysis pubis*. Also called the pubic symphysis, this joint has minimal movement in the normal, non-pregnant state. All mechanical movement at the pubic symphysis should otherwise be avoided. If injured, the symphysis pubis may become chronically hypermobile and troublesome, producing deep pain and chronic pelvic instability.

When one hipbone rotates forward and the other back, pelvic torsion can be produced. Precaution is necessary in closed-hip postures, especially extreme, counter-rotating hip poses, **Hanumanasana**, the Forward Split being most common. These postures create shearing forces on the pubic symphysis disc. To prevent damage, keep the hips square to the front of the mat and let the pose take place in the hip sockets exclusively without torque in the pelvis. Internally rotate the hip sockets and firmly scoop the tailbone forward. Engage muscular tension evenly through both hips and legs. Press out (inferior) through both inner heels.

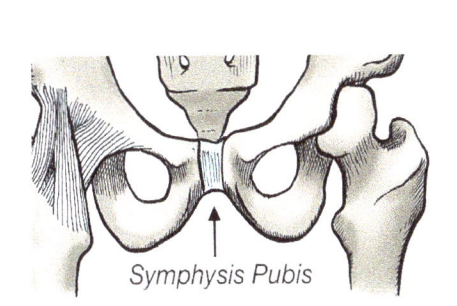

Symphysis Pubis

Sacroiliac joints: the posterior joints of the pelvis

The two iliac bones adjoin respective sides of the sacrum to form the sacroiliac joints. The sacroiliac joints (SI joints) are large and irregular in shape. They establish the foundation for the spine and torso and able to manage extensive weight bearing. Early anatomists had assumed the sacroiliac joints to be immovable. It is now well established that they do move, although only marginally.

Sacroiliac movement does not conform to the typical ranges of motion and designated by a unique name. Its two basic directions of movement are called *nutation* and *counter-nutation*, derived from the Greek word that means *to nod*. Nutation is an anterior, tip-and-glide movement. The posterior, reversed direction of nutation is counter-nutation. Experts have debated as to what the full range of nutation/counter-nutation is with most settling on approximately 6 mm. That range can increase to as much as 22 mm during a pregnant female's third trimester.

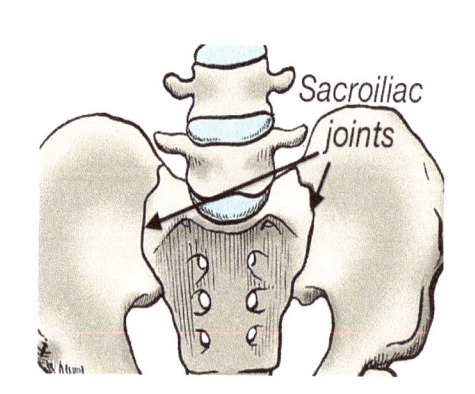

Sacroiliac joints are shock absorbers

The primary function of the sacroiliac joints is not to contribute gross movement but to absorb and distribute the various shocks and forces that the human frame constantly endures.

The strike of the heels creates upward forces that transfer through the legs and pelvis and continue to rise through the rest of the body. Heel strike forces are typically measured using a bite plate to record changes in pressure on the jaw. A vertical up-and-down, jump-in-place produces forces that can reach levels 10 times the weight of the body.

Gravity is the major downward force on the human frame. It compresses the body in a steady downward vector that directly impacts the pelvis where the sacroiliac joints effectively absorb the force of gravity.

As the forces of heel strike and gravity converge within the pelvis, the energy that is created reverberates and spreads around the pelvic rim. Acting as a pressure release valve for the build up of forces within the pelvic bowl, the sacroiliac joints subtly recoil to discharge the pressure.

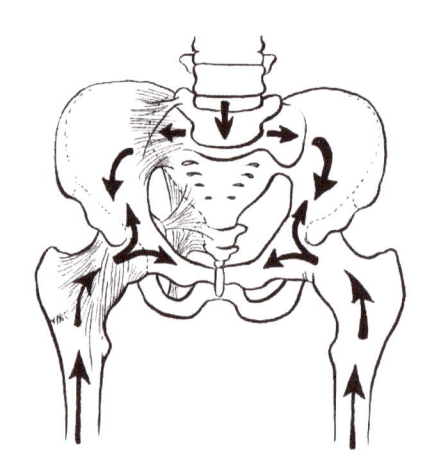

The forces around the pelvic brim circulate in a manner similar to a finger running around a glass goblet and emitting a vibratory hum.

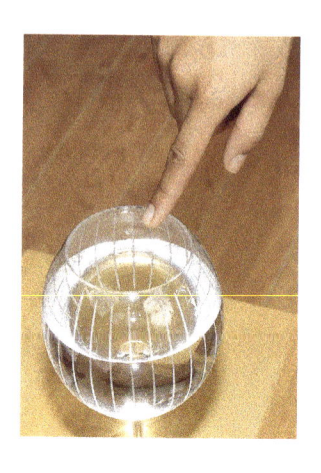

If one or both of the sacroiliac joints become immobilized, energy fails to be fully released. Instead, forces are unavoidably transmitted to the next most freely movable region, transferring up to the lower discs of the spine. If one sacroiliac joint becomes less mobile than the other, the result is an additional twist and torque on the lower discs with each pounding of shock they receive. Torsional stresses on the lumbar discs are a common cause of disc herniation. If stress on the disc becomes chronic, the long-term outlook is spinal deterioration.[4] Mechanical dysfunction of the sacroiliac joints is common and is a frequent cause of lower back pain.

Sacroiliac function: an open and shut case

The inner surfaces of both the sacrum and ilium are rough and irregular. With normal function, the small bumps and crevices on their interfaces match up and align, functioning similarly to the intermeshing teeth on a gear or sprocket. As the SI joints move, they open and close with a gear-like ratcheting and recoiling.

Iyengar-tradition yoga instructor and researcher Roger Cole equates the sacroiliac joint surfaces to a ceramic dish that is broken in two. The joints first open, separating their irregular surfaces and movement can occur. After the sacroiliac joints have opened to the desired degree, the joints close, locking into pre-set indentations and re-establish pelvic stability.

In making his analogy, Roger Cole also offers a hypothesis for injury:

"If you separate but misalign the two pieces (of the ceramic dish) in any direction, the bumps on one piece will be off set and clash with the bumps on the other. The surfaces of the sacrum and ilium have similar bumps and depressions that fit together but will clash with one another if you shift the bones out of place in any direction. The bump on bump pressure may be the source of sacroiliac pain. If it continues over long periods of time, it eventually causes cartilage and bone deterioration, causing deeper and chronic pain."[7]

How does the yoga practitioner control opening and closing of the sacroiliac joints? The next chapter describes the procedure in detail. Managing the sacroiliac joints is part of a broader methodology for alignment of the pelvis. This mechanical system is aptly called *pelvic integration*.

Sacroiliac function

One of the reasons that anatomists first assumed the sacroiliac joints were immovable is that they are purely ligamentous joints. No muscles directly move the sacrum in relation to the ilium. Instead, it is controlled through indirect actions by muscles such as the piriformis and the hamstrings. Research has uncovered that the biceps femoris muscle of the hamstrings can move the sacrum via a remnant of tendon tissue that connects to the sacrotuberous ligament. The legs are able to act as long levers that can efficiently utilize small muscular actions and create significant SI joint movements.[5] More details are provided in Chapter 20.

Sacrum as a trap door

Hormonal changes of later term pregnancy increase the elasticity of the ligaments to facilitate the processes of labor and delivery. The ligaments of the sacroiliac joints utilize this temporary elasticity by increasing nutation from the typical 6 mm to 22 mm. With this newfound freedom, the sacroiliac joints are able to separate widely, allowing the sacrum to drop open like a trap door, welcoming the baby's entrance to this world.

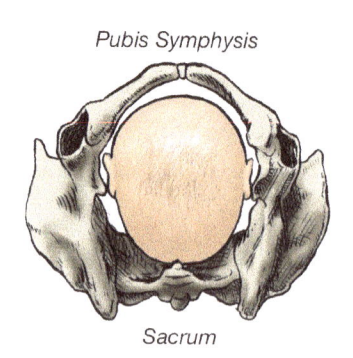

Pubis Symphysis

Sacrum

Insult and injury to the sacroiliac joints

The sacroiliac joints can become injured during yoga practice. Asana's constant transitioning between movement and stabilization places significant mechanical demand on the sacroiliac joints. Injuries can occur during either the open or closed phase of sacroiliac movement.
To illustrate this concept, the following examples are presented:

- If the sacroiliac joints are forced to move before first being opened, they cannot provide adequate mobility and strain the ligaments and other soft tissues affected by the movement.

- If the sacroiliac joints do not adequately spread open before the closing phase begins, the sacrum can be bruised as it jams against the closed joint.

- If the sacroiliac joints need to be locked and stable, such as in weight bearing postures, but are stuck their open position, injury can occur. If the joints are unable to fully close, they are unstable. This places excessive strain on the sacroiliac ligaments as they attempt to compensate for the instability of the joint.

A yoga practice that utilizes pelvic integrative alignment principles can prevent many injuries and rehabilitate existing ones. These principles, which are presented in the following chapter, also serve as essential tools used by professionals and in yoga therapy to address injured sacroiliac joints.

Sacral pump

The sacroiliac joints are designed to manage weight bearing and shock absorption and regulate their small, yet critical movements. Another subtle function of the sacroiliac joints is called the *boot mechanism*. This action is essential to the nervous system and, therefore, for the overall health of the body.

On the rough, inside surface of the sacroiliac joint is a small, smooth portion shaped in the form of a *boot*. The boot is situated at the level of the second sacral segment. This level is where the outer covering of the spinal cord, the dura mater, attaches to the spinal canal.

Sacral boot

Above this attachment point, a small, tapered channel is formed where a pooling of CSF, or cerebrospinal fluid, occurs. The cerebrospinal fluid is the vital liquid that nourishes the brain and nervous system. [8]

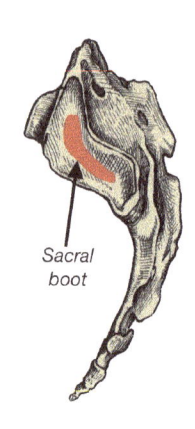

TWELVE: ANATOMY OF THE PELVIS AND SACROILIAC JOINTS | 79

Because the dural sleeve of the spinal cord attaches to the boot of the sacrum, movements of the sacroiliac joint tug the dura and pump the cerebrospinal fluid up the spine and back to the brain. This pumping action is called the *primary sacral respiratory mechanism*.[9] This action was discovered in the realm of osteopathic medicine. Its principles are fundamental in therapeutic modalities such as Sacro Occipital Technique, Cranial Release Therapy and Craniosacral Therapy. Arrays of bodily ailments that include headaches and back pain have been associated with inadequate sacral pumping.

Something else to chew on: the TMJ

Many holistic approaches to body therapy recognize the subtle but powerful neuro-mechanical relationship between the tempromandibular joints (TMJ) and the sacroiliac joints. Therapeutic approaches can be used to evaluate this functional relationship and restore imbalances that may develop between these two regions. The interconnection between the jaw and sacroiliac joints may seem like an unlikely association but it can be observed with weight bearing activities. As the sacroiliac joints engage when lifting a heavy object, the action is often accompanied by the clenching of the jaw.

Equator of the body – the sacrum/coccyx juncture

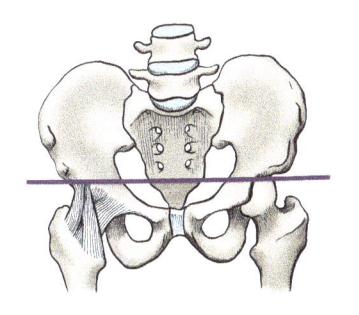

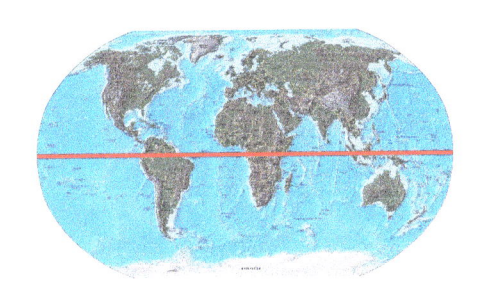

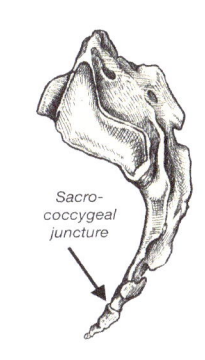

Sacro-coccygeal juncture

When passing over the equator in a boat or plane, no clear lines or signposts mark its location. But, from one side to the other, an invisible shift in polarized energy takes place, drawing subtly in opposite directions. As subtle as this demarcation is, it affects the earth as a whole. Conceptually, but not anatomically, the human body has a similar, energetic equator. It could be assumed that the location of the body's equator is somewhere in the middle region of the body, perhaps at the waist. This is not the case. The body's equator actually resides at the sacral-coccyx joint, the juncture where the bottom tip of the sacrum hinges with the top of the coccyx. Above this *sacro-coccygeal* juncture, muscular tension in the body energetically and isometrically draws upward in a superior direction toward the head. From the coccyx downward, all energy and muscular tension draws inferiorly toward the feet.[10]

The horizontal line of the equator aligns with the bottom fold of the gluteal muscles. At this level, the lower buttock rounds over the underlying hamstring muscles. In every yoga pose, the gluteal muscles contract and lift up toward the head to engage the energy of the body's northern equator.

The horizontal line of the equator bisects the center of each hip socket. As will be described in a later chapter that explores leg alignment, the principle of *lengthwise contraction* energetically "roots" down from the center of the hip sockets to the front of the heel bones. This line of action corresponds with the energy of the southern equator and two actions of the southern equator, the *forward tailbone scoop* and Mula Bandha.

In **Uttanasana**, the Forward Fold, the back of the sacrum is the highest point of the posture. A teacup, or perhaps for those less daring, a foam yoga block can be placed and balanced on top of the sacrum when practicing the pose.

In Forward Fold, the buttocks lift from the gluteal fold and draw upward toward the head. The tailbone scoops downward. From the center of the hip sockets, the pose "roots" down toward the feet, reducing strain on the hamstring tendons. The sacrum ideally occupies the highest point of the pose. This is the desired position for other forward bending postures as well, including **Parivrtta Trikonasana** (Revolved Triangle), **Prasarita Padottanasana** (Wide-angle Forward Fold), **Adho Mukha Svanasana** (Downward Facing Dog) and **Parsvottanasana** (Pyramid).

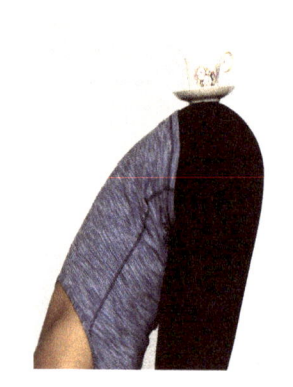

Not all students have the physical capability or have developed the skills needed to fully engage these directions of movement. As long as the student understands and can visualize the actions and has the intention to move in accordance with the body's equator, the alignment of the posture will be safe and energetically integrated. Most students can firmly root downward and scoop their tailbone, which will make forward folding postures safer and beneficial for the hamstrings. Equally, students must initiate the pose by first anteriorly tilting the pelvis, lifting from the gluteal fold. This helps to form the lumbar curve. If the curve is kept egg-sized, the lower back is safe in these postures.

Attention to the equator is fundamental to all postures and the key to safety in asana practice:

- From the lower tip of the sacrum and lower folds of the gluteal muscles, energetically lift upward
- Scoop the tailbone forward to engage Mula bandha
- From the center of the hip sockets, root the legs downward

As will be presented in the next two chapters, there are two steps of integrative alignment of the pelvis and lower back. These are essential steps necessary to engage in all forward fold postures. The first step will lift the gluteal muscles and deepen the lumbar curve. Step One corresponds to the northern equator. Step Two utilizes of the action of rooting down through the hip sockets along with Mula Bandha and corresponds to the southern equator.

13 Pelvis and Sacroiliac Joints Alignment Principles

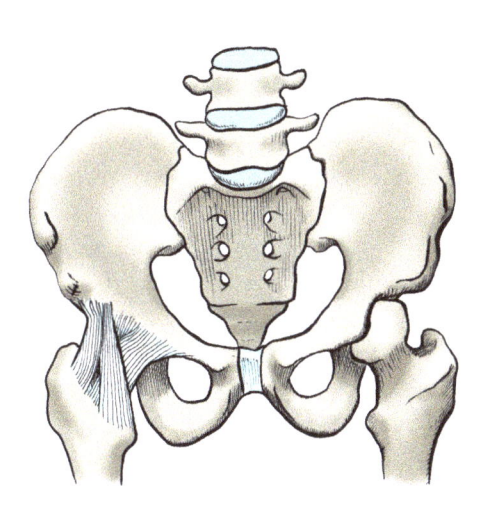

*As goes the pelvis, so goes the spine.
And with that, so goes everything else!*

The pelvis is the structural foundation of the spine. Its underside, the pelvic floor, is the termed the *perineum*. The perineum is a trampoline-like structure, wide and pliable and constructed from dense myofascia and muscle. It functions as an adjustable foundation for the pelvis and will shift the body's central axis when the pelvis tilts. The central axis rises vertically through the body's diaphragms and bandhas until exiting the posterior fontanelle of the skull. The perineum is believed to include remnants of muscles used to wag a long tail earlier along our evolutionary path. Engaging the perineal muscles during yoga practice may stir up the primordial energy of some ancient ancestors!

Move from the Mula

The Root chakra is called *Muladhara*. It is located at the center of the pelvic floor between the anal sphincter and urethra. Contraction of the muscles of the perineum engages *Mula Bandha*, the energy lock of the Root chakra. Although some yoga traditions may teach it differently, engaging the Mula bandha is not forceful nor does it tighten the anus or genitals; instead it is a gentle lift of the central perineum. *The action of scooping the tailbone forward engages Mula Bandha.*

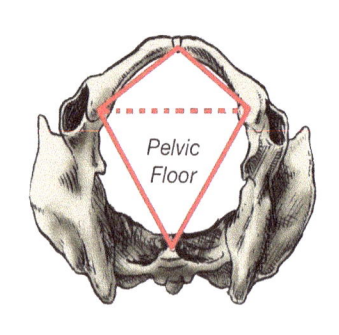

Some yoga traditions, such as Astaṅga Yoga, teach to first engage Mula Bandha before any asana begins. In fact, they suggest to keep Mula Bandha engaged 24/7 since we live life, in theory, always in a yogic state. The action is subtle, and does not alter the general position of the pelvis. The action of lifting the pelvic floor is integrated with lifting the thoracic diaphragm, which is the action of *Uddiyana Bandha*. Engaging Mula Bandha always coordinates Uddiyana Bandha.

Reduced hip mobility or tight hamstrings can limit the ability of the pelvis to tilt, which is necessary to shift the spine's axis as it adapts to gravity. Yoga postures that increase hip ranges of motion and stretch the hamstrings make Mula Bandha engagement more accessible and by that, promote freedom of movement throughout the body.

When initiating **Utthita Trikonasana**, the Triangle Pose, tilt from the Mula bandha. This maintains central axis alignment and helps lengthen the torso evenly. Moving from the Mula raises the front leg ilium away from the femur neck trochanter and prevents compression of the hip socket. A common instruction is to shift the hips posterior when bending into the pose; this is not recommended. This action fails to engage the Mula bandha and potentially produces shearing across the spinal discs.

Samasthiti, or *equal tension,* is an important quality in Triangle Pose. This is possible when the yogi focuses on lengthening the lower side body. This is more easily attained when initiating side bending from the Mula Bandha.

Utthita Trikonasana

Excluding the hip joints themselves, the only significant movement within the pelvis occurs at the sacroiliac joints. The sacroiliac joints move with a subtle, anterior-posterior nodding movement called *nutation*. The sacroiliac joints first open to allow mobility and then close to stabilize. The action of closing and stabilizing the sacroiliac joints is the same action as engaging Mula bandha. Mula bandha, however, must always follow sacroiliac joint spreading and opening before it is engaged.

A safe yoga practice always honors the mechanics of the sacroiliac joints. This is achieved through the principles of *Pelvic Integrative Alignment*. These principles manage the function of the sacroiliac joints but also affect the lumbar spine, the hips, pelvis, and legs. The discussion in the chapter specifically focuses on the sacroiliac joints even though the same actions address other regions of the body. The principles of pelvic integrative alignment will continue to be explored in following chapters as they apply to the other regions.

THIRTEEN: PELVIS AND SACROILIAC JOINTS ALIGNMENT PRINCIPLES

Pelvic integrative alignment

Pelvic integrative alignment consists of a two-step procedure. Both steps are used in every asana to operate the sacroiliac joints safely and efficiently. The first step is *inward hip release*, which opens the sacroiliac joints and creates space necessary for mobility. The second step is *forward tailbone scoop*, which closes and stabilizes the sacroiliac joints. With each of the two steps of pelvic alignment, there is a sub-set of actions that are engaged. For example, *inward hip release* represents five specific actions that occur together to fully make this first step effective. Fortunately, the separate actions within each step are mechanically linked together and function as one instinctive step. A full exploration of all the actions in pelvic integrative alignment and their effects will be presented in Chapter 14.

Step one: Inward hip release

- From the lesser trochanters of the femur bones located at the upper inner thighs (groin), the hips *roll in, draw back,* and *spread apart*. The legs act as long levers to initiate the action.
- *Inward hip release* slightly opens and widens the SI joints, which also creates space in the pelvic floor needed in step 2.[1]

Step 2: Forward tailbone scoop

- *Forward tailbone scoop* is performed exactly as its name describes: the tailbone scoops forward, under the pelvis and toward the pubic bones.[2]
- This step closes and stabilizes the sacroiliac joints. In some cases, *forward tailbone scoop* is a firm action. Other times, it occurs as the subtle engagement of Mula bandha. The physical action that always engages Mula bandha is essentially *forward tailbone scoop*.

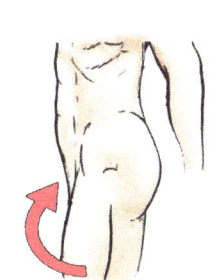

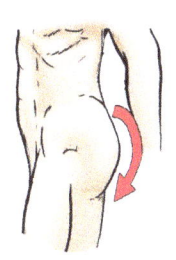

Baddha Konasana and the bruised sacroiliac joint

The outer form of **Baddha Konasana**, the Bound Angle Pose, places the hips in external rotation and abduction. These two directions of hip movement activate *forward tailbone scoop*, which causes the sacroiliac joints to close. It is important to first engage Step 1, *inward hip release*, with firm intention to open the joints and avoid a bruising injury that often produces pain in the buttock region.

Inward hip release may first seem counter-intuitive but its benefits are readily obvious.

To perform **Baddha Konasana** safely:

1. Engage *inward hip release* – roll upper thighs in, draw them back, and spread them apart. Lean off to one side to lift the opposite gluteal fold, rolling it outward and diagonal. Repeat.
2. Deepen the hip creases by pressing the front of the thighs down and away from the anterior spine of the hip.
3. Continue to actively internally rotate the thighs. Adduct the thighs toward the midline while the knees drop outward.
4. Scoop the tailbone forward into the now-open SI joints.

Causes of sacroiliac injuries

As the sacroiliac joints open and close when transitioning between mobility and stability, they are often in positions where they are vulnerable to injury. The sacroiliac joints can become stuck or fixated at any point between being fully opened and fully closed. A good analogy equates the sacroiliac joints getting stuck in their range of motion similar to a door getting stuck along the swing of its path.

Being stuck while the joint is fixated or closed is more common; perhaps because a closed joint is in its default, stable position and its movement is reduced or stopped. Asana practice and yoga therapy techniques that utilize step 1, *inward hip release*, frees open the sacroiliac joints.

Less often, the sacroiliac joints become "stuck" in the open position, making them unstable and unable to adequately support the torso. Therapeutically, joint instability is addressed by stabilizing the joints, which is more challenging to achieve than restoring mobility. Instability causes the surrounding musculature, predominantly the iliopsoas muscle, to respond in an effort to provide additional support. As a result, the muscles become tight, shortened and often fatigued as they attempt to support and stabilize the joints. The sacroiliac ligaments can sprain and inflame, causing them to overstretch and become chronically unstable. This is a common cause of lower back and pelvic pain. Professionals can test the relative strength and tension of the muscles of the pelvic region and use this as an indication of underlying joint instability.

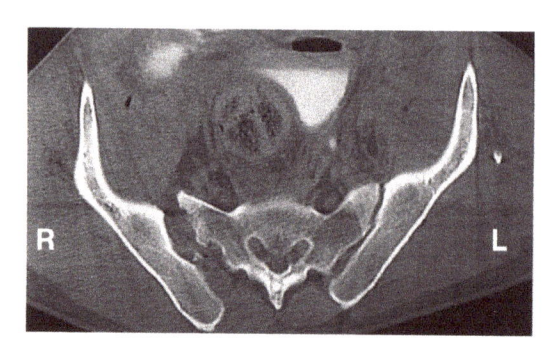

Radiographic image of the pelvis, viewed from head to toe with person on their back

A traumatic separation of the right Sacroiliac joint can be seen

Evaluating the function of the sacroiliac joints

Determining the mechanical state of the sacroiliac requires some skill and interpretation. The two tests presented can be performed by yogis and provide useful information to guide asana practice safely.

Marching in place

When the knee bends and lifts, the sacroiliac joint on that side normally opens. The ilium slightly drops posterior while sacrum remains stable.[3]

To test:
1. Assistant stands behind student, choosing one SI joint to test.
2. On tested side, one thumb is placed on the posterior superior iliac spine (PSIS) and the other thumb horizontally across on sacrum.
3. Student is directed to slowly bend and lift each leg at a time.

Normal observations:
- If the thumb on the ilium drops a few millimeters on the lifted leg side, this indicates normal movement and opening of the pelvis. Importantly, the thumb on sacrum does not move.
- When the opposite leg lifts, neither thumb moves; indicates the joint is stable and weight bearing.

Sacroiliac instability:
- As the leg lifts, the thumb on the sacrum drops, indicating that the sacroiliac joint cannot support the additional demand caused by the lifting of the leg.
- If the opposite, standing leg or sacroiliac collapses while testing the lifted leg side, the joint on the standing leg side is most likely unstable and its surrounding, supportive musculature weak.

Sacroiliac immobile and locked:
- As the leg lifts, the thumb on the ilium lifts: indicates the joint is fixated cannot nutate freely.

Test both sides. This test is not as complicated as it might appear. It is a general test but it is reliable. False interpretations are possible if there are muscular weaknesses or imbalances in the pelvis or legs unrelated to the sacroiliac joints. This is a not a common occurrence, however.

Observing the sway

A less predictable but useful method of sacroiliac evaluation is to observe the direction in which the pelvis sways when the yogi, with eyes closed, stands in the stillness of **Tadasana**, the Mountain Pose.

Side-to-side

- If the sacroiliac joints are locked, the pose may exhibit a subtle rocking forward and backward. This is surmised to be the method the body uses to generate sacral pumping of cerebrospinal fluid that becomes compromised when the sacrum is unable to move freely.[4]

- If the sway is side-to-side, this is indicative of one or both open and unstable sacroiliac joints. The sideways swaying is an attempt to re-establish contact with the weight-bearing portion of the joints.[5]

- A figure-eight motion combines the forward-and-back and side-to-side movements and signifies normal sacroiliac joint mobility.

Forward/back sway

- If no sway is observed, other possible conditions might be present, including lumbar disc herniation.

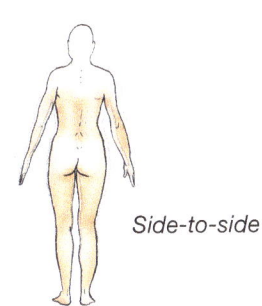

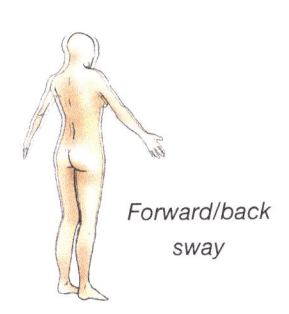

These unconscious swaying motions are controlled through a function of the nervous system called *proprioception*, which works with our sense of spatial orientation. The sacroiliac joints contain many proprioceptive sensors since it does not have direct muscle control.

In this chapter and subsequent ones, information is presented on the causes of injuries, evaluation of those injuries, and yoga therapeutic remedies. It must be emphasized that the therapeutic discussions in this book are educational. They offer guidance in administering self-assistance using yoga. They are not intended to negate, replace or discourage any evaluation or treatment from any qualified health care provider. Nor are they intended to be an absolute cure for anything.

That said, most health professionals are unaware of the therapeutic properties of yoga or how to utilize them. The techniques presented in this book, if carefully applied, are safe and may provide a breakthrough where improvement has been challenging. The appropriateness of any yoga posture or therapy is guided by the reduction of pain. If an asana or therapy is offering therapeutic value, pain will diminish during its administration and long-term ease and comfort will increase.

Sacroiliac Therapeutics

Imbalance between sacroiliac mobility and stability can be addressed using yoga asana to deliver a greater degree of either *inward hip release* or *forward tailbone scoop*. Once the type of sacroiliac imbalance is determined – either improperly open or closed - the yoga student can modify their practice and focus on asana that greater assist the action needed. Since every asana engages both *inward hip release* and *forward tailbone scoop,* doing postures that emphasize one step over the other is one way that yoga becomes therapeutic. A yoga practice that utilizes poses that emphasize both steps of pelvic integration will overall be therapeutic on its own merits.

Therapeutic Setu Bandha   Supportive Bridge Pose

Bridge Pose with the sacrum supported on a block is both therapeutic and diagnostic.

- Care is taken to place the yoga block directly under the sacrum and not on the lumbar spine.

- If one or both sacroiliac joints are tight, use the block's narrowest side to support the sacrum. This position stabilizes the sacrum while gravity slowly moves the ilia and opens the sacroiliac joints. This is a good restorative position and can be held for extended time. An add-on to the pose is to cross the leg of a tight sacroiliac joint over opposite knee, perhaps in an Eagle Pose's leg position.

- If one or both the sacroiliac joints are unstable, place the block horizontally. In this position, the block fully supports and stabilizes the sacroiliac joints. By holding the pose, the ligaments are proprioceptively stimulated and the musculature can strengthen and provide greater stabilization. The Figure-4 leg position can be used to further stabilize the sacroiliac on the side of the bent leg.

Supported Bridge Pose is also an evaluation tool for SI mobility. The relative comfort between the two positions can suggest whether the sacroiliac joints are locked or unstable.

- Choose in the horizontal block position. Hold approximately 30 seconds. Repeat the pose with the block in the vertical position. The correct position for the yoga block produces comfort, while an incorrect block placement can cause discomfort or pain. The pain indicates that the block position is either increasing the mobility of unstable joints or compressing joints that are already locked. Once the more comfortable position is determined, use only that block position when performing the pose. If no pain is produced in either position, the sacroiliac joints are normal.

If it is difficult to lift onto the highest positions of the block, its lower counterparts can instead be used with equal effectiveness.

Horizontal block stabilizes sacroiliac joints

Vertical block increases mobility

Asana that open fixated sacroiliac joints

Gomukhasana — Cow Face Pose

In **Gomukhasana**, the leg position places one knee nestled over the other. This leverages the legs to spread open the sacroiliac joints with the effect more pronounced in the SI joint of the top leg. *Inward hip release* initiates the posture and *Forward tailbone scoop* is engaged to complete the pose. Incorporating **Gomukhasana** into a regular practice helps keep the sacroiliac joints open.

Garudasana — Eagle Pose

The wrap-around leg configuration of **Garudasana** is highly effective for releasing the sacroiliac joints. It delivers powerful leverage that opens the joints and increases their mobility. Here again, the effect is greater in the sacroiliac joint of the front or top leg. The therapeutic focus in Eagle pose derives from *inward hip release,* however, as in all asana, *forward tailbone scoop* is also engaged to complete both steps of pelvic integrative alignment.

"Eagle legs" is most effective when performed as a restorative pose as lying on the back (supine) avoids the added balance challenge. Although no longer weight bearing, the feet stay active and press out through all four corners, especially through the inner heels.

A spinal twist can be added to the pose. To keep the central axis aligned, shift the hips 10-12 inches toward the side of the top leg before dropping the legs across to the opposite side. This reduces back strain. An egg-sized curve in the lumbar spine is maintained. The thoracic spine and the upper rib cage rotate toward the floor in the opposite direction of the legs. Focus on dropping the upper knee down to open the sacroiliac joints rather than overly forcing the top shoulder to flatten to the mat.

Twisting poses when the sacroiliac joints are unstable

Placing a block between the knees in a supine lumbar twist eliminates any torque created by the legs on the sacroiliac joints. The block between the knees keeps the sacroiliac joints in a neutral position. The hips cannot significantly internally rotate, which would cause the sacroiliac joints to spread farther. This procedure allows the yogi to receive the benefits of twists when the sacroiliac joints are weak. It is a protective strategy but not a therapy for either locked or unstable SI joints.

Asana and yoga therapy for stabilizing open sacroiliac joints

When the sacroiliac joints are stuck in an open position, they are unstable and cannot offer adequate weight-bearing support. Asana that bring the sacroiliac joints together will be therapeutic. The second step of pelvic integration, *forward tailbone scoop*, becomes the primary focus in these poses to close the joint and assist the healing and rehabilitation of the ligaments and associated soft tissues. *Inward hip release* is still slightly or energetically engaged at the start of each asana and therapy. It ensures that adequate SI joint space is created and that the hips, pelvis, and spine are correctly positioned.

> Every yoga posture or therapy, even when sacroiliac movement is clearly limited, engages *forward tailbone scoop,* even if only energetically, to balance the mechanical actions of pelvic integration

Sacroiliac strap stabilization

Support unstable sacroiliac joints by tightening a yoga strap across the center of the SI joints and encircling the pelvis. The strap horizontally crosses the level of the greater trochanters and measures approximately two finger distances above the pubic bones on the front of the pelvis.

Baddha Konasana   Bound Angle Pose

External rotation and abduction are engaged to create the outer form of this pose. Both actions close and stabilize the sacroiliac joints. The pose can be performed sitting or as supine, **Supta Baddha Konasana**, which may be easier and more effective when the joints are very weak.

As previously noted, although the final form of Bound Angle Pose closes the sacroiliac joints, it is important to initially engage *inward hip release* before deeply abducting and externally rotating the hips. *Forward tailbone scoop* is held in the final pose if seated or naturally engaged by the sacrum of the floor if supine.

The feet can help control the pose: the soles press together as the pose initiates to prompt *inward hip release.* The soles open like a book in the final pose to assist *forward tailbone scoop.*

Baddha Konasana

Virasana and Supta Virasana with blanket or sacral block

In the traditional forms of **Virasana** (Hero Pose) and **Supta Virasana** (Reclined Hero Pose), grounding the sacrum with the floor is important to allow the nervous system to calm and the musculature to safely release. If the sacrum is not in contact with the floor or a prop, *inward hip release* may become over exaggerated. This causes the lower back to overarch and the sacroiliac joints to open farther and become unstable. However, when a prop is positioned under the sacrum and body weight firmly rests upon it, the joints are stabilized. The block or blanket supports *forward tailbone scoop*. Any sized block can be used to provide a solid foundation. A blanket or block is always recommended anytime there is lumbar spine overarching and strain in these poses.

Even without injuries, Supta Virasana can be challenging. Tight quadriceps and weak hamstrings often contribute to its difficulty. Sacral support in Virasana and Supta Virasana is always recommended if the sacrum does not fully reach the floor. Comfort in the lower back strain is a good indicator that a support is beneficial. If the sacrum does not ground, the poses will become a sacroiliac joint opener. This may not be the yogi's desired intention. As always, to protect the lower back and sacroiliac joints in this and all asana, *forward tailbone scoop* must be firmly engaged.

Supta Virasana with sacrum on floor

Virasana with block

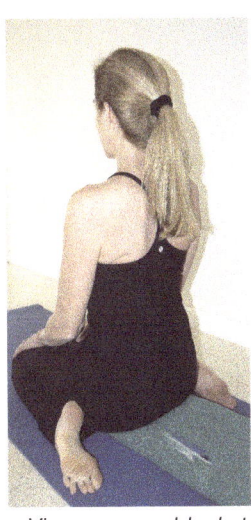

Virasana on a blanket

14 Integrative Alignment of the Pelvis

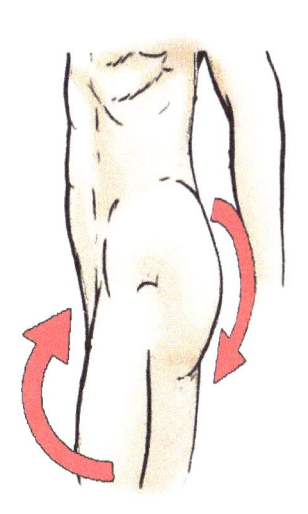

In Chapter 13, the basic principles of *pelvic integrative alignment* were introduced, focusing on their application specifically to sacroiliac joint function. This chapter expands on its effect on the pelvis and lower torso as a whole. Pelvic integrative alignment expands upon many teachings pervading through yoga studies on structural alignment, particularly embracing the work of B.K.S. Iyengar and the subsequent innovations and modifications developed in the Anusara yoga system. It offers a sophisticated and precise, yet user-friendly approach to full pelvic alignment.

YOGA ALIGNMENT PRINCIPLES AND PRACTICE

Pelvic integrative alignment applies a two-step process for the function of the pelvis. These steps are followed in every posture, even when the outer form of the asana over exaggerates one step over the other. In those cases, in may be appropriate to only energetically engage the over emphasized step.

1. Step one: Inward hip release
2. Step two: Forward tailbone scoop

Inward Hip Release – Step one

Inward hip release is the first step of pelvic integrative alignment. It consists of three interdependent actions that are activated from the upper thigh (groin), specifically engaging their movement from the lesser trochanter.

- Upper, inner thighs roll in [1]
- Front of the thighs draw back [2]
- Upper, inner thighs spread apart

Inward hip release lifts and spreads each buttock outward, widening in an oblique, posterior direction. This normally causes the buttock to stick out. Care is given to keep the knees forward facing; they do not roll inward as the hip joints engage. With practice, moving the upper femur without disturbing the position of the knee and lower leg will become easily achievable.

Effects of Inward Hip Release

- Loosens the primary hips ligaments, causing them to micro-pleat and unwrap and enable increased hip mobility.
- Sacroiliac joints slightly spread apart to allow for mobility.
- Pelvic floor widens, broadening the perineum to create space that will be needed for step two, *forward tailbone scoop.*
- Anterior pelvic flexion, which causes the top of the sacrum to tilt forward. This tips forward the foundation on which the spine rests, causing the lumbar spinal curve to deepen.
- The psoas muscles tone, helping maintain the lumbar curve [3].
- The upper, inner thighs spread apart. This produces hip adduction, which may at first seem counter-intuitive to picture. *Thighs apart* will release the hip ligaments and align the hamstring muscle fibers lengthwise, increasing both their flexibility and strength.
- Spreading the thighs apart reduces compression of the femur heads into the labral collars that secure them to the hip sockets. *Thighs apart* enables freer joint motion. It is therapeutically valuable in the rehab of labral cartilage tears. It is also a primary yoga therapy for knee rehabilitation when paired with its counter movement, *shins in.*

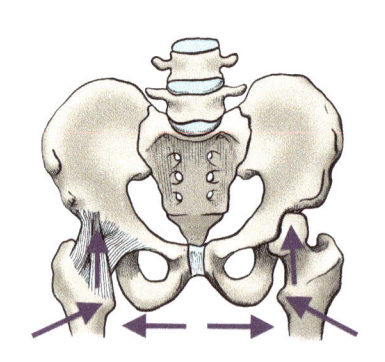

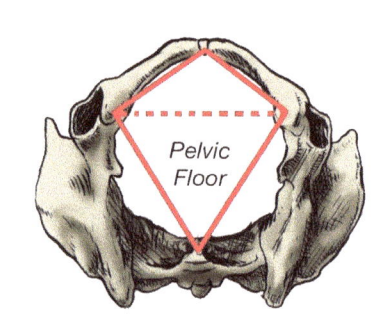

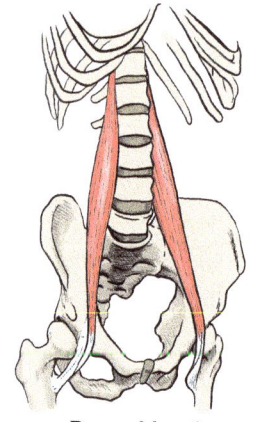

Pelvic Floor

Psoas Muscle

> "Mantra" for inward hip release:
> Roll-in - Back - Apart

Anterior Pelvic Tilt    *Posterior Pelvic tilt*

Forward tailbone scoop – Step two

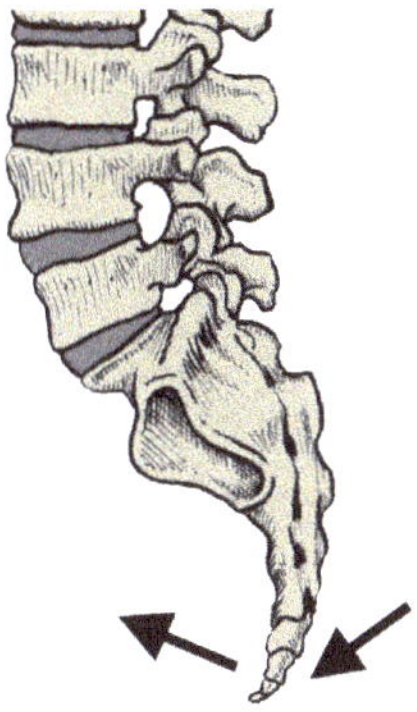

Forward tailbone scoop, the essential second step of pelvic integration, presses the coccyx forward toward the pubis symphysis. This action engages *Mula Bandha* and tones the pelvic floor.

After *inward hip release* creates a general opening and mobility to the pelvis, *forward tailbone scoop* stabilizes the pelvis. The sacroiliac joints close. *Inward hip release* externally rotates, extends, and abducts the hips, which tightens its ligaments and stabilizes the hips.

Forward tailbone scoop prevents hyperextension and compression of the lumbar discs [4]. It helps stabilize the lumbar curve in extension poses.

Forward tailbone scoop references the coccyx, which is formed by the lowest three vertebrae of the spine that fuse together by adulthood. The action of *forward tailbone scoop* represents the "intention" to move the coccyx. To be clear, intentional movement of the coccyx alone is not possible and is always joined by the sacrum, which in itself is a limited and difficult motion. The two bones basically act as one solid, non-separated unit. The amount of gross physical movement that actually occurs in *forward tailbone scoop* is minimal and is mostly an energetic action or an isometric muscle contraction.

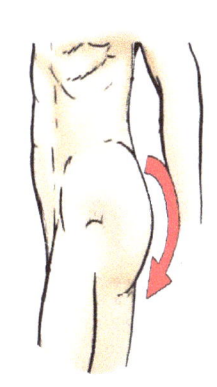

Effects of Forward Tailbone Scoop

- Closes and stabilizes the sacroiliac joints
- Externally rotates and stabilizes the hip joints
- Limits degree of lumbar curve, preventing over-arching and compression of lumbar spinal discs
- When performed subtly, the perineum gently lifts and tones to engage *Mula bandha*
- Tilts the pelvis posterior, what is also called pelvic extension.

Finding the sweet spot

In every posture, both steps of pelvic integration are utilized The balance of effort between *inward hip release* and *forward tailbone scoop* is determined by what produces the "sweet spot" for ideal pelvic alignment. Both steps are deliberate but never forced.

In all straight-legged postures, there are two determinants that indicate balance between both steps. The first is when the greater trochanters of the hips align directly over the ankles (same as over the anterior calcaneus bones). The knees are never forced into alignment but will naturally find their position somewhat close to the same vertical line.

A second gauge for balanced pelvic integration is when the combined action of both steps creates an egg-sized curve in the lower lumbar spine at the level of L4-5. Yogis should avoid allowing the lower thoracic spine and lower ribs to jut forward as an effort to create the desired curve but instead move primarily from the inner thighs and tailbone.

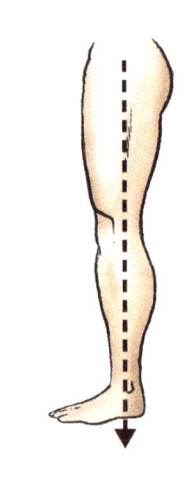

The greater trochanter aligns over the ankle when the pelvis is balanced.

Each student finds the sweet spot of balance between *inward hip release* and *forward tailbone scoop* differently. Some yoga students may naturally possess a lifted, protruding buttock while others have a flattened posterior. Those yogis with a "big booty" already have an abundance of *inward hip release* in their normal posture. They need to barely engage *inward hip release*, focusing most of their efforts on the singular action of *spreading thighs apart*. Their greater efforts and actions are needed in step two with the *forward tailbone scoop*.

Conversely, flat-butted yogis need to more deeply engage *inward hip release*, especially the action of *thighs back*. These yogis will still engage step two, the *forward tailbone scoop* but only slightly because it is already well pronounced in their normal posture.

Students with tight hips and have limited ability in hip internal rotation. Their focus would be on *inward hip release* in every asana, giving additional attention to the action of *rolling in* the upper, inner thighs.

Putting on tight pants – how to engage pelvic integration

First attempts at engaging pelvic integration might be confusing and elusive. As a visual tool, imagine the feeling of pulling up and zippering a pair of tight pants.
- When first putting on tight pants, the thighs draw back, the pelvis tilts forward, and the buttocks sticks out and broadens posterior. This is the action of *inward hip release*
- As the pants are pulled up, the thighs and pelvis remain drawn back. Finally, the front zipper is zipped from bottom up. This action corresponds with *forward tailbone scoop*

Scoop vs. Tuck

Yoga teacher Betsey Downing, PhD., asserts that the difference between the terms "scoop" and "tuck" is significant. Tucking implies a tightening and contraction where scooping suggests a more lengthened and spacious action

Scooping also allows the action of Mula bandha to be better engaged

> When alignment is applied to yoga practice with confidence and clarity, it makes our time on the mat not only more therapeutic but also much more fun!

The Human Pez® dispenser

Another effective learning tool for discovering pelvic integration is to place a yoga block between the legs at the level of the upper groin. This may conjure the image of a human Pez® dispenser.

Procedure:

1. To experience step one of *inward hip release*, place a block the narrowest side between the upper thighs, directly against the lesser trochanter of each femur. The block creates a spacer that keeps the thighs spread apart. Using the musculature of the inner, upper legs and pelvis, roll the block in and back. This action can feel like the defensive response to a feigned kick to the groin. It produces the *roll-in and drawing back* of the thighs while the position of the block maintains the spreading apart aspect of *inward hip release*.

2. To experience step two, *forward tailbone scoop*, press the fingers firmly against the front of the block. Engage *forward tailbone scoop* while the fingers prevent the block from being extruded forward (like the Pez® candy), which would lose the rolling inward and thigh back actions previously created.

3. Balance between the two actions is complete when the *forward tailbone scoop* is engaged as fully as possible while the outer hips (greater trochanters) remain in line over the ankles or front of heels. The thighs remain posterior and an egg-shaped curve in not lost the lumbar spine. The lower rib cage does not jut forward.

4. Once in balance, there is a definite sense of comfort in the lower back, pelvis, and legs. Eventually, balancing *inward hip release* and *forward tailbone scoop* becomes second nature. The mechanics of engagement will become easier, seamless and subtle, requiring less step-by-step concentration.

5. To increase the intensity of the spreading apart action of *inward hip release*, place the block at its middle width. Do not squeeze into the block; instead, press the outer shins toward the midline. The block will naturally spread the inner groin apart. This position will be appreciatively less comfortable but is an excellent therapy for limited hip range issues and especially useful for labral tears.

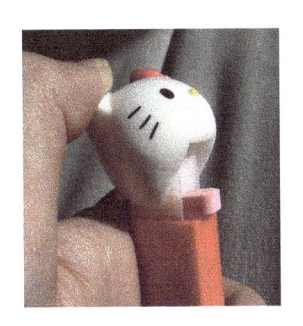

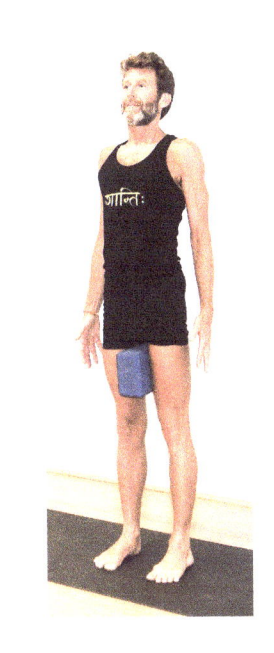

Pelvic positions

There are various terms that describe the position of the pelvis. Some terms are scientific and others have been adopted from varies exercise regimens. The anterior and posterior spines of the ilium are often used as guides to determine the pelvis' position.

Some teachers actually forego the pelvic alignment principles altogether and reference the anterior spines of the pelvis, or the ASIS, to instruct pelvic alignment. They may refer to them as "hip points" and describe moving them closer together or farther apart as the method to balance the pelvis. Some students find this easier to visualize and manage.

In time, it all becomes easier to understand and utilize.

- Neutral pelvis: ASIS is slightly higher than PSIS
- Anterior pelvis: ASIS drops below than PSIS
- Posterior pelvis: PSIS drops further below ASIS

    ASIS: Anterior superior iliac spine
    PSIS: Posterior superior iliac spine

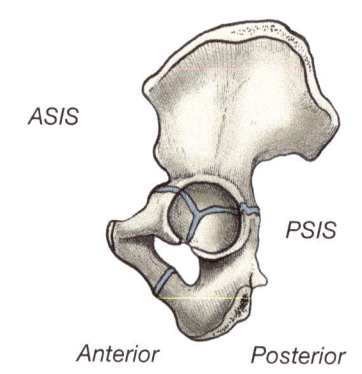

*An additional value in visualizing pelvic alignment from the point-of-view of the tailbone (coccyx) is that the physical action of *forward tailbone scoop* is linked with engagement of *Mula bandha*. By moving from the tailbone, the powerful effects that Mula Bandha provide can be utilized both energetically and in its protection of the sacroiliac joints and lower spinal discs.

Backward "Butt Walking"

This is another procedure that teaches students how to engage *inward hip release*. It can also be used therapeutically for limited hip rotation.

- Sit in **Dandasana** at the front of the yoga mat with both hips squared. Lift one hip off the mat and, from the upper inner groin, internally rotate the thigh, placing the ischial tuberosity (sitz bone) down a few inches further behind its original position. Shift your weight over to the opposite pelvis and repeat the procedure. Slowly continue the backward walking until the back of the mat is reached.
- This procedure engages all three actions of *inward hip release*. The thigh draws back and internally rotates as the hipbone moves back a few inches. As weight is shifted from one hip to the other, the thighs spread apart.
- It is also a fun teaching tool!

15 The Lower Thorax

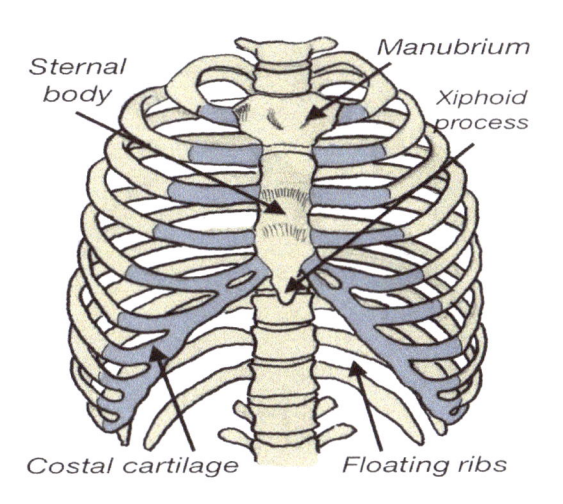

Sternal body — *Manubrium* — *Xiphoid process* — *Costal cartilage* — *Floating ribs*

Although this chapter is short and its focus specific, alignment of the lower rib cage is instrumental in establishing structural integrity in the torso. It is also required for optimum function of the extremities. The lower rib cage may seem an unlikely place to focus attention in the overall approach to alignment yet it is a pivotal region to engage in every asana. Without aligning and stabilizing the lower rib cage, the pelvis and shoulder girdle cannot integrate and function in sync with each other.

Lower rib cage alignment is simple: draw the lower floating ribs posterior and widen across the back.

Anatomically, the lower ribs have no anterior attachments to bone or cartilage on the front body. Their only structural attachment is posteriorly to the spinal column. The vertebrae that attach to the floating ribs occupy the strenuous transition between lumbar and thoracic spines. The twelfth thoracic vertebra acts as the fulcrum between the upper and lower torso and serves as the spine's primary *swivel point*.

The lower rib cage is where the diaphragm and abdominal muscles, the primary muscles of respiration attach. Drawing the ribs of the lower thoracic cage back is an important aspect of respiration.

It is challenging to move the lower portion of the rib cage interdependently, including for those yogis considered most limber. Often and incorrectly, when drawing the lower ribs back, the shoulders roll forward and the chest drops down.

98 YOGA ALIGNMENT PRINCIPLES AND PRACTICE

Conversely, the lower rib cage tends to jut forward as a seesaw type of response when drawing the shoulders back. If the lower ribs jut forward, the entire lumbar spinal curve deepens too far and the lower lumbar spinal discs can compress.[1]

Aligning the lower thoracic region reduces spinal compression by gently lengthening the spine at this dynamic transition point. Learning to isolate, align, and interdependently move the lower rib cage from other parts of the body take both practice and patience.

Method to engage lower rib cage integration

1. Lift and lengthen both lateral sides of the body, from the iliac crests to the armpits, creating copious space between the ribs. This region of the body is sometimes called the side-body ribs
2. The bottom tip of the breastbone (*xiphoid process*) scoops toward the navel and continues energetically toward the back body[2]
3. The lower, floating ribs draw back, lift, and spread across the back horizontally
4. Drawing the navel into the abdomen is not performed with a highly forceful contraction of the abdominal musculature
5. When aligning the lower rib cage, the front of the body remains lengthened and does not shorten or collapse

Visualizing lower rib cage integration

"Zip" the bottom tip of the breastbone (xiphoid process) down towards the navel while gently drawing the navel into the abdomen. This zipping action, as previously presented in Chapter 11, is used in conjunction with the upward zipping action of the fly of a pair of pants as described when pulling up a tight pair of pants.

Scoop the breastbone, Scoop the tailbone, Draw in navel

This principle is presented repetitively throughout the book, as it is an essential, fundamental action for integrating the upper and lower torso. To keep the lumbar spine safe in back bending postures, it is necessary to keep the lower rib cage from jutting forward. The actions below work in conjunction with each other to integrate the lower torso:

- **Scoop the tailbone**: stabilizes the sacroiliac joints
- **Scoop the breastbone**: prevents overarching of thoraco-lumbar and upper-lumbar spine; also resists forward jutting of the lower rib cage
- **Draw in the navel**: stabilizes the entire lumbar spine

A useful method for learning to engage these perhaps foreign and elusive actions is to perform a few, extremely slow rounds of *Kabalabhati* breathing (forced exhalation). All the steps are instinctively engaged and can be isolated and practiced until they become accessible. If you feel that, due to time restraints, only one, quick action is possible, draw inward from two inches below the navel.

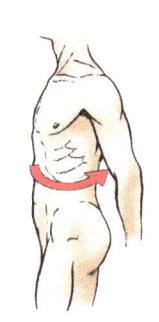

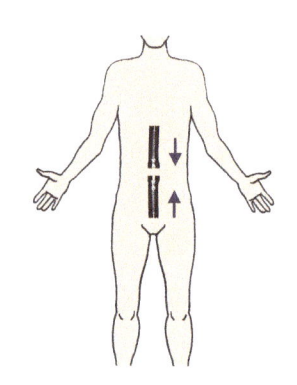

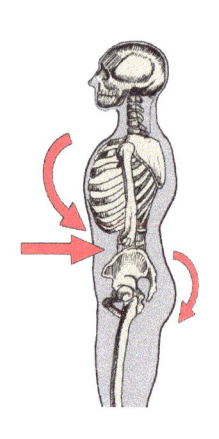

FIFTEEN: THE LOWER THORAX | 99

As mentioned in the previous chapters, *scoop the tailbone, scoop the breastbone,* and *draw in the navel* corresponds with engaging the bandhas and musculature contraction of the diaphragms.

- Scoop tailbone: Mula Bandha   Perineum
- Scoop breastbone: Uddiyana Bandha   Diaphragm
- Draw in navel: Manipura chakra   Diaphragm assist from abdominals

Another set of instructions for aligning the lower thoracic region comes from yoga teacher Jaye Martin who instructs with these simple steps:

1. Drop the front ribs
2. Draw back the side ribs
3. Lift up the back ribs

Urdhva Dhanurasana   Upward Bow Pose

The following instructions relate only to the thoracic spine:

1. Press the bottom tips of the shoulder blades anterior, located at the level of the T-7 vertebra
2. Extend the middle thoracic spine anterior to expand the chest
3. Draw the lower portion of the rib cage posterior, scooping the breastbone toward the navel
4. Draw the navel in toward the spine
5. Firmly scoop the tailbone forward to resist lower back arching

Back bend poses are essentially thoracic extension poses. The goal is not to exploit the already curved lower spine. If the above instructions are not utilized, the lumbar spine is exploited in the backbend, angled sharply and compressed instead of lengthening with a smooth arc.

Urdhva Dhanurasana, Upward Facing Bow, is also called **Chakrāsana**, the Wheel Pose. To use either term more accurately, the pose is a "Half Wheel" with an imagined lower portion of the wheel continuing below the plane of the floor. Ideally, the spine will curve evenly along the outer rim of the wheel, keeping the spokes equal in length.

Ustrasana   Camel Pose

1. The same instructions outlined for the Wheel Pose are essentially followed to open and expand the chest and thoracic spine.
2. The lower rib cage firmly draws back, preventing the lower ribs from jutting forward.
3. The lower ribs widen on the back and the navel firmly draws in.
4. *Forward tailbone scoop* firmly engages to stop the lumbar spine from over extending.
5. The hips align directly over the knees, vertically centered from the greater trochanter to the kneecaps. The hips and upper thighs do not press forward but instead engage *inward hip release* to create space in the sacroiliac joints before scooping the tailbone.

Ustrasana

Bhujangasana   Cobra Pose

1. Engage *Scoop, Scoop, Draw in Navel*.
2. The lower rib cage draws back as the tailbone scoops down toward the floor. This is a less common method but a highly effective action to align the asana.
3. The navel is firmly drawn in.

Imagine that an assistant places their hands clutched along the lower ribs. Initiate the pose by "feeling" the imaginary hands firmly drawing back of the lower ribs to make the torso lift.

Additional alignment instructions for Cobra Pose:
1. Keep the elbows in line with the side body rib cage.
2. Draw the throat back without lifting the chin. Instead, subtly lift the back of the skull.
3. When the throat draws back, the lower ribs and thighs will also draw back, naturally.

An assistant can use a strap to draw the lower ribs back with his knees stabilizing the sacroiliac joints.

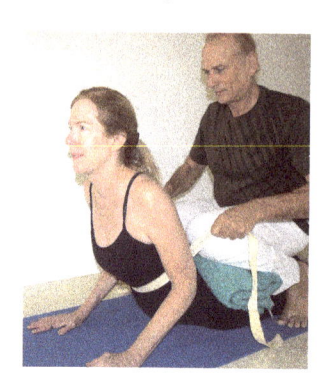

Bhujangasana

Adho Mukha Vrksasana   Handstand   Upward Facing Tree

1. From All Fours Pose, drop the chest anterior and draw the heads of arm bones back into their joints.
2. Draw the lower rib cage posterior.
3. Form an egg-sized curve in the lower lumbar spine.
4. Lift into Downward Facing Dog, maintaining all prior actions.
5. Walk the hips toward the shoulders, bringing them as forward and vertical, as possible.
6. Keep the shoulders vertical over wrists or knuckles
7. Keep lower rib cage posterior to prevent overarching the spine
8. The hips remain in anterior flexion and the lumbar spine maintains its egg-sized curve *
9. Lead to lift from the pelvis, not by extending the straight, lifted leg toward the wall. The leg does not lead the lift but behaves as one solid, direct continuation of the hip.
10. The power to hop or press up comes from the second, bent leg.
11. Once in the pose, the lower ribs draw back to avoid overarching into a "banana back". Scoop tailbone to support a straight torso.

* Overarching the lumbar spine is common misalignment in Handstand.

16 The Alignment of Sitting

"Where we start is where we're going!"

Yoga classes often begin with students being instructed to take a basic seated posture, either cross-legged in **Sukhasana** or aligning heel-to-heel-to-pubis in **Siddhasana**. Some students quickly begin to fidget, barely able to sit still while restlessly waiting for class "to start". Others are fully engaged and focused, aware that their practice began long before their mat was unrolled. For them, it is understood that the seemingly insignificant posture of sitting represents the ultimate intention of yoga asana.

The ability to sit with effortless precision is an essential prerequisite for performing the more advanced forms of yoga, such as Pranayama and the meditative practices. Sitting skillfully prevents the body from creating distraction to the mind as it ascends into greater control of consciousness. The seated posture is where we start our yoga practice and ultimately where we are going.

The rigors of asana practice put the body through a series of complex and ever challenging postures. As the student becomes more proficient with the demands of asana, the simple act of sitting becomes easier to maintain for longer periods of time.

The following list offers detailed alignment cues used in seated posture. Because these instructions are universal alignment principles, they apply, not only to the sitting poses, but also to every posture and position that the body may take, both in yoga classes and in our everyday life activities. This is one way that our lives become the constant practice of yoga.

Although the number of instructions may seem extensive, each is simple and significantly adds to and will improve overall alignment. An aligned posture enables sitting for long periods to be comfortable.

- Sit at a height where the knees do not elevate above the hips
- Close the knee joints as simple hinges, facing as much of the anterior shin inferior onto the mat. Placing the outer side of the shin down causes knee strain called tibial torsion (see Chapter 21 for more information)
- Roll the inner thighs inward, drawn them back, and spread them apart
- Lean off to each side to lift each opposite buttock posterior and oblique
- Rock forward on the ischial tuberosities until the lower lumbar spine forms a small egg-sized curve. The closer one sits toward the pubic bone on the front pelvis, the deeper the lumbar curve
- Sit broadly on the pelvic floor, enabling the perineum (Mula bandha) to make complete contact with the sitting surface
- The thoracic diaphragm and soft palate draw back to align over the perineum
- The perineum, thoracic diaphragm, and soft palate align vertically, forming the central axis of the body that rises through the posterior fontanelle of the skull
- The back of the sacrum, shoulder blades, and skull align vertically
- The 3rd lumbar vertebra and the base of the heart also align horizontally with the floor
- The throat draws back and lifts in a line toward the lower occiput at the back of the ears
- Roof of the mouth remains level and horizontal, in line with the center of the ears
- Bottom tip of the breastbone scoops inward as the chest remains lifted
- Side-body lifts from the hips to the armpits; armpits deepen without the shoulders shrugging
- Collarbones square and widen to the outer edges of the shoulders
- Shoulders move posterior, drawing back simultaneously from both the inner armpits and outer edges of the shoulders
- Shoulder blades glide onto the back toward the spine
- The inferior tips of the shoulder blades gently press forward onto the back ribs
- The head and neck balance effortlessly on the broad foundation formed by the shoulders
- Triceps muscles roll inward to bring the inner elbows in line with the side body rib cage; hands can be kept supinated or turned over since the forearms can rotate independent of shoulder alignment

While the seated posture is held, the principle of *Samasthiti* is engaged. Samasthiti is measured and is engaged along every body surface, creating subtle balance in muscular and energetic tension between each opposing part of the body.

17 The Hip Joint

Are you *hip* to this?

When a yoga teacher instructs a class to stand with their feet hip-width apart, some students slap the sides of their pelvis and shudder a silent, "Oh no!" while reluctantly widening their stance. The good news is that our "nightmare hips" are not what the teacher has in mind. The intended distance is not the broad outer crests of the iliac bones but instead, a more narrow position found two to four inches further in from each outer surface. Anatomically, *hip-width apart* refers to the distance between the hip sockets, measured from where the head of each femur comes into contact with the inner surface of each hip socket (acetabulum). The acetabulum is the point of convergence of the three bones that form each hipbone. When the legs align directly below this point, a stable standing foundation is established for the pelvis and heel-strike forces are able to travel through the center of the acetabulum and spread throughout the pelvic bowl.

How far apart are the feet when spaced hip-width apart?

The short answer is 4-6 inches, or *fist-width* to *head-width* apart.

A longer answer relies on hip anatomy to determine a mechanically desirable distance. Since the deep, inner surfaces of the acetabula are where forces of heel strike most concentrate, measuring the width from the distance between the two is appropriate. This distance corresponds to the width of the head, without the ears and is approximately the width of a standard yoga block: six inches across.

Another distance to consider is the width between the ischial tuberosities. Mechanically, when the heels fully draw back toward the buttocks, their natural points of contact are the ischial tuberosities. This distance begins at approximately a fist-width distance, or four inches apart.

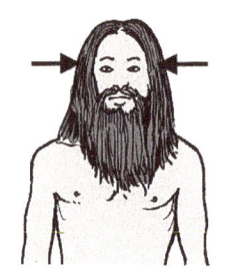

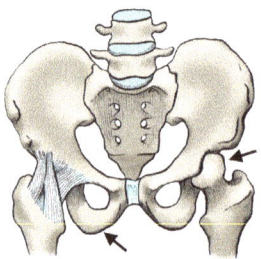

Ischial tuberosity

Some yoga traditions instruct students to touch the inner borders of the feet together. Anatomically, there are factors to consider before following this or any stylized form. Narrow stances may produce a refined appearance but are problematic for the knees. Wider stances are more stable and therapeutic and best manage the forces and tensions transferring through the lower extremities. Wider stances also reduce the actual distance to the floor that the hamstrings must negotiate in a forward fold poses, which lessens muscle strain and allows easier hip flexion without any resistance from tight hamstrings.

Anatomically, a stance of 4-6 inches is the sweet spot. However, some students with narrow hips and are very flexible may benefit from using a narrower stance in closed hip, revolved postures such as **Parivrtta Utthita Trikonasana**, Revolved Triangle.

A snug fit

The femur head fits securely in the acetabulum. A dense ligamentous capsule and a thick cartilage collar/gasket called the labrum surround the joint.[1] It creates a snug fit with a suctioning, vacuum effect on the femur head, which makes dislocating the hip difficult without tearing the cartilage first to release the suction. According to orthopedist and yoga anatomist Ray Long, MD, labral tears occur at nearly epidemic levels. By age 35, more than 70% of people have torn labrums. That number increases to 90% by end of life.[2] Labral tears are often a pre-cursor to hip osteoarthritis and degeneration. Proper hip alignment and moving according to the design of the joint and ligaments are essential for injury prevention and the rehabilitation of existing labral tears.

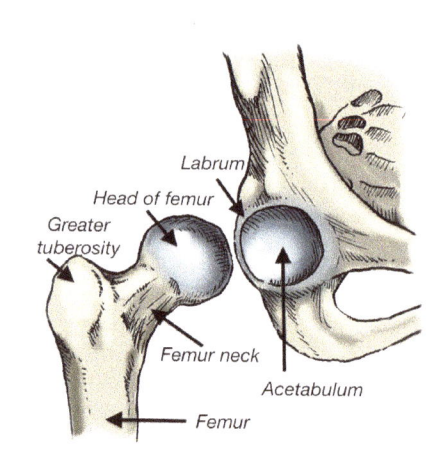

Labrum
Head of femur
Greater tuberosity
Femur neck
Acetabulum
Femur

As first introduced in Chapter 13, one alignment action engaged in pelvic integrative alignment is to widen the inner hips from side-to-side. This action increases the hip socket's "joint play", enabling freedom for easier mobility. Importantly, this action reduces compression on the hip socket and torsional tension on the labrum. The action of widening the hip joint does not alter the alignment of the femur head on the central axis of the acetabulum. Widening the hips is an essential tool when asana is used therapeutically to rehabilitate a hip with a labral tear.

The swing of things

The hips are designed to pivot between the two femur bones in a fashion similar to the swinging of an upside-down bucket between its handles. This configuration is beneficial for the knees, swinging toward the midline of the body as they flex, an action necessary for walking and body balance. The ideal angle for the neck of the femur bone to resist hip dislocation and provide proper swing is approximately 120-125°. A wider pelvis as often seen with women may bring the angle as small as 90°, producing forcible stress on the hips and the knees.

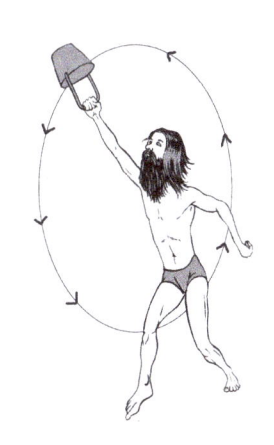

Hips move in three axes of motion

Hip ranges of motion occur along axes in three planes - *sagittal*, *frontal* and *vertical*. In the sagittal plane (forward-back), the hips flex and extend. In the frontal or coronal plane (side-to-side), the hips move either toward the body's midline (*adduction*) or away from the midline (*abduction*). Along the vertical axis (superior-inferior), the hips rotate internally and externally.

"Mind the gap" – deepen the hip crease

The region of the front hip and upper thigh is called the *femoral triangle*. It is a dense location with massive muscles and tendons. To effectively create movement from this area, a deep, spacious *hip crease* must be formed. A deep hip crease is formed by enlarging the "gap" between the anterior femur and the anterior spine of the pelvis (ASIS).

In **Uttanasana**, Forward Fold Pose, deepen the hip creases by lifting the ASIS up and away from the thigh and drawing the femurs back.

For greater precision, deepen the hip creases more fully at their lateral, outer aspects. This area is where greater tissue compression can occur and can compromise the anterior fibers of the *tensor fascia latae*.

Therapeutically, a partially rolled blanket or yoga mat is wedged deeply into the anterior space between the femur and the ASIS. This is most easily performed in a standing forward fold. The blanket creates a physical opening of the gap and compresses the myofascia of the upper thighs. It also produces tactile stimulation to the body that assists the sensory nervous system in "remembering" to produce the gap space.

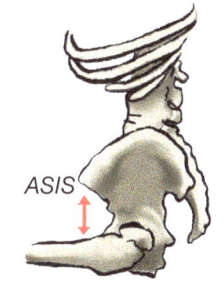

Femur

Uttanasana

Avoid the pinch

Hip abduction, bringing the leg out to the side while in a forward-facing position, causes the greater trochanter of the femur to jam into both the upper rim of the acetabulum and the lower ridge of the iliac crest. This irritates the bone and surrounding soft tissue. To avoid abrasive injury, the greater trochanter must be diverted to avoid the pinch. External rotation, turning the femur head outward, is necessary when abducting the hip. For the front foot forward in most *open hip* poses such as **Utthita Trikonasana** (Triangle Pose), deepen the outer part of the hip crease, then externally rotate the femur head.

When the front leg turns outward (external rotation) and abducts, the hip ligaments tighten, stabilizing the joint. If the hips externally rotate before the hips have fully aligned, the hips cannot fully and safely open.

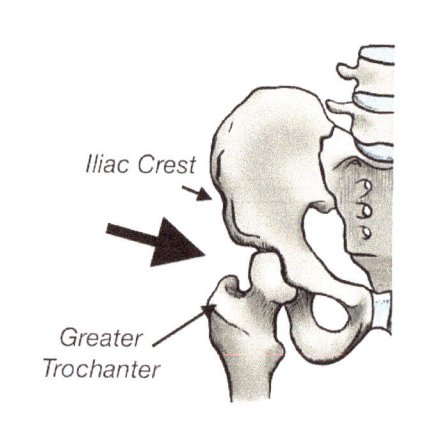

To remedy:
1. Slightly bend the front, straight knee and shift the knee medial
2. Deepen the outer hip crease and draw the femur head back and wide to release the hip
3. Glide the knee lateral until kneecap aligns over small toe side of the foot
4. Re-straighten the leg

External rotation of the hip also occurs in the standing leg hip in Wild Dog Pose. The hip of the upper, raised leg internally rotates while forming Three-legged Dog Pose and is maintained as the hips stack vertically and the top knee bends into Wild Dog; the same action occurs in Nataraja the Dancer's Pose.

Prasarita Padottanasana   Wide-Angle Forward Bend

In this pose, the hip joints are almost fully abducted, a direction that tightens hip ligaments and limits joint mobility. To loosen the ligaments, fully engage *inward hip release*.
1. Align the feet forward and parallel to each other, which helps resist external rotation
2. Once forward hip flexion begins, avoid a lateral pinch at the greater trochanter by widening from the outer portion of the hip crease
3. *Forward tailbone scoop* is added when forward flexion reaches 90°. This reduces stress on the sacroiliac joints and hamstring muscles where they attach on the ischial tuberosities

Prasarita Padottanasana

In most *closed-hip* postures, the hip joint of the rear leg must internally rotate to allow the ligaments to loosen to enable the hips to square forward.[2] Warrior One and Revolved Triangle are examples of *closed-hip* poses where joint mobility of the rear hip determines how well both hips can square to the front of the mat. All of the actions of *Inward hip release* are engaged in the rear hip of most asana.

In the final position of most standing poses, the front leg externally rotates and abducts, causing the ligaments to become taut to provide stability. These are actions that occur with *Forward tailbone scoop*, which is often engaged with greater effort for the front leg in these poses.

The ligaments' micro-pleating system plays an important role in asana alignment and in preventing injuries (details on ligament function found in Chapter 7). This chart below reviews the directions of movements and their effect on the ligaments.

Ligaments Loosen
- Flexion
- Internal rotation
- Adduction

Ligaments Tighten
- Extension
- External rotation
- Abduction

Baddha Konasana   Bound Angle

The outer appearance of Bound Angle widens the hips toward the floor, bringing them into external rotation and abduction, two directions that limit mobility. To counteract, apply *Inward hip release* and deepen the hip creases. Rock the pelvis forward toward the pubic bones. Rotate the hips internally and lift the gluteal muscles up, back, and oblique.

The Pinwheel Pose

Pinwheel Pose is challenging for yoga students who have limited hip internal rotation. For the pose to be effective, the hips, knees, and ankle joints all form 90° angles. The ultimate position is to sit squarely and evenly on both ischial tuberosities with the spine vertical. It may be difficult to bring both ischial tuberosities to the floor but it is important not to force the legs down. Support can be placed under the pelvis or knees to minimize strain on the joints. Explore hip joint mobility and ligament and tendon release by gently "pistoning" the internally rotated leg into its hip socket with added side-to-side, figure eight movements. The effects of Pinwheel Pose are more pronounced on the internally rotated hip. Used gently, it can be a valuable therapy for hip arthritis.

Agnistambhasana   Fire logs pose

Fire Logs Pose is a challenging hip-opening pose. The shin of one leg is placed over the other, as if stacking firewood. To be effective, the ankle of the top foot rests on top of the kneecap of the lower leg with the sole extending beyond the knee's outer surface. Both Achilles' tendons, as viewed from behind the ankles, remain smooth and "wrinkle-free".[13]

The outer form of Fire Logs Pose has a significant degree of hip external rotation, which makes this pose difficult. By firmly engaging *Inward hip release,* a greater, safer hip opening can be achieved. Sitting forward on the pelvis toward the pubic bone also assists hip mobility. To protect the knees and ankles from torsional strain, press the inner heels outward and smooth out the Achilles tendons. Pressing inferiorly through the inner heels assists hip internal rotation and further releases the legs.

Baddha Konasana

Pinwheel Pose

Agnistambhasana

In an externally biased world, internal rotation is highly valued

Many of our daily activities tend to splay our legs apart, externally rotating and abducting the hips and thighs. Sinking into a car seat or collapsing on a soft sofa causes the legs to spread apart. The now illegal "man-spreading" once seen on New York subways is a good example. These sitting postures cause the hip flexor muscles to become taut, particularly the iliopsoas group.

Standing with the knees locked in hyperextension and feet pointed outward causes hip external rotation with the thighs to rolling forward and out. This stance often appears when students are fatigued or in their lazy posture. It requires less muscular effort to be "somewhat" erect. The calf muscles are very powerful and in theory, when contracted just by themselves, there is enough strength to hold the entire skeleton upright if all the joints are held stable. Postural habits that over-emphasize external rotation may overdevelop the musculature in the calves. This is seen with students of ballet, martial arts, and athletics that require turned out stances.

An additional challenge to internal rotation of the hips comes from the myofascia of the upper thighs. It possesses a naturally occurring, externally rotating torque. The more the upper leg muscles develop, the greater the degree of torque

Virasana (Hero Pose) and Supta Virasana

With all of the external rotation existing in our lives, internal rotation gets little opportunity to develop. Few yoga asana emphasize internal rotation of the hips. The Pinwheel pose is an excellent one. A variation sometimes called Supine Figure-4 Pose can be performed supine, dropping one bent knee to the midline and placing other ankle on top of its lateral side. **Virasana** (Hero Pose) and **Supta Virasana** (Reclined Hero Pose) are two additional postures that increase internal rotation, as long as the knees remain less than hip-width apart. Supta Virasana adds the challenge of stretching the quadriceps muscles, which tend to be short and tight. Effort is made to resist overarching the lower back, which is a compensation for the limited flexibility of the quadriceps muscles when maximally stretched.

Virasana

Supta Virasana

> *Inward hip release* is engaged whenever mobility is required,
> especially in the rear hip in standing asana

18 Hip Extension

As presented in the previous chapter, healthy functioning hips are essential for a safe, advancing asana practice and overall physical wellbeing. Anyone challenged by pain or arthritis in the hip joints needs no reminder of the fundamental role that the hips play in almost all of our daily activities.

More than seventeen muscles move each hip joint through a total of six different directions.[1] One of these directions is extension: movement of the femur posterior from the center of the hipbone.

Hip extension might seem to be a relatively simple concept but clearly there is something significant about hip extension to warrant its own discussion!

The hip's limited range of motion

It is important to know the distinction between moving into extension and moving in the *direction* of extension. A straight joint, which forms a 180° angle between its two bones, is in its neutral position. Beyond straight or neutral, or greater than 180°, the movement is not extension but is hyperextension, a physiologically unsafe position.

Anatomists once argued that the hip joints are not capable of extension, following the observation that the knees and elbows do not extend. Hip extension is deceptively limited. In most asana, attempts at hip extension mostly extend the lumbar spine, barely bringing the hip into its neutral position. If observing the typical yoga student, true hip extension is rarely achieved, despite the many poses that call for it.

The physiological range of hip extension is:

- 10° with knee in full flexion
- 20° with knee straight
- 30° in a passive (assisted) stretch

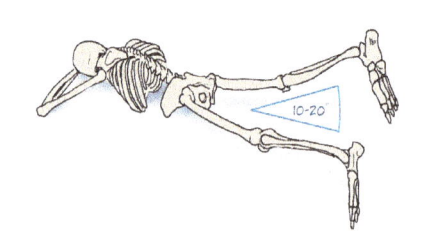

The degree of hip extension is determined by the following unique factors:

- The shape of the acetabulum. A shallow, flatter articular surface with a less developed brim allows greater range of motion. See details in Chapter 11 for an anatomical review of the acetabulum.
- The genetically determined percentage of elastic-type collagen fibers in the hip ligaments.
- Strength of the hip extensors, particularly the gluteus maximus and hamstring muscles.
- Resistance created by the *antagonist* hip flexor muscles. The hip flexor muscles, the quadriceps and iliopsoas, play a significant role in the degree of available extension. When the flexors are taut and tight, hip extension becomes restricted. The further the knee closes into full flexion, the greater the tension on the quadriceps and the less hip extension, reducing porportionally from 20° the 10°.

Experiencing the limits of hip extension

Lie on your stomach (prone) with both legs straight. Reach under the front of the hip joint and observe the hip crease space between the hip joint and the floor. The space reveals that, even at rest, the hip maintains a slight degree of flexion.

With the anterior iliac crests in contact with the floor, lift both legs while avoiding arching of the lower back. Most students find it difficult to raise the legs into extension without arching the back. One-leg extensions use resistance provided by the floor against the non-lifted leg to make the maneuver easier. With either pose, care is taken not to allow the hip(s) to lift from the floor, which will halt hip extension and immediately engage arching of the lumbar spine.

> *Antagonist* muscles are muscles that oppose each other's action. They are often paired and located opposite one another along a joint.
>
> *Synergists* are muscles that work together. They are generally found on the same side of the joint, above or below.
>
> (More details will follow in Chapter 19)

Limited hip extension can injure the lumbar spine

Without skillful action, hip extension can force the joint to move beyond its normal limitations. Combined with a student's exuberance to create deeper postures, injury can occur. Most often, strain and any resulting injuries are to the lumbar spine.

Many asana involve hip extension, an actual movement in some postures and a less obvious, isometric muscular engagement in others, such as the rear leg in Lunge Pose and Warrior One. Because the lumbar spine is already in extension and hip extension is so limited, it is easiest for the body to engage a pose with extension from the lumbar spine before any hip extension begins. Lumbar hyperextension occurs easily but results in compression of the discs and the nerves of the lower back. It can bruise the vertebrae and, worse, result in disc herniation. Chronic back pain often results from repetitive lumbar hyperextension. Flexible yogis who routinely hyperextend the lower back may not notice pain for many years. Unfortunately, spinal degeneration can start long before any symptoms begin.

Hip hyperextension trauma can also occur at the front of the hip socket. Repetitive overstretching in hip extension can weaken the ilio-femoral ligament that stabilizes the head of the femur from slipping forward in the socket. Once this ligament is overstretched, the hip may become permanently unstable. Degenerative hip joint disease and eventually hip replacement is a common result of chronic hip instability.

Natarajasana (Dancing Shiva) challenges the limits of hip extension and put the lumbar spine at risk of hyperextension. The lower back is often hyperextended without any significant hip extension engagement.

The integrative alignment principles of *Scoop the tailbone, Scoop the breastbone, Draw in the navel* are essential to apply in Dancer's Pose to reduce the likelihood of lumbar spinal trauma.

Hip flexion, by comparison, is significantly more accessible. Although this imbalance is part of the our natural gait, it nevertheless contributes to imbalance in the leg muscles. The quadriceps on the anterior thigh build strength needed for forward flexion, but their flexibility is limited. The hamstrings on the posterior thigh are less activated due to limited extension and weaken, despite their extensive use in walking.

Natarajasana

Eka Pada Urdhva Dhanurasana

Comparison: hip flexion and extension

Hip flexion ranges:
- 90° with straight leg
- 120° with bent knee
- 120° passive straight leg stretch
- 140° passive with knee bent

Hip extension ranges:
- 10° with knee bent
- 20° with knee straight
- 30° in a passive stretch

Muscles of hip extension

The gluteus maximus and hamstring muscles are the major muscles involved in hip extension. When demand is low, such as during the back swing of the leg in the normal gait (walking), the hamstring muscles provide most of the power for hip extension. As demand for hip extension increases, such as when running and climbing stairs, the gluteus maximus contraction increases. The quadriceps muscles oppose hip flexion, making them antagonists to the hamstring muscles. The iliopsoas is considered to be the primary antagonist to the gluteus maximus.

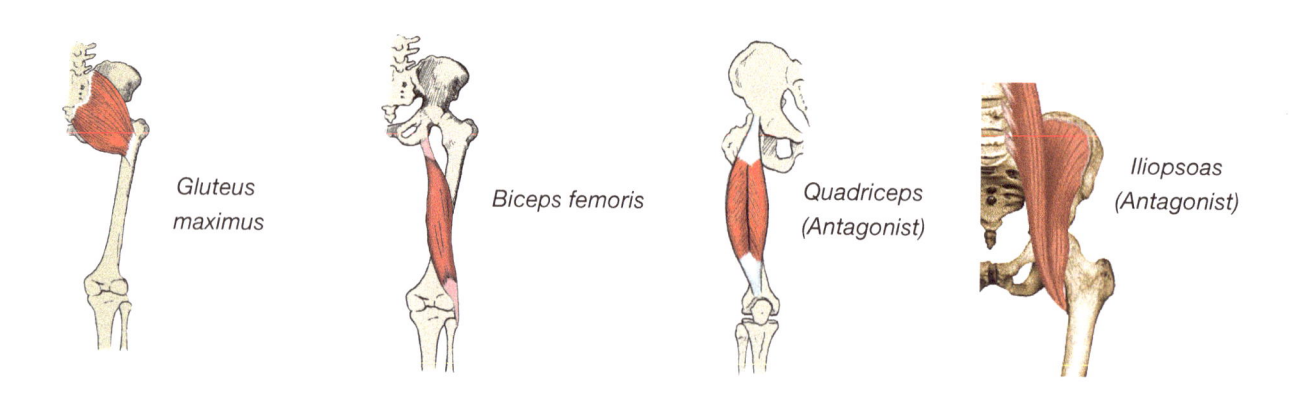

Gluteus maximus *Biceps femoris* *Quadriceps (Antagonist)* *Iliopsoas (Antagonist)*

Quadriceps stretch

One consequence of limited hip extension is that the quadriceps muscle group has limited opportunity to deeply stretch, causing the muscles to become tight, short, and ultimately inflexible. Conversely, tight quadriceps limits the degree of hip extension that is available. Stretching the quadriceps directly increases hip extension. It is also an important practice for improving hamstring flexibility.

To stretch the quadriceps:

1. Bend one knee, placing kneecap as close to the wall as possible.
2. The front leg lunges with its knee directly over the ankle.
3. The rear foot plantar flexes (points) and aligns along the outside of hip, as in **Virasana** (Hero's Pose). The foot turns out, slightly lateral to reduce sickling of the foot.
4. Slightly tilt the pelvis anterior to slightly flex the hip. This releases the hip ligaments and relaxes the quadriceps. Further discussion on flexion before extension is found at the end of this chapter.
5. Straighten the torso toward into **Tadasana**, attempting to place the three "S"s against the wall: the sacrum, shoulders, and skull. As the ability to stretch in the quadriceps progresses, the sacrum, shoulder blades, and skull may eventually all touch the wall.
6. Engage *Scoop the tailbone, Scoop the breastbone, Draw in the navel* to reduce lumbar spine hyperextension and stabilize sacroiliac joints.
7. For added stretch, slide the pelvis forward and bend the front knee more deeply to engage **Anjaneyasana** (Crescent Lunge). The rear hip will experience additional extension in this position
8. Arms can be lifted overhead and contact the wall, if desired.

"Intense" quadriceps stretch

EIGHTEEN: HIP EXTENSION | 113

Iliopsoas muscle stretch

Because hip extension is limited, the iliopsoas, which is the major hip flexor and an antagonist to the hip extensors, has less opportunity to stretch. Iliopsoas contraction is required in most postures and plays an important role on the weight-bearing side in balancing poses. But an opportunity to deeply stretch the iliopsoas is rare. Flexibility of the iliopsoas is an important factor to increase the range of hip extension. Runner's Lunges poses are a valuable way to stretch this muscle group.

For a deeper stretch:
From a supine position, drop one leg over the edge of a massage table or a set-up of raised yoga blocks. The dropped leg extends the hip and its iliopsoas stretches. The sacrum and opposite hip are held stable by the table or blocks. To prevent lumbar hyperextension and sacroiliac strain, engage *Scoop the tailbone, Scoop the breastbone, Draw in the navel*. The dropped leg is considered the rear leg of the posture and actively engages *inward hip release* with the thigh rolled-in and drawn back.[2]
This stretch can also be performed simply with one block placed under the sacrum and both legs dropping and stretching toward the floor.

Extenuating circumstances

Two major challenges in hip extension asana: resisting lumbar spine hyperextension and keeping the thighs firmly drawn back. Engaging *Inward Hip Release* throughout extension draws the thighs back and loosens the hip ligaments for greater extension. Engaging *Scoop the tailbone, Scoop breastbone, Draw in the navel* throughout the posture protects the sacroiliac joints and lumbar vertebrae.

When carefully observing the degree of extension the hips achieve in extension postures, it is evident that the thighs barely move from a straight position, despite the deceiving outward appearance that a deeper action is occurring. Limitation of hip extension is obvious in the poses below and in additional postures such as **Virabhadrasana I** (Warrior One) and **Anjaneyasana** (Crescent Lunge). The postures below provide an excellent way to develop greater quadriceps flexibility and hip extension.

Dhanurasana

Supta Virasana

Setu Bandha Sarvangasana

Urdhva Dhanurasana

Bi-articular muscles: Why thigh muscles are massive and less flexible

Another name for a joint is an *articulation*. When a muscle crosses over two joints, it is called a *bi-articular* muscle. The role of most bi-articular muscles is to transfer the power of muscle contraction across two joints. When a muscle's primary duty is to deliver power, it is most efficient when it is short, massive, and remains under tension (tone). The side effect of this is reduced flexibility.

The muscles of the thighs are bi-articular muscles. They transfer the powerful, contractile forces of locomotion from the hips to the knees. Because of the demand for power transfer on the quadriceps and hamstring muscles, flexibility is often compromised. Cyclists, runners and those engaged in sports that require surges of explosive movement such as football and basketball are often challenged in their flexibility. Power lifting can also cause the thigh muscles to become massive and less flexible although some recent studies have produced contradictory but interesting findings.[3]

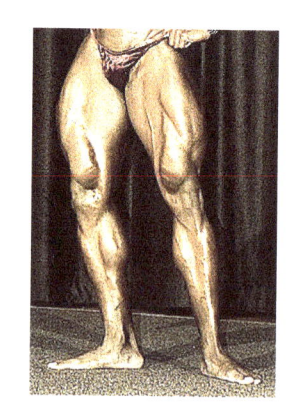

Safe stretching of massive muscles should be slow, in small increments of release, and originate from the muscle bellies. Hugging the muscles to the bone while stretching or eccentric stretching is also effective.

Flexion before extension– the anterior pelvis

When the pelvis tilts forward (anterior), the hip slightly flexes. Those few degrees of flexion generated from anterior pelvic tilt can be "borrowed" and applied in hip extension poses. Flexion releases the ligaments and relaxes tension on the quadriceps, adding extra mobility before the limitations of actual extension are reached. Tilting the pelvis anterior, however, cannot be wildly exploited. Although it allows the pose's outer form to go deeper, it also increases the base angle of the sacrum and with that, an increase in the lumbar curve and potential compression.

The "extra" degrees of flexion can be used poses such as **Dhanurasana** and **Salabhasana**, Locust Pose, High Lunge, and Quad Stretch at the wall. In **Eka Pada Adho Mukha Svanasana**, One-legged Downward Facing Dog, the lifted leg slightly flexes anterior at the hip. This adjustment does not increase true hip extension but will make it easier to lift the leg while keeping the pelvis square. If the pose opens up to Wild Dog, the hips stack vertically. The lower hip must externally rotate, an action that tightens the hip joint. Anterior pelvic tilt loosens the hip ligaments. Anterior tilt also helps form and support the lumbar curve. As discussed earlier, anterior pelvic tilt is one of the components of *inward hip release*. In almost all postures, the rear leg engages all components. Widening the inner groin is also a good tip to help increase extension.

Anterior pelvic tilt

Eka Pada Adho Mukha Svanasana

Some degree of *Forward tailbone scoop* is always important to engage and necessary to balance the action of anterior pelvic tilt. In all postures, *Forward tailbone scoop* protects the sacroiliac joints from strain and lumbar spine from hyperextension injury.

19 Alignment of the Legs

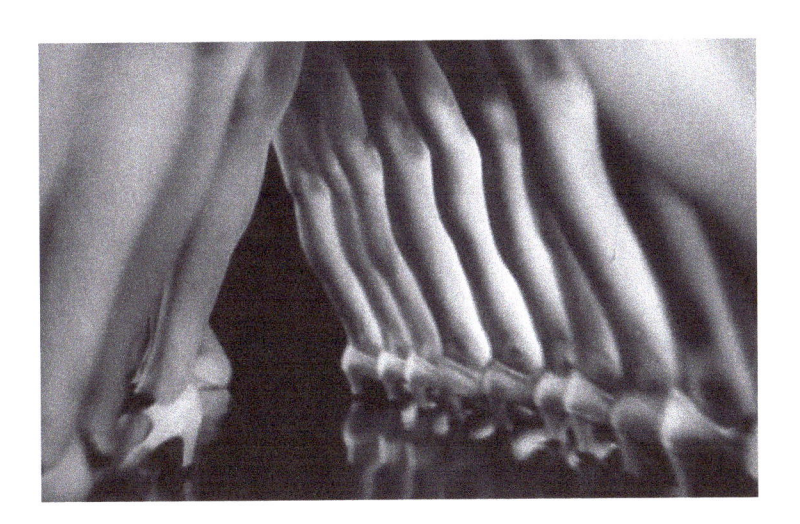

The legs are the masters of the pelvis

Perhaps you have already heard this statement. Expanding this concept, the legs provide not only the foundation for the hips and pelvis, but having a direct effect on alignment throughout the body.

The legs provide stability that is essential in many postures, yet they are not stiff and static. The legs are constantly active, balancing muscular tension with the forces of gravity and heel strike. Alignment of the legs, especially through the leg's vertical axis, manages the powerful mechanical demands that transfer through the hip, knee, and ankle. Leg alignment and integration is essential for maximizing it all: flexibility, strength, mobility, and muscle efficiency. All therapeutic approaches to the joints and muscles of the legs rely on precise alignment to produce successful healing.

Benefits of leg alignment in yoga practice

- Protects the knees from hyperextension and its resultant injury
- Prevents collapsing of the ankles and arches of the feet
- Increases the efficiency, flexibility, and rehabilitation of leg muscles by aligning the muscle fibers, especially in the hamstring muscles

General alignment of the legs

From the side view of the leg:

- The greater trochanter of the femur aligns vertically over the lateral malleolus (ankle bone)
- The knee does is not forced to be in line with the hip and ankle
- The femur aligns vertically with the tibia. The substantial difference in muscular thickness between the anterior thigh and anterior shin may make the thigh appear more anterior when it is actually aligned
- The knee maintains a micro-bend to prevent hyperextension

From the front view of the body:

- Center of the hip socket aligns with second toe and center of ankle
- Here, the knee aligns with the hip and ankle. If it deviates medially or laterally, specific corrections are utilized (instructions discussed later in this chapter)

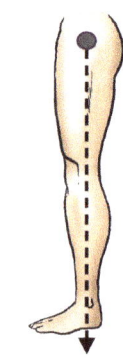

Greater trochanter aligns vertically over ankle

Center of hip socket aligns over center of knee and second toe

In Open Hip postures such as **Utthita Trikonasana**, the Extended Triangle, the ankles align vertically with the center hip socket. Typically, students are instructed to position the front heel in line with the center of the inner arch of the back foot. Should a student have fully flexible and open hips, it is mechanically correct to align heel-to-heel. For most students, however, their hip range of motion is adequate but not fully open, making the first, more common instruction appropriate.

The Q-angle

Viewing the leg from the front, the shaft of the femur does not descend along a perfect vertical line. Instead, it forms an angle 3° lateral of true vertical. This angle can vary, based on individual hip-width differences. Students with wider hips often possess wider angles.

The Quadriceps angle, or *Q-angle* is a measurement used to assess the potential for knee injury. It determines if the angles of force through the leg are excessive, particularly if the forces of muscle contraction are balanced above and below the knee.

The quadriceps muscle tendon is angled as it crosses over the kneecap to its attachment on the tibial tuberosity. A desirable Q-angle does not exceed 15° in men or 20° in women. The lower the angle, the safer is the knee and the higher the efficiency of the quadriceps.

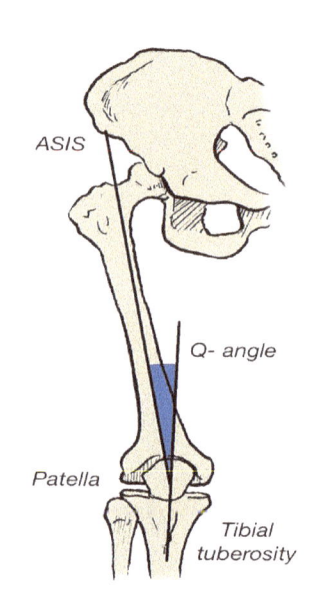

> **Technical details for the Q-angle**
>
> The Q-angle compares the contractile tension of the rectus femoris muscle with tension created by the patellar tendon on the tibia. The Q-angle is measured in the frontal plane by establishing two ascending lines - one from the tibial tubercle to the middle of the patella and the other from the middle of the patella to the ASIS.
>
> If the angle increases beyond the normal range, the patella will track improperly in the groove formed between the femoral condyles where it is seated. This can result in pain and eventual damage to the knee joint. *Patellofemoral arthralgia* and degenerative joint disease are complications arising from a chronically increased Q-angle.
>
> The Q-angle often increases as a result of collapsed arches and excessive foot pronation. This causes the tibia to medially rotate and the tibia to twist, what is called *tibial torsion*. The result is excessive strain on the knee and the quadriceps tendon. If pronation of the foot is severe, orthotic correction may be appropriate.

Compensation for a large Q-angle

1. The simplest way to compensate for a large Q-angle is to take a wider stance. Adding 1-3 inches of distance between the feet significantly reduces the angle
2. Slightly bend the knees; widen the distance between them. Re-straighten knees with new width
3. Engage the gluteus medius and tensa fascia lata (TFL) muscles by "squeezing" outer hip, which will isometrically abduct the femur head. This action helps to reduce deviation of the knees toward the midline (knock-knees)
4. Spread the thighs apart from the inner groin
5. Press firmly through the front, outer foot

Specifics of leg alignment

Alignment utilizes either actual movements or is simply isometric or energetic actions. This applies to the legs and the whole body, in general. And, the same alignment instructions apply to all asana, not only for standing postures. Exceptions, where they apply, will be noted.

The legs align in all three planes of the body - sagittal, frontal, and axial:

- Sagittal plane: Anterior/posterior (front-to-back)
- Frontal plane   Medial/lateral (side-to-side; also called the coronal plane)
- Axial plane: Central (vertical, or top-to-bottom)

In each plane, the shins and thighs counter-balance with each other. This makes the legs stronger and more stable, yet also more flexible. Each alignment principle is named simply by the action it performs.

The actions are:

- Sagittal plane: Shins forward-thighs back
- Frontal plane: Shins in-thighs apart
- Axial plane: Lengthwise contraction (from hip to heel and back)

Shins forward-thighs back

In the sagittal plane, the shins press forward and the thighs draw back. This action is engaged in every asana, whether standing or seated; legs bent or straight. Follow these steps to better facilitate this action:

1. Slightly bend the knee by pressing the shin forward
2. Stabilize the shin in its forward position
3. Draw the thigh back while resisting the shin to remain forward
4. If straightening the knee, stop with a slight micro-bend
5. Squeeze into the inner and outer side of each knee (collateral) to stabilize the micro-bend position

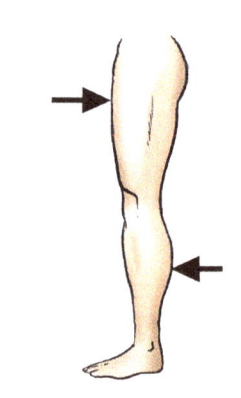

Shins forward-Thighs back

Engaging *Shins forward-thighs back* can be an elusive action for both the beginner and experienced student. In fact, some practitioners habitually reverse these directions. This reversed misalignment can affect the entire posture, causing the lumbo-thoracic spine to overarch and upper back muscles to tense. This reversal of correct alignment- drawing the shins back instead of forward- unfortunately "feels" natural because it exploits the extraordinary power of the calf muscles to support the entire body, especially when the legs fatigue.

Shins forward-thighs back prevents hyperextension of the knees. Hyperextension overstretches the posterior cruciate ligaments and compresses the knee cartilage (meniscus). *Shins forward-thighs back* is a fundamental tool in yoga therapy to rehab many injuries or dysfunctions of the knees. More details are presented in Chapter 21.

When standing, eliminate any visual confusion caused by a greater thickness of the quadriceps by aligning the greater trochanter of the hip vertical over the ankle. This action brings an imaginary seam of the pants in line vertically. Throughout this effort, knee hyperextension is resisted.

Alignment Tip: When straightening the legs, resist (press) the shins forward and draw the thighs back

Shins in-thighs apart [1]

Shins in-thighs apart aligns the legs from side-to-side, or across the frontal plane. This action helps stabilize the knees, increases hamstring flexibility, and can improve flattened foot arches.

The *Shins in* portion of the action contracts the peroneus muscles (Chapter 33) located on the outer shin, which directly affect the arches.

Thighs apart contracts the muscles located on the outer hip and thigh, particularly the tensor fascia lata and the gluteus medius. When the feet are weight bearing, the adductor muscles also are engaged.

Thighs apart is best engaged from the upper inner thighs, spreading apart the lesser trochanters found on each upper, inner humeral shaft. *Thighs apart* is one of three actions engaged in the pelvic integrative alignment principle of *Inward hip release.* [2]

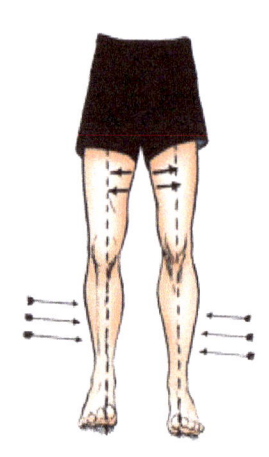

Shins in-Thighs apart

Lengthwise contraction

The long axis of the leg runs vertically, from the center of the hip socket through the ankle and directly anterior of the heel bone (calcaneus). The musculature of the leg engages isometrically along this axis in both directions. This is referred to as *lengthwise contraction*. It both "roots" down from the hip to the foot and equally draws up from the foot to hip.

Although both directions are always engaged, one may be more accessible than the other in certain postures. Contracting upwards is more easily activated in non-weight-bearing and seated postures such as **Paschimottanasana**, the Seated Forward Bend. To engage in Happy Baby Pose, **Balasana**, press from the bent kneecaps downward into the hips before bringing the knees toward the armpits.

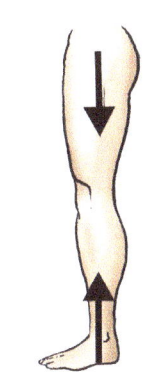

Lengthwise contraction

Equator of the body: *Lengthwise contraction* engages the lower equator of the body. As described in Chapter 12, the center of each hip socket is on the same horizontal plane as the sacral-coccyx juncture, the body's energetic equator. *Forward tailbone scoop*, which engages Mula bandha, also coordinates with the action of *lengthwise contraction* and contracts the leg's musculature.

> Lengthwise Contraction - Southern Equator - Mula Bandha - Forward Tailbone Scoop: All function together

Are you pulling my leg?

These instructions may seem somewhat abstract or confusing! If so, go back and slowly re-read. Also, try to visualize each step of alignment and then engage each, one at a time. Eventually, using the concepts in practice will make them fully understandable and obtainable. Mastering the principles of leg alignment will change the very nature of your entire asana practice, not only standing poses!

Asana and yoga therapies presented on the next few pages put each of the three principles of leg alignment into practice. Practicing each alignment cue and integrating each one into a few basic postures is the best way to master these principles.

Roman sandal strapping

Shin forward-Thigh back

1. Wrap a strap over the front of the heel bone and behind the calf. This keeps the shin forward while assisting *lengthwise contraction* from heel to hip.
2. Slightly bend the knee by pressing the shin forward. Re-straighten the leg by drawing the thigh back while keeping the shin forward. This strap modification is beneficial to assist students who habitually hyperextend their knees.

Various asana explorations and assists

Vrksasana   Tree Pose

Shins in-thighs apart

1. Press the heel of the bent leg firmly into the upper, inner femur of the standing leg. If possible, press directly into the *lesser trochanter*, the hip flexors' insertion point, located high on the groin. The outward pressure of the heel provides *thighs apart* action on the standing leg.
2. The standing leg instinctively engages *shins in* as it attempts to balance by contracting the peroneus muscle group, located on the lateral shin. This may be observed as mild fluttering on the outer shin.

Students unable to press their lifted heel into the thigh should avoid heel pressure directly on the knee and place the foot on the inner shin. In this case, the shin presses into the sole of the lifted foot, effectively engaging *shins in*.

Vrksasana

Trikonasana with block   Front leg support

Shins in-thighs apart

1. Front leg: The forearm or a block is placed and pressed firmly against the outside of the front shin. This reinforces the action of *shins in*. The thigh widens laterally for *thighs apart*.
2. Rear leg: Lift the outer edge of the rear foot to engage *shin in*, which also reduces inward sickling of the ankle. Firmly widen the upper, inner thigh, *thigh apart,* until the foot returns to rest fully on the floor.

In **Trikonasana,** some yoga traditions place the front hand to the inside of the leg. Placing the hand outside the shin with forearm or a block pressing into the outer shin reinforces *shins in* and one of the many reasons why this is the better aligned form of the pose.

Trikonasana

Tadasana   Mountain Pose with props

Shins in-thighs apart

1. Place a block, if possible at the second widest surface, between the upper, inner groin to spread the upper thighs apart. The block is not squeezed but acts to spread the space wide apart.
2. Squeeze a block or tighten a yoga strap between the shins to create the action of *shins in*.
3. This pose can also be performed with only the upper groin block. Use the activation of the feet to engage the peroneus muscles and *shins in*.

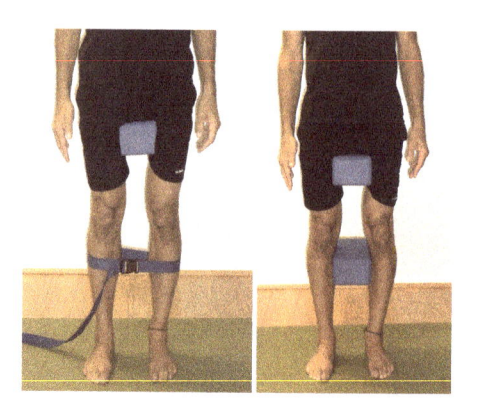

Trikonasana with rear leg assist

Shins in-thighs apart

1. The student sets their rear leg in the pose, aligning the ankle in line with the greater trochanter.
2. The assistant stands in tandem behind the student, using his rear foot to stabilize the student's outer rear foot.
3. Standing behind the student's rear leg, the assistant places one flat hand securely on the musculature of the student's inner, upper thigh.
4. The assistant presses his rear hand directly into the outer, mid shin, avoiding the knee joint. The assistant's elbow presses against his back knee for leverage.
5. The two hands press toward each other, through the midline plane of the leg, providing a firm counter-resistance between the top and bottom bones of the leg. Avoid adding rotation to this assist.
6. The student will experience greater stability in their rear leg.

Utkatasana   Active Chair Pose

Shins forward-thighs back

1. Move into Chair Pose, pressing the *shins forward*. The knees do not move beyond the front of the toes.
2. Draw the *thighs back* and deepen the anterior hip creases.
3. Draw the heads of the femur bones draw deep into their sockets.
4. Lift the front of the pelvis away from the anterior femur, creating additional space at the hip crease.

Tadasana stabilization assist

Shins forward-thighs back

1. Student slightly bends one knee while keeping weight on both feet.
2. Assist one leg at a time. Support behind one calf, holding it forward in a *shins forward* position.
3. The student slowly straightens their assisted leg by engaging *thighs back* against the firm resistance provided by the assistant.
4. Once the student can "sense" the position of the shin being forward, the assistant's hand is slowly removed as the student maintains the forward position of the shin. Repeat with other side.

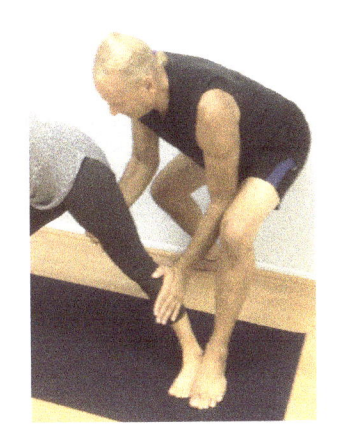

Trikonasana   Triangle Pose with block

Shins forward-thighs back

1. Angle a block behind the fleshy portion of front calf. This position stabilizes the shin forward and prevents calf from moving back.
2. Firmly draw the *thigh back*. The knee will be unable to lock.

This procedure is beneficial for rehabilitation of knee hyperextension injuries. Also, injured hamstring muscles can be safely stretched in this supported position. The supported calf position enables the quadriceps to be more easily engaged to build strength in an injured leg.

If there is a torn anterior cruciate ligament, this procedure is avoided or adapted by simply stabilizing the leg with the block in position and isometrically contracting the quadriceps. Additionally, isometrically contract the accessory muscles (side-to-side muscles of the knee).

Reciprocal inhibition

The quadriceps and hamstring muscle groups are *antagonists*, working in opposition to one another. When one contracts, the other is inhibited from contracting. If contracted simultaneously, there would be no actual stretching or movement. This is called *reciprocal inhibition*. Similar oppositional relationships between major muscles exist throughout the body. Reciprocal inhibition is necessary for all major joint movements to occur. It is controlled by the nervous system through feedback loops that control muscle contraction and relaxation.

Synergists are muscles that contract or stretch in synchrony with each other. Stretching is best achieved when the synergists work together and antagonists contract together. Stretching the hamstring muscles, their synergists also stretch, that being the gastrocnemius and soleus (calf muscles). Their antagonists, the quadriceps muscles, contract to provide reciprocal inhibition.

Lift your kneecaps!

The quadriceps muscles are antagonists to the hamstring muscles. To stretch the hamstrings effectively in a forward bend pose such as **Uttanasana**, actively contract the quadriceps by firmly lifting the kneecaps, utilizing the neuromuscular principle of reciprocal inhibition.

Adho Mukha Svanasana (Downward Facing Dog) stretches the hamstring muscles. As the heels press toward the floor, the calf muscles also stretch as synergists to the hamstring muscles. The quadriceps, the antagonist muscles to the hamstrings, contract, lifting the kneecaps.

A well-outlined, "fried egg" look to the muscles and tissues around the kneecap is an indication the quadriceps have effectively contracted.

Uttanasana

Co-activation

Should this chapter not be completely confusing, there is another action that oppositional muscles do called *co-activation*. Co-activation occurs when a muscle switching its performance from being an antagonist to a synergist that contributes to the overall strength to a muscle. Co-activation enables large, powerful muscles to move with refinement and not overpower fine motor movements produced by smaller muscles.[3]

All three hamstring muscles co-activate with the quadriceps muscles beginning at 9° before full extension of the knee. The hamstring muscles contribute up to 20 percent additional strength to quadriceps extension (measured at the knee joint).[4] Co-activation by the hamstring muscles helps to stabilize the knee and prevent hyperextension. Co-activation distributes pressure across the knee joint more evenly and reduces strain on the anterior cruciate ligaments. Co-activation protects the tibia from dislocating should a quick, severe contraction of the quadriceps muscles occur during the end stage of knee extension.[5]

Eka Pada Padangusthasana

Concentric vs. eccentric contraction

Chapter 9 has explored the physiology of these two muscle actions where the details can be reviewed.

By definition, a *concentric* muscle contraction reduces the angle within a joint. A familiar example of concentric contraction occurs with flexion of the elbow by the biceps brachaii. The muscle contracts and the arm bones come closer together as the joint angle decreases. Concentric contraction is more common than eccentric contraction.

When *eccentric* contraction is engaged at the elbow, the opposite action occurs. The biceps brachaii contracts, but instead, the joint angle increases and the arm bones move away from each other. *Isometric* contraction is similar in sensation to eccentric contraction except that isometric contraction does not change the joint angle or cause any movement of the arm bones.

Eccentric contraction is an important musculo-skeletal mechanical action and a valuable aid in muscle injury rehabilitation. It provides slow and mindful lengthening, protecting injured muscle fibers as they stretch. Healing of injured muscle benefits from eccentric contraction while stretching and is utilized in various rehab approaches, such as Iso-Kinetics and PNF stretching.

Eccentric contraction for stretching the hamstring muscles:

1. Bend the knees.
2. Contract both the quadriceps and hamstring muscle groups simultaneously
3. Maintain both contractions while slowly straightening the knees as far as possible
4. If discomfort occurs, slightly back off contraction for a few breaths, then slowly continue
5. Hold the intensity of hamstring contraction throughout the entire pose

When straightening the legs by stretching the hamstrings in this fashion, the quadriceps *concentrically* contact and the hamstrings *eccentrically* contract.[6]

Walking Well   Avoid forward hip thrust

Upon taking our first baby steps amidst cheers from loving observers, it is unlikely that we had coaching in the finer details of locomotion. Early in life, we mimic postural habits that become the patterns that we follow throughout our lives. In a significant way, early postural and movement patterns influence how our bones form and musculature develops.

Engaging our body mechanics correctly when standing, walking, and running utilize the principles of Samasthiti and Brahmacharya - even tension and balance and muscle efficiency. Agility and coordination are also essential skills.

A commonly observed, but unfavorable habit when walking is to first lift and then thrust the hip forward. This waddling-type gait fails to properly engage the hip flexors (iliopsoas and quadriceps). It externally rotates the thighs and flattens the pelvis and lumbar spinal curve. This habit is common for a few reasons: when the quadriceps muscles become weak or fatigued, instead of lifting the knee to lead the step, it is easier to use side-to-side momentum to swing the hip forward. This action incorrectly uses the quadratus lumborum muscles.

Habitually standing with uneven hips may over-encourage quadratus lumborum activation when initial steps are taken. Hip thrusting may also be the result of weak, shortened iliopsoas muscles. Another cause of failure to fully lift the knee is hamstring inflexibility that will restrict the quadriceps muscles when they contract and attempt to raise the knees.

Step-by-step to walking well:

1. Draw the femur back into the socket and create space in the inner, upper groin.
2. Deepen the hip crease. This increases quadriceps efficiency and help the hips remain square to the front and aligned.
3. Lift the knee from the kneecap, as if a puppet string draws is up.
4. Place the heel to the floor directly below the knee and roll through the bottom of the foot to access the *propulsion* for the next step.
5. The kneecap does not drift inward past the big toe. The more the kneecap aligns over the smaller two toes, the more stable will be the leg and more power transferred across the joints.
6. *Shins in-thighs apart* is engaged with every step to protect the knee from twisting and better transferring the shock of the heel strike force.
7. Roll through the bottom of the foot until the knee is slightly forward of the toes before the hamstrings engage to extend the hip. This is a similar concept to that of automobile engine mechanics in setting the release of power at 5° past top dead center.
8. Maintain an egg-sized curve in the lumbar spine. Movement initiates from the inner groin, not from the swinging of the hips. The hips and spine naturally follow the action of the legs and not inhibited but free to go with the flow.

20 The Hamstring Muscles

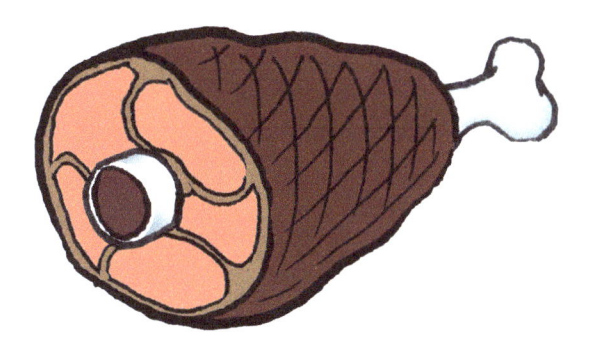

In Wikipedia's definition of hamstring, *ham* refers to a cut of meat taken from the thigh of the back leg of an animal, a pig in particular.[1] The *string* refers to the tendon from which the hindquarter is hung on a hook while the ham cures. Non meat-eaters, such as myself, may find the idea interesting, at best. Being "hamstrung" describes the rendering of someone or something powerless, ineffective, crippled or thwarted.[2] Yogis students with tight hamstring muscles will find this second definition something to which they can relate. From my additional point of view as a long-distance runner, not only can I relate to the second meaning of hamstring but also after a challenging road race, my legs have sometimes felt like the first!

For less flexible yoga practitioners, tight hamstring muscles tend to be a primary culprit thwarting their aspiring practices. Hamstring limitation is most evident in postures that have straight-legged, forward bending. Limited forward hip flexion will also restrict pelvic tilt, the formation of the lumbar curve, and mobility of the sacroiliac joints.

More ham, less string – muscle to tendon ratio

Muscle fibers can stretch to nearly 200% of their resting length, while tendons fibers can only safely elongate 4-8% stretch. One predictor of hamstring flexibility is the ratio between the lengths of a muscle vs. its tendon. Long, thick, cable-like tendons are ideal for transferring power across the knees and hips. And, smaller muscle bellies have smaller, more compact muscle fibers that are less available for stretching.

To better stretch the hamstrings with this limited ratio, first engage all alignment principles. Then, focus the stretch from behind knees instead of the pelvic, ischial tuberosity attachments to reduce muscle strain and utilize the muscle belly more effectively.

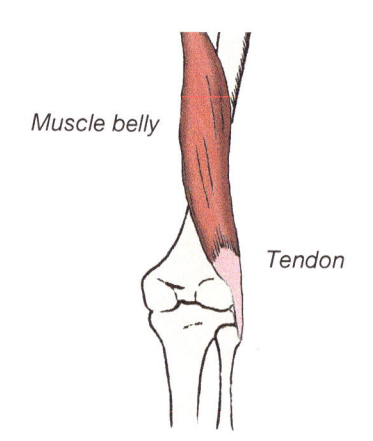

Basic anatomy of the hamstring muscles

Found at the posterior thigh, the hamstrings are a three-muscle group, although referred to singularly. They comprise the *semimembranosus, semitendinosus*, and the *biceps femoris* with its two parts, the *long head* and the *short head*. All muscles originate at the ischial tuberosity, except the biceps femoris short head that originates from the lower femur and inserts at the lateral head of the fibula and tibia.

The function of the hamstrings muscles

The hamstring muscles are bi-articular muscles, crossing both the hip and knee joints. Their function is to extend the hip and flex the knee, two somewhat opposing actions. The knee's position determines the hamstring's strength, efficiency, and flexibility - performing best when the knee is flexed. As the leg straightens, the hamstrings myofibrils lengthen beyond a 20% stretch and limit the muscle's function.

Semimembranosus

The semimembranosus originates at the ischial tuberosity and inserts medially on the posterior tibia. It flexes the knee and medially rotates the tibia. At the hip joint, the muscle extends, adducts and medially rotates the femur. The semimembranosus provides important stabilization of the posterior and medial aspects of the knee. When the leg is stationary, the semimembranosus stabilizes the pelvis posteriorly and extends the hip.

Semitendinosus

Along with its companion, the semitendinosus originates at the ischial tuberosity but inserts medially on the tibial condyle, sharing a common insertion with the gracilis and sartorius muscles into a three-pronged mound called the Goose's foot, or *pes anserinus*. The semitendinosus also flexes the knee and rotates the tibia medially. At the hip, it extends, adducts, and medially rotates the leg. It protects the knee from torque and shearing forces during knee rotation. It is an important stabilizer and prevents *valgus deviation*, commonly known as knock-knees.

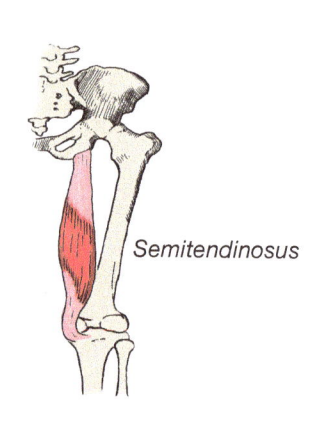

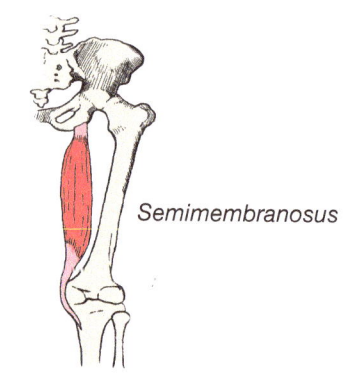

Biceps femoris   Long head

The long head of the biceps femoris muscle also originates at the ischial tuberosity. It crosses the rear thigh to insert laterally on the fibula. The biceps femoris flexes and externally rotates the knee. It extends and adducts the hip, but unlike the medial hamstring muscles, it laterally rotates the femur. Fibers from the biceps femoris embed into the lateral collateral ligaments, TFL, and fascia surrounding the knee. This enables the knee to be squeezed laterally to assist the ligaments in stabilization.

Biceps femoris- long head

Biceps femoris   Short head

The short head of the biceps femoris originates at the lower shaft of the femur and with the long head, inserts with a common tendon to the lateral tibia. It is not designated as a fourth hamstring muscle since it is not bi-articular, only crossing the knee and not the hip. It also has a different nerve supply than the rest of the hamstring muscle group. The tibial nerve innervates all hamstring muscles except for the short head of the biceps femoris, which is supplied by the common perineal nerve.

Biceps femoris- short head

When the knee is fully extended, the short head acts as a "key" muscle, unlocking the knee and allowing it to flex. The long head of the biceps femoris and other hamstring muscles are too inefficient in the position of full knee extension and depend upon the short head to initiate flexion.

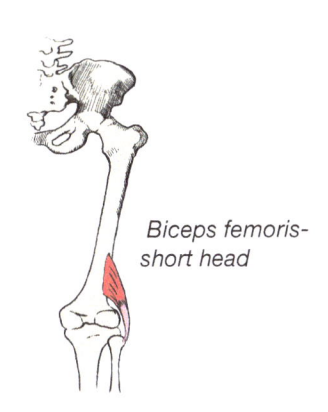

> **Technically Speaking**
>
> As introduced in Chapter 12, the sacroiliac joints are stabilized to resist forward flexion (nutation) by an obscure anatomical feature of the biceps femoris muscle.[3] In our human evolutionary past, the biceps femoris tendon once connected directly to the sacrum. The *sacrotuberous ligament* was once the upper portion of the biceps femoris tendon before evolving into a new, separate tissue. The two structures remain interconnected. Because of this, the legs can act as long levers and affect the motion of the sacroiliac joint. Although many muscles indirectly move the sacroiliac joint, this remote connection is the only instance of direct muscle movement.

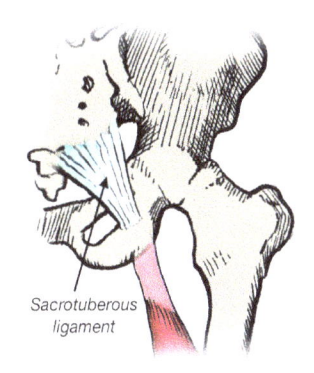

Sacrotuberous ligament

Why do tight hamstring muscles cause back pain

Limited hamstring muscle flexibility is a common pre-condition for lower back pain. The three hamstring muscles each attach to the ischial tuberosity of the pelvis. When the hamstring muscles are tight and short, the pelvis is pulled down and held in a posteriorly tilted position that restricts forward flexion of the pelvis.

128 YOGA ALIGNMENT PRINCIPLES AND PRACTICE

If the pelvis cannot flex, the top of the sacrum (sacral base) cannot tip, and the lumbar curve cannot deepen. Tight hamstring muscles make it more difficult to engage *inward hip release*, resulting in inadequately open and less mobile sacroiliac joints. The mechanical limitations caused by tight hamstring muscles are most evident in forward-bending poses. Injury to the hamstring tendons and surrounding soft tissue can easily occur when forward-bending asana are forced beyond safe limits.

As discussed in Chapter 18, hip extension is more limited when as flexion of the knee increases. This is partially due to increased tension at the front of the knee from the quadriceps muscles. It is another reminder of the inter-relationship between the various components of each extremity.

Tips and refinements for the hamstring muscles

- Practice hip opening poses prior to deep hamstring stretching.
- In forward-bending postures, use a wider stance to reduce hamstring strain. Resist the tendency in the wide stance for the knees to roll inward by keeping the kneecap aligned with the outer toes.
- Engage *Shins In-Thighs Apart*, which helps align the fibers of the hamstring muscles. It increases flexibility, muscle efficiency, and safety. This is a critical alignment needed for hamstring therapy and the healing of upper leg myofascial injuries.
- Engage *Lengthwise Contraction-* from the hips to the heels. This reduces hamstring strain. Couple this action with both *Forward Tailbone Scoop* and posterior pelvis tilt. All three actions function together to reduce the tug on the hamstring attachments at the ischial tuberosity, protecting the muscle from tearing at this highly vulnerable spot. These actions also inhibit the stretch reflex, restraining the contraction response that would otherwise tighten the muscle and limit stretching.
- Prevent overstretching and unsafe knee hyperextension by "hugging" the hamstrings firmly to the femur, as if being Ace™ bandaged and squeezing the muscle insertions into the sides of the knee.
- Contract the quadriceps, the hamstring's antagonist, to activate reciprocal inhibition (see Chapter 19 for details).

> Failing to activate *forward tailbone scoop* can result in inflammation or tearing of the hamstring tendons or muscles!

Extend thighs back, not straighten knees in these poses

Urdhva Dhanurasana (Upward Bow) and **Setu Bandha Sarvangasana** (Bridge Pose) require firm hip extension to lift into the poses.

Hip extension is how the pelvis elevates. This is accomplished by contracting primarily the hamstrings and partly the gluteus maximus. The quadriceps' are the antagonists. They flex the hips and extend the knees. Although their contraction provides important stability in the hips, legs and knees, the quadriceps are not specifically used to lift the pelvis.

In **Eka Pada Urdhva Dhanurasana** (One-legged Upward Bow), the lifted leg is in hip flexion, where the hamstring muscles stretch and the quadriceps contract. In the leg that is pushing up is in hip extension, the hamstrings contract and the quadriceps stretch.

Flexion

Extension

Eka Pada Urdhva Dhanurasana

Uttanasana: a two-stage procedure

Forward bending postures, such as **Uttanasana** (Standing Forward Fold), are essentially poses of hip flexion. In theory, the torso remains in Tadasana as it flexes forward over the legs, hinging from the hip joints. Hamstring inflexibility is the primary limiting factor in Uttanasana. It restricts the hips from flexing and prevents the natural spinal curves to be held. The combination of the hips in flexion and the legs fully extended make the hamstrings vulnerable to injury.

Forward bending is divided into two distinct steps:

Step one flexes the hips and pelvis forward while the torso remains in Tadasana. The spine's natural curves are maintained. The shoulders do not round forward. In stage one, *inward hip release* directs initial movement into the asana.

Inward hip release tones the psoas muscle, assisting in its maintenance of the lumbar curve. The tendency of the femurs in this position is to externally rotate. This is due to the pull of the biceps femoris and the natural external torque of the myofascia of the upper thigh. Because of this, *inward hip release* provides an important counter-balance action. *Thighs apart*, part of *inward hip release*, widens the hamstring muscle attachments and vertically aligns the muscle fibers. It increases flexibility and safety when stretching.

Step 1

Step two engages once the lumbar curve can no longer be maintained. Step two activates *forward tailbone scoop* and *lengthwise contraction*, both actions of the southern equator of the body. Scooping the tailbone forward reduces tension on the hamstring attachments by preventing the ischial tuberosities from lifting up to where they can be torn; instead, it essentially shortens the distance the muscle and tendon are being stretched. *Forward tailbone scoop* also stabilizes the sacroiliac joints to prevent their injury.

Yoga teacher Jaye Martin instructs a useful determinant, if the lumbar spine does not curve anterior of the sacral base in step one but instead, rounds to the posterior, then it is not appropriate to continue into step two. Step one strengthens the lumbar musculature if the yogi focuses on forming the curve. This strength is a valuable benefit of the pose and necessary to have before deepening into the full folded position. This concept applies to all forward fold postures, **Uttanasana, Upavistha Konasana,** and **Paschimottanasana**. The straight-legged seated posture, **Dandasana**, is a valuable asana to practice Step One.

Step 2

A frequent error when going into a full forward flexion posture and a common cause of hamstring injury is only engaging Step One

Dandasana

Sitting hamstring stretch

The yoga poses **Paschimottanasana** (Sitting Forward Fold) and **Janu Sirsasana** (Knee Head Pose) are popular, seated hip-flexion postures. Hamstring inflexibility is the limiting factor that challenges most yoga students when attempting to deepen these poses.

As in the standing hip-flexion poses, the same two-step strategy for forward flexion is followed. Additionally, when forward-folding asana are performed on the floor, the heel of the straight leg presses down into the floor while the leg's muscular tension is drawn up, from the heel to the hip socket. The entire leg medially rotates so that the bulk of the hamstring and calf muscles make firm contact with the floor. These procedures should not cause the knees to internally rotate; the kneecaps continue facing upward. The muscles and flesh of the buttocks are manually spread back and apart, making contact of the ischial tuberosities with the floor. Broader contact of the muscles and bone with the floor "calms" (attenuates) the nervous system its spinal reflexes, allowing a more effective stretch. *Inward hip release* and floor contact produce a visible toning of the quadriceps, especially around the knee with the classic "fried egg" appearance to the kneecap.

When the hamstring muscles are tight and short, sitting on blankets assists forward flexion of the pelvis. Sitting on props, however, may compromise full contact of the hamstrings with the floor. If the leg lifts significantly off the floor, a blanket can be placed under the thigh to allow better contact. This set-up is similar to the hamstring rehabilitation set-up that can be viewed on the next page.

*Paschimottanasana
Rounding the upper back and dropping the shoulders forward is counterproductive, increasing strain on the lumbar spine and its musculature.*

Rehabilitative stretching of the hamstrings

Should the hamstring muscles become torn or injured at their ischial tuberosity attachment, a yoga strap can be used therapeutically to prevent further injury. The strap or belt is tightened around the upper thigh, as high up as comfortable. The strap posteriorly crosses the hamstring attachment, overriding the injured tendon and anchoring the hamstrings to the femur from a lower, uninjured part of the muscle. This procedure reduces the repetitive stress of muscle contraction on the actual attachment. It prevents overstretching and potential re-injury of the healing tissue. The strap can be used not only during yoga practice but also during daily activities. The excess end of the strap can be tucked into the wrap. To avoid the pinching off of blood circulation, it is important not to cinch the strap too tightly and to release its tension throughout the day.

Restorative hamstring rehabilitation is effective, particularly for hamstring muscle-belly injuries. It requires an assortment of props and maintaining the posture for 5-15 minutes. The strategy is to compress the muscles to the femur bone while engaging *lengthwise contraction* through the leg. Muscle compression is a valuable therapeutic tool for rehabilitating injured muscle and myofascia. Compression flattens the myofascia, widens the muscle fibers, and creates a neurologically calming effect similar to 'hugging" the muscles to the bone.

To set-up this therapy, sit in **Janu Sirsasana** (Knee Head Pose). Loop a yoga strap behind the hips at the level of the hip socket, around the sacroiliac joints and down to the front of the heel. This step alone can be used alone as a complete, effective therapy. To go further, place a sandbag or weight on the top of the thigh to compress the muscles into the femur bone and the leg to the floor. The amount of weight will vary, from one or two yoga sandbags to using weightlifting plates equaling one hundred pounds or more. If necessary, cushion the top of the thigh before placing weights on it, eliminating any hard edges of the weights pressing into the bone.

An option is to sit against a wall for support and on blankets to better form the lumbar curve. If the straight leg does not touch the floor or if the knee is vulnerable to hyperextension injury, place blankets or a folded yoga mat under the leg to provide firm contact from below.

Points to consider with yoga therapeutics

- Integrative alignment is the fundamental principle for all therapeutic and restorative postures.
- Long-held postures are especially effective for healing damaged connective tissue.
- Postural alignment allows tissues to rebuild according to their anatomical design and not forced to follow forces and tensions that are distorted and compensatory.
- Not aligning muscles and joints before stretching or strengthening reinforces misalignment and distortions and the result will be poor rehabilitation.

Another effective option for hamstring therapeutic stretching is to make use of a vertical pole, sometimes found in many yoga spaces. Once set up, the pose can be held for five minutes or more and repeated with each leg. If this therapy is utilized daily, the myofascia of the posterior thigh will slowly and affectively lengthen and hamstring flexibility will increase.

Hamstring pole stretch therapy

1. Lie supine. Lift one leg to 90° and place it against the pole.
2. Bring the sacrum as close to the pole as possible; maintain the egg-sized, lumbar curve.
3. If necessary, a thin blanket can be place under the pelvis to initiate forward pelvic tilt.
4. Maintain the lumbar curve and keep the shoulders and head on the floor throughout the therapy.
5. Place a strap around the upper thigh, high up to the hip crease, as possible. Tighten the strap, pulling it medially around the pole to promote internal rotation of the upper thigh.
6. A second strap, or the end of the first one if it is long enough, is placed around the lower thigh, two or three finger distances above the kneecap. Pull and tighten in a similar fashion as the first strap.*
7. A weight or sandbag can be placed on top of the thigh of the leg resting on the floor. Attempt to flatten and medially roll the hamstring muscle of the leg on the floor.
8. Both feet are actively engaged, pressing through the inner heels. A sandbag can be placed on upper foot but with caution not to let it drop (not shown).

Placing a strap below the knee is not advised. That placement would reverse the *shins forward* alignment that is normal in postural alignment of the legs and potentially strain the ligaments of the knee.

When the hamstring muscles are chronically short and tight, this therapy is very effective in increasing their flexibility if used consistently. The angle or space behind the knee can be used to monitor the progress gained with the therapy. The space should diminish over time until the knee is flat to the post. In a gym, the post of a weightlifting rack can be used if a freestanding pole is not available. Plate weights can be used in place of a sandbag.

"The poison is the cure" [4]

This mantra of homeopathy also serves as an underlying principle for yoga therapy. It applies directly to hamstring injuries and the prevention of potential traumas. The hamstring muscles are most often injured while stretching. In spite of its possible hazardous consequence, stretching is essential for successful rehabilitation of the hamstrings. It realigns the myofibrils, improves blood supply and promotes the removal of ill-suited chemicals for healing, such as lactic aid.

Stretching, like poison, depends upon precision in its use for it to be safe and effective. Successful stretching will become a cure if performed respect for the healing process and a uncompromising attention to precise alignment.

21 Knee Alignment Principles

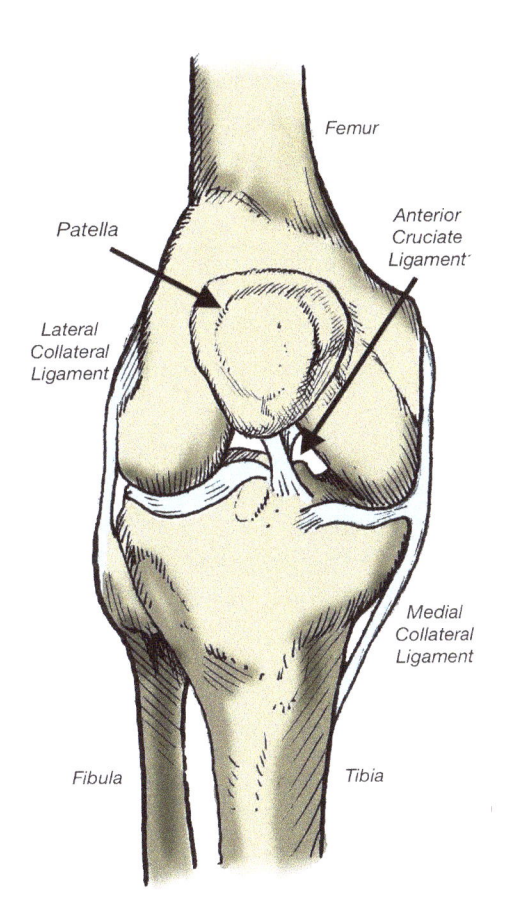

Visit any gym or yoga studio and you will see braced knees and hear frustrating accounts of injury by athletes, young and old, lamenting their intractable knees. The knees are an almost universal area of complaint among those who live a highly physically active life. A greater understanding of the knee can prevent their injury and offer therapeutic options for knees that have already been injured.

The knee is the largest joint in the body. It is also one of the most complex and faces a number of anatomical challenges. The knee must lock into position when stability and support are needed yet be able to release quickly to manage rapidly applied motion delivered by the body's most powerful muscles. To handle these diverse, mechanical demands, the knee has evolved an intricate design. In order to securely lock, the knee's joint surfaces and cartilage, the menisci, have shapes and sizes that are irregular and asymmetrical. Its ligaments wrap extensively around the joint to add support. These design features stabilize the joint in the extended and fully flexed positions. The movements between those two positions are where the knees are pre-disposed to sprains, tears, and dislocations.

The knee is primarily a hinge joint. It performs most safely and will rehabilitate from injury most effectively when it stays in the sagittal plane and is limited to flexion and extension. Knee flexion, the movement that brings the heel toward the buttock, is the knee's primary action and an important component of gait. Extension, the knees' other major motion is enabled by quadriceps contraction but can only occur once the joint has flexed. Technically, the knee joint does not have extension but will move from a flexed position by using extension. When the leg is straight (extended), the knee is not in extension but at its neutral angle. Extending beyond this neutral position, called hyperextension, can cause injury.

The knee has limited ability to pivot but only when it is partially flexed (bent). This is called internal and external rotation. When the knee is fully extended (straight leg), rotation is not possible. The ability of the knee to rotate increases proportionally as the degree of flexion increases, stopping at full flexion.

Knee movements are coordinated with the hip and the ankle during walking and running. They function together to allow the foot to land safely on irregular surfaces. By the same reasoning, issues with the hip or foot usually cause compromise and injury to the knee.

Muscles that flex the knees

The hamstring muscles, the three-muscle group located on the back of the thigh, are the primary flexors of the knee. As detailed in Chapter 20, the hamstring muscles originate at the pelvis on the ischial tuberosity and insert below the knee on the tibia and fibula.

Other muscles participating in knee flexion include the *sartorius*, a long, thin muscle located along the inner thigh. It crosses the front of the thigh to the inner knee, acting with the *gracilis*. When the leg is straight, the hamstring muscles are inefficient and require the assistance of "key" muscles to unlock its power and action. Assistance comes from the *short head of the biceps femoris* and the short, powerful *popliteus*, located behind the knee joints, unlocking the knee from the straight-legged, fully "extended" position.

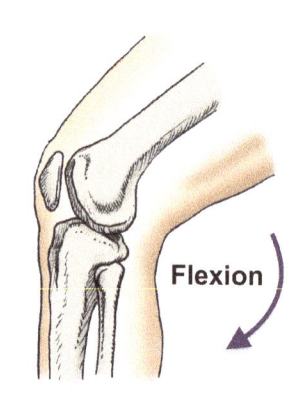

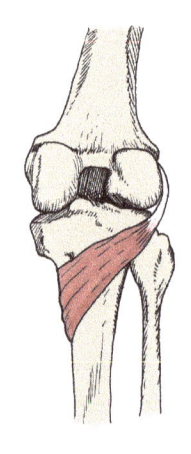

Flexion

Popliteus

Extension

The four-part quadriceps muscles are the primary muscles that extend the knee. Located on the anterior thigh, the quadriceps is comprised of the three vastus muscles: *vastus lateralis*, *medialis*, and *intermedius*. They originate on the femur and insert on the knee. Their dedicated function is knee extension.

The fourth and most centrally positioned muscle is the *rectus femoris*. It originates on the ilium, not the femur, which enables it to flex the hip as well as extending the knee along with the rest of the quadriceps.

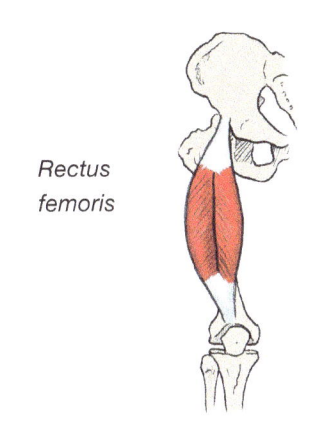

Rectus femoris

Rotation

Rotation of the knee is only possible when the knee is in flexion. The internal rotator muscles are the *sartorius, gracilis,* and *popliteus*, along with the two hamstring muscles, *semitendinosus* and *semimembranosus*. The knee external rotators are the *biceps femoris*, part of the hamstring muscle group, and the *tensor fascia lata*, on the lateral thigh.[1]

Knee injury statistics

Between 1999 and 2008, there were 6.6 million emergency room knee injuries.[2] In 2003 alone, there were approximately 19.4 million visits to individual physicians' offices due to knee problems. Reports indicate these statistics have continued to increase yearly. Knee problems are reported as the most common complaint to an orthopedic surgeon.[3]

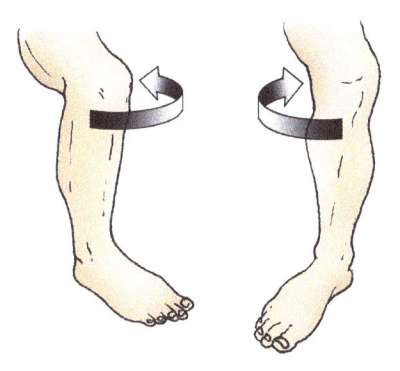

Internal Rotation   External Rotation

Vulnerable knee positions

The knee is most stable in full extension or full flexion. It also responds best to yoga therapy when utilizing postures where it is fully extended or fully flexed. In between these extreme positions, what is sometimes referred to as "no man's land", the knee is unstable and vulnerable to injury.
External rotation is a more stable position for the knee due to ligament function and anatomical design. Internal rotation offers little stability. As the degree of internal rotation increases, the knee is more prone to damage.

Movements happen quickly for the knee. Forcefully straightening the leg or a sudden shift of position from internal rotation torques the knee and can tear the meniscus cartilage. This is a common sports injury that occurs especially when kicking a ball.

Concussive forces that vector across the knee from lateral to medial are particularly damaging, especially when the knee is in flexion. Such a force is common in football tackles, martial arts, and side-impacting skiing injuries.

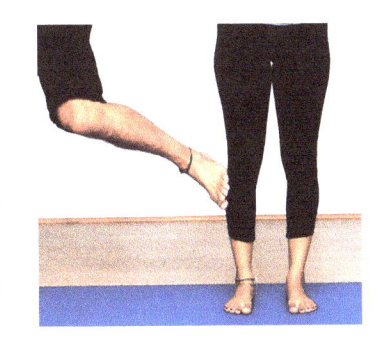

Meniscus   The cartilage of the knee

The *menisci* (plural) are two irregularly shaped discs made of fibro-cartilage. The menisci attach to the tibial surface of each knee, following the contour of the inner surfaces of the tibial condyles. *Condyles* are the round prominences that form the ends of most long bones, what are often exaggerated in cartoon depictions of bones. The menisci cushion the forces of the femur bone as they impact the joint. The menisci attach only to the anterior portion of the tibia, allowing them to float and adapt to the movements of the knee. This configuration, however, makes the knee cartilage susceptible to tearing.[4]

Medial meniscus

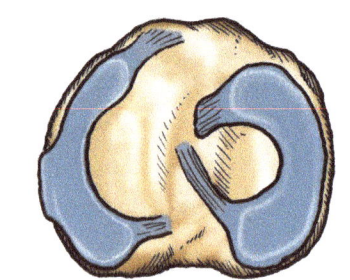

Lateral meniscus

Roll and glide

The knee joint moves like a mortar and pestle. As the knee flexes, the femoral condyles roll and glide, from back to front, over the top of the menisci. During this movement, the unattached, posterior flap of the menisci is vulnerable to being lifted and getting caught and torn by the femoral condyles as they glide across the joint.

When the knee flexes, it compresses the menisci, forcing them to shift laterally in the joint toward the outer edges. In extension, the menisci reverse directions, moving medial to their original position.

Because the medial portion of the knee joint has less space, medial meniscus is more easily injured than the lateral. The medial meniscus can only displace one half the distance available to the lateral meniscus.

Knee injury is often accompanied by inflammation. *Synovial fluid* is the viscous liquid that lubricates the knee, as well as most joints. If the knee joint becomes inflamed, the synovial fluid stagnates and adheres to the surfaces of the menisci, making them more susceptible to being torn.

To avoid injury, full closure of the knee joint should be in a straight-line hinge, without twist or torque. If rotation occurs at final closure, the torque created can tear the menisci. Avoid rotation in the final degrees of full knee flexion. **Virasana** positions the ankles lateral of the knees but can be performed without menisci torsion if the hips are first internally rotated.

Ligaments of the knee   Crosses and strips

Two sets of ligaments connect the femur with the two bones below, the tibia and fibula. They serve as the primary stabilizers of the knee.[5]

The *cruciate* ligaments, anterior and posterior, cross within the knee's central, interior surfaces respective to their names.

The *collateral* ligaments are long, thick, vertical strips located along the inner and outer sides of the knee that stabilize side-to-side movement of the joint.

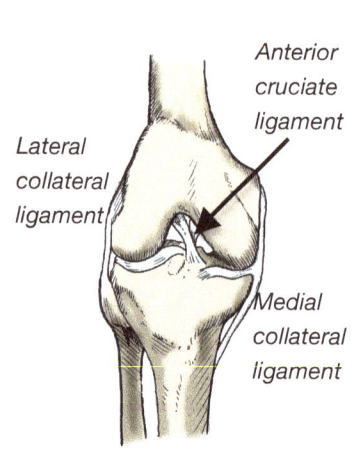

Anterior cruciate ligament

Lateral collateral ligament

Medial collateral ligament

Fibula   *Tibia*

A small number of fibers from the leg muscles that course alongside the knee joint embed into the collateral ligaments and contribute to knee stabilization.[6] This anatomical feature offers an additional method to help stabilize the knee when the collateral ligaments are injured or other instabilities are present. By squeezing, or contracting the knee from side-to-side, the accessory musculature provides valuable support.

The anterior and posterior cruciate ligaments are deeply recessed in the joint. The anterior cruciate ligament (ACL) criss-crosses toward the front of the knee and the posterior cruciate ligaments criss-cross toward the back. The cruciate ligaments provide stability to the front and rear of the joint. Injury to the anterior cruciate ligament (ACL) is the most frequent ligament injury of the knee.

Shins in – thighs apart

The following are common causes of ACL injury:

- Direct trauma (e.g., injuries sustained in athletics)
- Quick change in direction or speed
- An abrupt, hard jump down into a squat
- Landing from a jump with the body weight surging forward

Studies have shown that in certain sports, female athletes have a higher incidence of ACL injury than males. Possible explanations are a woman's increased Q-angle, differences in strength, or hormonal effects on ligament flexibility.

> Muscle fibers embedded in the ligaments enable the ability to actively assist knee stability. With the action of *Shins in-Thighs apart*, the embedded accessory muscle fibers are engaged to stabilize the knee during asana. This action can be used as yoga therapy to support weakened or damaged cartilage or ligaments.

Hyperextension of the knees

As stated previously, the knee moves in the direction of extension but in itself does not extend. Once in the straight-legged position (180°) the angle of the knee is neutral. Extension would require the knee to go beyond neutral and would be referred to as *hyperextension*, which is not desirable or mechanically safe. Habitually pressing the knee beyond the neutral, straight-legged position will overtime, overstretch the posterior cruciate ligaments, which are the major stabilizers of the posterior knee. Once overstretched, the ligaments will be unable to support the knee from its posterior aspect, shifting additional stress onto the muscle tendons of the posterior leg and compressing the menisci posteriorly.

Hyperextension

> Always straighten the legs by pressing or resisting the shins forward as the thighs draw back

The following actions can reduce habitual hyperextension.

- Micro-bend the knee and press the shin forward.
- Always keep the shin pressed forward when straightening the leg.
- Squeeze into the knee side-to-side to engage the embedded accessory muscles of the knee. This helps habit keeps the collateral ligaments strong and healthy.
- Squeeze the mid calf around its circumference to contract the soleus muscle, a foot extensor muscle responsible for plantar flexion, which helps press the shin forward.
- Engage *shins-in* to activate the peroneus muscles. Located on the lateral shin, the peroneus resembles a racing stripe that flattens against the outer leg.

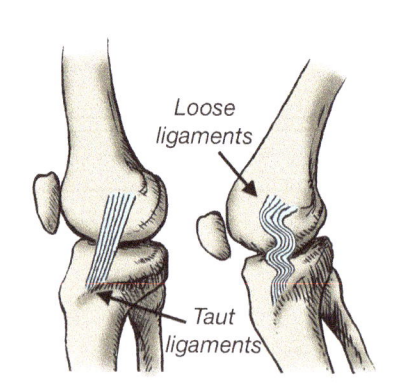

Into the fold

The knee's collateral and cruciate ligaments are the *least* elastic ligaments in the body, a design that best fosters stability. To provide flexibility, these ligaments use their micro-pleating mechanism and wrap-around-the-bone arrangement to modulate their tension and relaxation. Utilizing the ligaments correctly creates stable postures and prevents injuries. (See Chapter 8 for details)

In **Virabhadrasana Two**, Warrior Two pose, the front knee is in flexion. Since flexion loosens the ligaments and makes the joint less stable, other available actions are engaged to compensate and stabilize the knee:

- Should the knee be in a deeply flexed position, slightly release to allow the kneecap to track laterally, re-positioning toward the foot's outer edge. Instead of tracking over the 2-3rd toes, track over the 4-5th toes.
- Re-bend knee toward 90°. Align the outer knee with the outer edge of the hip and isometrically, externally twist the knee and squeeze both the outer hip and outer knee musculature.

These actions engage *external rotation* and *abduction*, two of the three directions that tighten the ligaments and stabilize the knee.

Virabhadrasana Two

Detailed instructions for knee alignment in asana

Knee injuries occur in yoga practice on a regular basis. Yet, injuries are preventable when correct alignment is integrated into every asana. Poor alignment results in faulty mechanics that place strain on the ligaments, cartilage, and tendons of the knee and overtime, leads to degenerative damage. Following proper alignment protocol should create a positive outcome with reduced pain, greater strength, and greater stability.

The following are specific actions that align the knees. Many of these actions will be engaged with only isometric muscular contraction:

- Move the shins forward and the thighs back (*shins forward-thighs back*)
- Draw the shins in and the thighs apart (*shins in-thighs apart*)
- If the tendency is to hyperextend, follow the procedure outlined for hyperextension
- Squeeze both sides of the knee toward the center of the joint
- Hug the quadriceps muscle onto the femur bone and firmly lift the kneecap towards the hip
- Draw the hamstring muscles down the back of the thighs and into the knee creases
- Lift up the calf muscles up toward the knee from below
- Viewed from the front, the knee aligns along a vertical line formed between the center of the hip and the center of the ankle, terminating through the second toe. The knee itself is not forced into alignment; instead, the previous actions usually help bring the knee into its correct position
- Viewing from side-to-side, the femur and tibia are essentially in the same vertical line. The larger, forward-protruding muscle mass of the thigh must be considered when making this observation

The Squat

The half-squat, common throughout Western culture, causes the lower femur to forcefully compress into the superior condyles of the tibia. This forward driving force compresses the menisci and strain on the anterior cruciate ligaments.

In deep squats, the form of squat practiced in yoga and within many indigenous cultures - the femur does not compress the joint. Instead, as the hips drop below the knees, the femurs *draw away* from the joint, leaving the tibia and menisci free of downward, compressive pressure.

Half-squat

In **Anjaneyasana**, the Low Lunge, the hips drop below the knees. This allows the front knee to move forward of the ankle safely with little risk of injury. Deep knee bends are therapeutic for knee injuries.

Utkatasana, the Chair Pose, brings the knees to a half-squat position, which could be problematic. It is relatively safe if the knees remain over, but not past the toes. To avoid injury, deepen the pose by focusing on increased hip flexion and tracking the kneecaps over the 4-5th toes.

Deep-squat

Meniscus therapy

The knee cartilage (menisci) function most safely when space is maintained behind the knee and the femur is lifted away from the tibia. This is especially important after a meniscus injury to avoid cartilage compression. This procedure can be used whenever possible in many seated asana:

1. Spread the two bellies of each calf muscle apart
2. In posterior knee crease, place a soft dowel (a rolled towel or piece of a yoga mat). This action lifts the femur away from the cartilage. This can be used during most flexed-knee postures.
3. A yoga strap can secure the dowel into the deeply flexed knee, increasing the effect of the yoga therapy delivered. The strap is placed well past the knee joint but on the thigh and shin

140 | YOGA ALIGNMENT PRINCIPLES AND PRACTICE

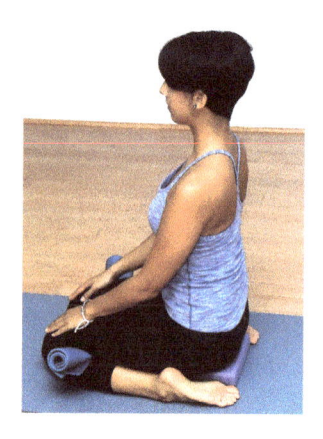

*This procedure may be contra-indicated if an unresolved, torn anterior cruciate ligament is present.

Tibial torsion

Tibial torsion is a rotational misalignment of the shinbone, or tibia. The tibia twists either inward or outward, yet the femur remains facing forward. Inward torsion is more common. The term tibial torsion is more often associated with infant leg development but is also used with adults when rotation of the tibia is greater than 25°.[7] It becomes a cause of concern when transitioning from bent to straight legs in weight-bearing postures, causing a forceful "cork-screwing" action into the menisci.

To reduce tibial torsion, stand in **Tadasana** with the feet parallel. Press the inner edges of the feet into a block placed in its middle-width position between them. Slightly bend the knees and align the kneecaps outer edge of the foot. Firmly straighten the knees by drawing back the thighs while keep the shins forward, avoiding hyperextension.

To protect the knee from both types of tibial torsion in non-weight bearing postures, press through the inner heel. The Achilles' tendon, located behind the ankle, should be straight, not sickled, and without wrinkles. **Padmasana**, the Lotus Pose can readily create tibial torsion. It is a good example of where applying these techniques is valuable.

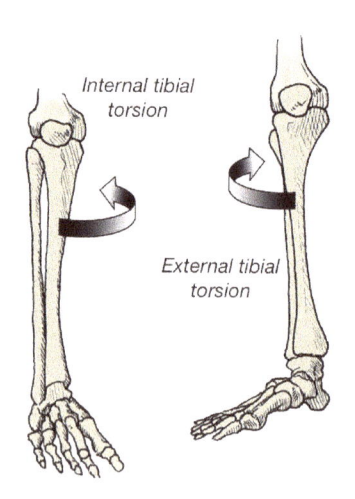

Internal tibial torsion

External tibial torsion

Avoid tibial torsion   Keep knee as a simple hinge

The knee can rotate when positioned between full flexion and straight, however it is most vulnerable to tibial torsion. The following suggestions for knee positioning can minimize stress in these postures:

- Seated and Child's Pose: Close the knee from pure flexion and place as much of the anterior shin inferior onto the mat. Simply placing the outer shin down can create tibial torsion, especially when twisting from that position

- Squats and Chair Pose: Keep knee cap tracking over the outer toes to allow external rotation stabilize the knee and prevent tibial torsion

Oh my Goddess Pose!

A popular yoga pose is the Goddess Pose. It calls for the legs to widen, the hips to externally rotate, and the knees to flex. Most students will allow their knees to drop medially and internally rotate. They often will press their thighs forward, using adduction, while holding the posture. Flexion, internal rotation, and adduction are all actions that de-stabilize the knees by loosening their ligaments.

Additionally, teachers often instruct students to repetitively bend and straighten the knees while in the pose. This action creates tibial torsion into the joint and directly compresses the menisci. For Goddess Pose to be safe, the outer edges of the knees must draw posterior and align over the back, outer edges of the feet. External rotation and abduction must be applied to the knee by squeezing their outer musculature while holding a position. The knees must soften, feet move toward the center and point forward when returning to a standing position.

Eka Pada Rajakapotasana   Pigeon Prep

The full version and true intention of Pigeon Pose is to be a deep closed hip, back extension posture. Pigeon Prep, the pose more commonly, is a challenging hip opener. It stretches the hip external rotator muscles such as the piriformis (front leg) and stretches the hip flexors such as the iliopsoas (rear leg). A more accurate name for the pose may be **Eka Pada Agnistambhasana**, One-legged Fire Logs Pose. Knee injury can be avoided by developing greater hip flexibility.

Pigeon Prep pose conceals many hidden dangers for the knee, creating damaging counter-strain on the menisci. The safest version of the posture calls for the front knee to maintain a full 90° angle. If the front shin is not parallel to the front of the mat, tibial torsion is created. Since this is challenging, there are two options for the front knee to take in order to be safe in the pose:

The first option is fully closing the front knee as a pure hinge without any rotation. The front of the shin, not the side of the shin, is placed down on the mat. The thigh is tucked under the pelvis. This keeps the knee positioned in one of its safest positions and eliminates tibial torsion.

The other option is to bring the front knee to 90° with both hips squaring toward the front of the mat. This requires the front hip to fully achieve the position of external rotation. Although external rotation is the outer form of the pose, the necessary actions within the hip are internal rotation and *Inward hip release*.

Neither the knee nor the hip should be forced to the floor. Instead, a support, block or blanket is placed under the front hip or upper thigh. The knee does not rotate but attempts to remain an open but simple, flexion/extension hinge. Pressing inferior through the inner heel, and removing wrinkles in the Achilles tendon protects the knee from torsion. It is also a remote action that encourages the hip to internally rotate farther.

"X" and "O"

Two common knee misalignments are *valgus* and *varus*.

Valgus comes from the Latin word for knock-kneed. With a valgus knee configuration, the legs resemble the shape of an "X". The valgus knee misalignment increases the Q-angle in the leg and can also contribute to bunion formation.

Varus misalignments cause the knees to bow away from the midline, creating an "O" appearance to the legs. Disorders of the hips often accompany the presence varus knees.

The causes for both types of misalignment include muscle imbalances, faulty foot mechanics, hip disorders, and knee trauma. Both varus and valgus misalignments damage the collateral ligaments and cause degeneration of the menisci.

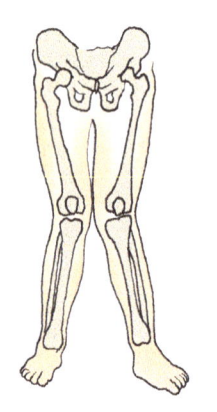

X - Valgus

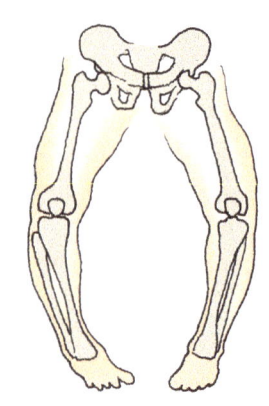

O - Varus

Valgus/Varus therapy

Yoga therapy is very effective in addressing knee misalignments. Therapy procedures are performed with yoga straps and blocks in either standing or supine positions.

For valgus knees (X), place a block between the knees and a strap across the shins. The knee block stabilizes the knee and the belt creates *shins in*. This set-up draws the lower legs toward the midline, reducing lateral displacement of the tibia. Should the ankles and feet shift to the midline when the knee block is positioned, a second block can be snugly wedged between the ankles.

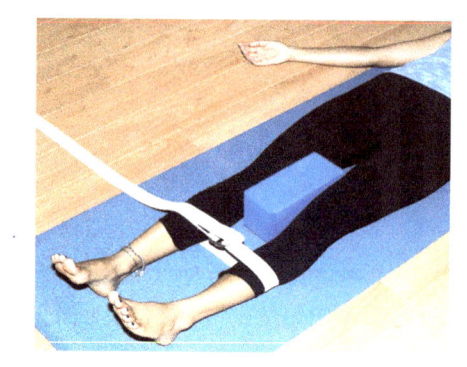

Valgus correction

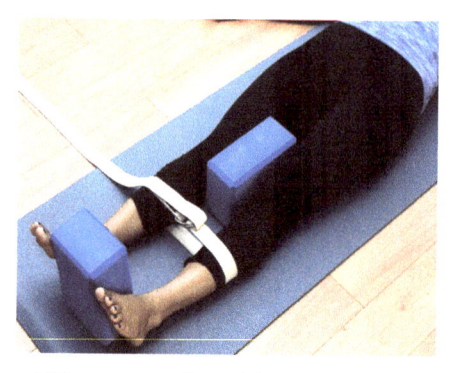

Valgus correction with ankle block

For varus knees (O), place a block between the inner ankles and tighten a strap around the shins. This will bring the legs inline with the action of *shins-in*. When a student has a significant degree of varus knees, the results can be dramatic, both in the immediate application and more importantly, over time. During asana practice, a block can easily be placed between the knees or ankles and the muscular contraction of *shins-in* applied.

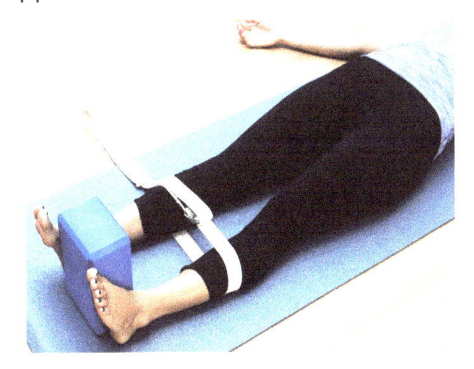

Varus correction

The Patella

The patella, or kneecap, is a small, boney plate located at the front of the knee. One function is as a protective shield for the knee. It sits in a special groove between the femoral condyles, riding above and slightly lateral of the center of the knee joint. This is the best location to rest on the knee. The patella bears no weight. It is held to the tibia by the patellar ligament. The patella embeds into the quadriceps tendon where, acting as a pulley, it increases the efficiency of the quadriceps muscle.

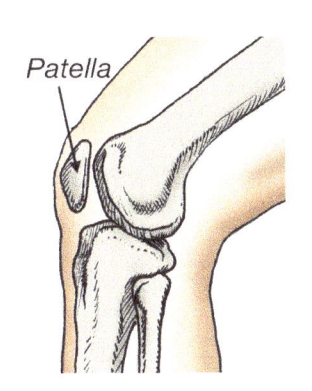

Baker's cyst

A Baker's cyst is a swelling that can form behind the knee. It is often painful and limits knee flexion. It can result from intense muscle activity and is often mistaken for a hamstring tendon tear. Named after its discoverer, a Baker's cyst occurs when a portion of the joint lining, the *synovium*, gets trapped and bulges into the popliteal fossa, the space behind the knee. Baker's cysts often occur together with inflammation of the bursa on the semimembranosus tendon (outer hamstring muscle) where it overlaps with the gastrocnemius, the calf muscle.[8]

Knee hyperextension is frequently causes a Baker's cyst. If the femoral condyles glide excessively and repetitively posterior, the synovium will become compressed and irritated. This scenario triggers cyst formation. Reducing hyperextension can prevent, or at least, reduce the size of the existing cyst and the pain that results from it.

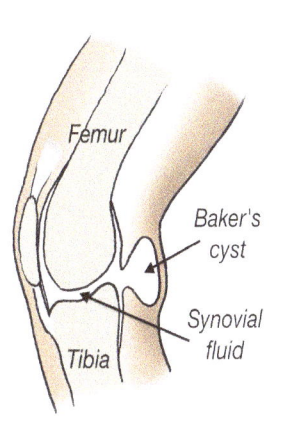

"Don't it always seem to go, that you don't know what you've got 'til its gone…"

Joni Mitchell's insight is true not only for parking lots but also for healthy knees. Our knees are taken for granted, abused, and forced to meet all sorts of unrealistic demands. If they become injured, a negative effect on our ease and confidence with practically everything physical is immediate. Yoga and most of our daily tasks are compromised or often stopped entirely.

By using precise alignment in asana practice, students have reported significant improvement in knee conditions - from acute strains to meniscus tears to degenerative arthritis.

Meniscus tears are very common and surgery is a popular option. Although surgical procedures for menisci continually improve, studies have shown that a positive, long-term outcome for patients receiving surgery is not guaranteed.[9] Furthermore, no studies have researched the effects of yoga therapy on meniscus tears and its effectiveness may be significant.

The use of yoga therapy for knee injuries, however, is not an either/or situation. Yoga can complement medical and surgical procedures and serve as excellent rehabilitation. Of course, as with all use of therapy, precise alignment must be essential!

As with all advice in this book, the health care choices that anyone makes are personal. Hopefully, decisions are made using a thoughtful, yogic mind. Conscious awareness, intelligence, and wisdom are the ingredients for any successful outcome. *Pain* and *injury* can be our greatest teachers. If we respect their presence and listen to what they tell us, our yoga practice will serve us gracefully!

Joni Mitchell

22 The Ankle

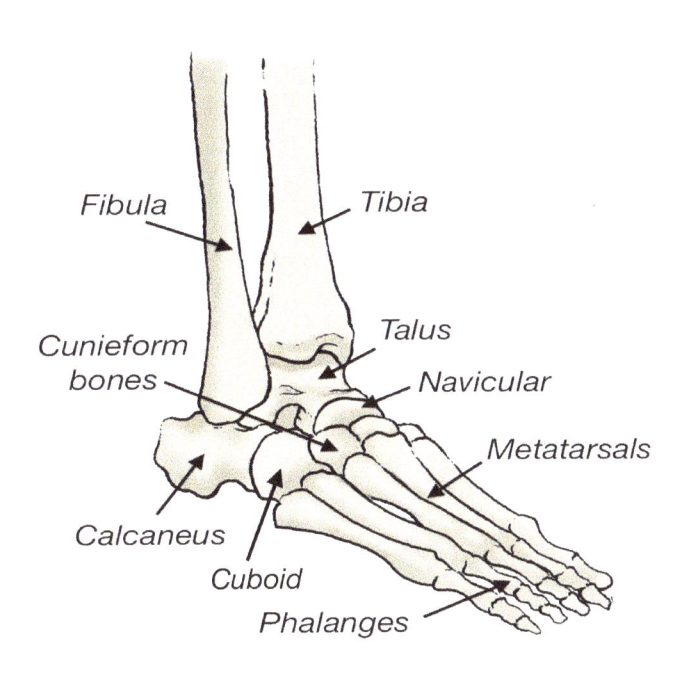

Hardly an athlete has not experienced an ankle sprain. Many sports careers have been compromised or ended by a sprain or break to this complex joint. Yoga asana and yoga therapy offer effective tools for stabilizing the ankle, preventing injuries, and supporting the healing process.

The movement of the ankle is fully integrated into the mechanics of the entire lower extremity. The finesse and grace expressed in the simple act of foot placement cannot occur without full coordination of the actions of the hip, knee and ankle. Alignment, as is true for all joints of the body, is essential for healthy function of the ankle.

The talus joint

The ankle, also called the *talus joint*, consists of three bones: the *tibia*, *fibula* and *talus*. The ankle is a large hinge capable only of flexion and extension. Technically, the ankle is a rolling mortise joint, gliding and rolling in a fashion similar to the knee. The ankle's design allows it to absorb vertical shocks to the body. Seventy percent of the up-and-down, heel-strike forces pass through the tibia and the remaining thirty percent through the fibula.[1] The "bump" on the inside of the ankle is the distal end of the tibia, named the *medial malleolus*. The outer ankle protrusion located on the outer end of the lower fibula is named the *lateral malleolus*. The lateral malleolus is larger than the medial malleolus and is set more posterior.

Below the talus joint are two sub-talar joints that contribute to the function of the ankle, the *talar-calcaneus* and *talar-navicular* joints. These joints prevent excessive side-to-side motion that attempts to shift the anklebones as the feet accommodate for irregularities in ground surfaces.

Inverted "T"

Alignment of the ankle with the foot follows the form of a capital T. Looking down at your own feet, the T is upside down; it positions correctly when looking across at the feet of someone else. The top of the T cuts across the anterior crease of the ankle, the outer landmark for the talus joint. The stem of the T runs through the calcaneus (heel bone) and between the navicular and cuboid and continues through the 2nd metatarsal to the proximal aspect (knuckle) of the second toe (proximal phalange). Measuring from the actual toe is not recommended because toes often deviate due to fractures, shoe wearing, or other injuries.

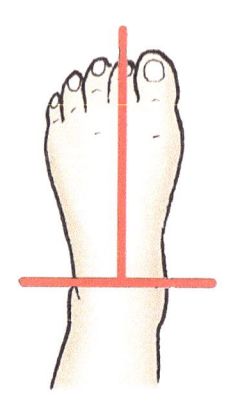

Details of ankle alignment while standing

1. Keep the Achilles' tendon straight and smooth with no inward sickling (turning in or inversion)
2. Lift the talus bone, moving it medial to lateral to a position that is horizontal to the floor
3. Keep the ankle crease (known as the talus joint) perpendicular to the calcaneus
4. If the foot inverts (sickles), press the down into the inner heel while keeping the talus joint horizontal. This will balance any uneven muscle tension, helping to strengthen muscles on the outer, elongated side
5. Deepen the ankle crease
6. Contract the center of the calf muscles, using slight pressure to soften the stretch on the Achilles' tendon and remove any visible wrinkles

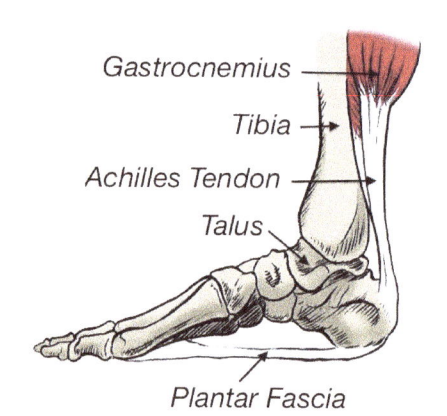

No wrinkles

Padmasana (Lotus) and **Ardha Baddha Padmasana** (Half-bound Lotus) are postures that require the Achilles' tendons to be straight, smooth, and without wrinkles visible in its ropey, surrounding skin. To keep the ankles neutral and balanced, also draw the inner talus bones toward the knee and press them inferior through the inner heels that are on top in these poses.

The powerful calf muscles

The calf consists of two muscles: the *gastrocnemius* and the *soleus*. The gastrocnemius has two unevenly sized heads that form the bulge that is visible at the calf. Underlying the outer gastrocnemius is the single-headed soleus muscle. Together, the two muscles are referred to as the *triceps surae*.

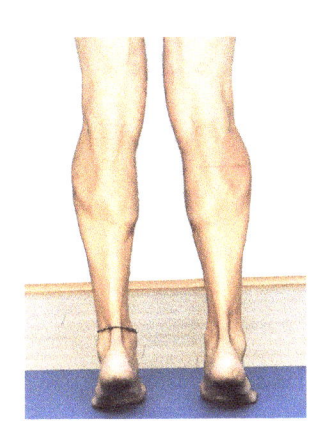

Although the gastrocnemius is one of the body's more powerful muscles, it increasingly weakens as the knee bends into flexion. Rising from a fully bent knee position, as in a squat, must be initiated by the soleus because the gastrocnemius does not have enough efficiency until the knee angle is more open.

Because of asymmetry of the calf muscles, the ankles tend to sickle inward (inversion) when non-weight bearing or forcefully contracted. Along with the ankle, the foot will also invert and supinates. (details and definitions follow in chapter 23)

Using accessory muscles to reduce foot sickling

Foot sickling, in general, is undesirable in yoga postures. Feet are best kept in their neutral position in what is often referred to as "active feet". (See Chapter 23 for details).

In order to limit foot inversion, keep the calf muscles more passive and engage the accessory, or secondary muscles of the joint by squeezing the joint from side-to-side. This reduces foot deviation and allows for more pure flexion and extension at the ankle.

The accessory muscles of the ankle are:

- Peroneus longus and brevis
- Tibialis anterior and tibialis posterior
- Flexor digitalis longus [3]

In **Paschimottanasana** (Seated Forward Bend), keep the feet active by engaging *shins in* and pressing out (inferior) through the inner heel. This produces neutral, balanced alignment through the ankles.

Motions of the ankle

- Flexion   Dorsi-flexion (top of foot flexes superior)
- Extension   Plantar-flexion (bottom of foot presses inferior)
- Adduction   Sole of foot turns inward - same as medial rotation
- Abduction   Sole of foot turns outward - same as lateral rotation
- Supination   Sole of the foot lifts medially (inward)
- Pronation   Sole of the foot lifts laterally (outward)

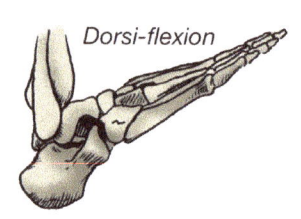

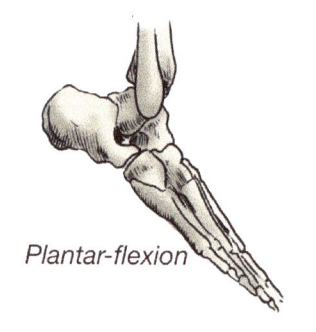

Dorsi-flexion

Plantar-flexion

Ankle ligaments

Transferring heel strike forces up and through the legs and managing the downward, weight-bearing forces of gravity all pass through the ankles. This requires an extensive system of thick ligaments to support the joint structure. These ligaments have only minimal elasticity. The major ligaments of the ankle are called the *collateral ligaments*, providing side-to-side ankle stability. The medial collateral ligament of the ankle is also called the *deltoid* ligament.[4]

Sprain has sprung

Visualize a cool, spring morning. A runner steps through his front door and is ready to finally get back into shape. But as he reaches the curb, he steps unknowingly on an uneven crack in the cement and turns his ankle. The foot gives way and the runner immediately collapses to the ground. In a matter of moments, the ankle is swollen and signs of blood appear under the skin, especially below the joint.

Across North America, sprained ankles occur at a frequency of 25,000 each day.[5] Once the ankle is sprained, the ligaments become overstretched and fibrous scar tissue infiltrates into their original make-up of collagen and elastin fibers. This tissue reconstruction is a permanent change. Multiple ankle sprains lead to overall permanent instability in the ankle joints.

Sprains are graded on a scale of one to four. Grade-1 is minor ligament damage. A grade-4 injury is a complete tear, rupturing the ligament's collagen fibers.

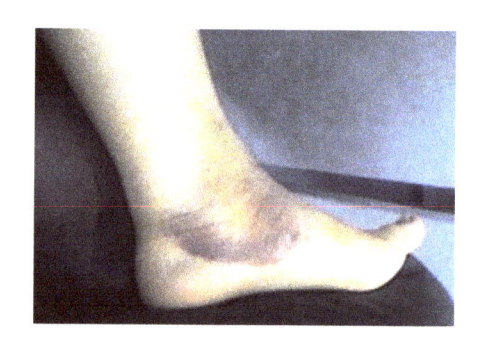

> Any severe injury should be professionally evaluated as soon as possible to be graded and to rule out fractures or other possible complications

RICE, plus

The most accepted approach to soft tissue injury is known by the acronym RICE: Rest, Ice, Compress, and Elevate. This regimen is followed through the initial healing process and beyond, depending on the severity of the injury. In most situations, 24 to 48 hours is recommended with ice being applied for a 15-20 minute period each hour, and repeated as often as possible.

During the rehabilitation process, alignment is critical. Alignment enables ligaments and surrounding tissues to heal with the least amount of scar tissue. As weight bearing re-introduced, neglecting to stand and walk without alignment can re-injure the tissues and cause a quick return of pain and instability. Precisely aligned asana can provide excellent rehabilitation. Establishing precise alignment, not only in the injured ankle but also throughout the lower limbs and pelvis will be immediately apparent by a decrease in pain and more stability in the ankle. Pain becomes a masterful teacher!

General guidelines for soft tissue healing response

The list below is derived from research and clinical experiences. It presents the approximate soft-tissue recovery time needed to reach minimal pain and reduced risk of re-injury. Of course, many extraneous conditions cause variances in these estimates.

Muscle injury	4 - 6 days
Myofascia	2 - 4 weeks
Ligaments	3 - 6 months
Cartilage	6 - 12 months

Ankle support

The alignment principles of the body call for the creases of the ankles to draw back (posterior). If an ankle becomes injured, its ligaments have minimal tolerance for the ankle to shift anterior without creating significant discomfort. **Virasana**, the Hero's Pose requires the foot to bend back (plantar-flexion), which can force an unstable ankle anterior. Place a dowel, a rolled-up cloth, or the edge of a blanket in front of the ankle crease to provide support to keep the ankle from collapsing anterior.

Virasana is an excellent posture for developing strength in the arches of the feet. Engage the peroneus muscles located along the sides of the shins with *shins in* and press through the inner heels, keeping the small toes flat to the floor. Discussion of the arches of the feet is presented in Chapter 23.

Virasana

Supta Virasana

Collapsed talus joint and foot pronation

The following set-up can help in a variety of foot and ankle misalignments. This therapy will strengthen the shin muscles, best accomplished while the feet are in their neutral position. It provides excellent therapy for medially dropped talus bones or excessive foot pronation.

1. Place a block firmly between the ankles.
2. Lift and maintain the longitudinal foot arches. Bring the inner feet as close to the block, as possible.
3. Engage all four corners of the feet and press evenly to the floor.
4. Attach a yoga strap around the shins (as shown) to provide counter-resistance to the black and to allow the pose to be more passive.
5. The pose can also be performed without a strap. Firmly engage *shins in* to produce the desired muscular effect. In many cases, muscular contraction without the strap is a more effective therapy for strengthening the accessory muscles of the ankle.
6. If the knees rotate or collapse medially, an additional block can be placed between the knees (not shown).

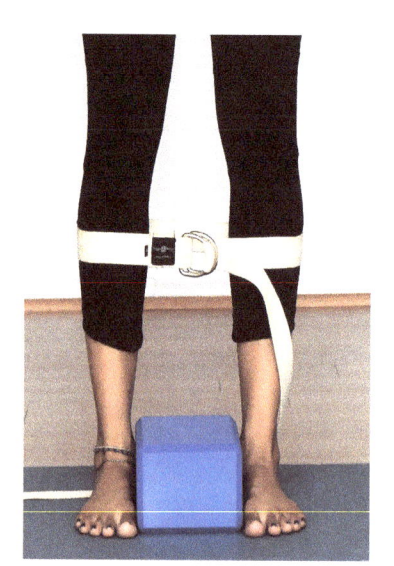

> Alignment is a powerful tool for rehabilitation. It speeds recovery time, reduces pain and sets the state for high-quality healing

23 The Feet

Yoga students might begin their morning practice by standing at the front of their mat, staring down contemplatively at their curious appendages below. They may sense that the feet play an important role in establishing the foundation for the upcoming sun salutation series. But how do the feet engage? Do the toes spread apart, or extend out, or is there any difference? How do the feet align to support and balance the weight and length of the body?

Nearly twenty-five percent of the bones of the skeleton are found in the feet. The alignment of the feet and their foundational role in asana practice is often an afterthought or simply taken for granted. But, to be sure, the feet work very hard in our practice and in everyday life!

No wheelies, please!

Ideally, body weight is distributed evenly on the four corners of each foot, supported like an automobile on its four tires. Additionally, each foot aligns with its counterpart as if they were two toy cars, traveling side-by-side on a track. They remain in line without colliding into each other or diverting to the side of the road.

The foot is neither a square nor a rectangle. Its front and back widths are different. Technically, each foot is an obtuse trapezoid. Anatomical alignment favors that the heels are kept slightly separated, even though some yoga traditions instruct students to bring the rear inner heels together, displaying a more stylized position.

As noted in the previous chapter, alignment through each foot forms the alignment of capital "T" through the second metatarsal to the ankle. Because the toes are often crooked, naturally or due to injury or footwear, it is best to disregard the entire second toe when aligning the foot. The stem of each T is kept parallel to the other, which will keep the two "cars" perpetually in line.

With a single up-and-down jump, the force of heel strike can reach nearly ten times body weight. The trampoline-like design of the feet provides shock absorption to manage the concussive forces of foot strike. Forces are managed at the soles of the feet by the three arches that cross the plantar (bottom) surface of the foot. The arches lift the plantar fascia to increase the spring tension of the tissues and to best absorb the shock.

As mentioned in the previous chapter, the arches of the feet regulate side-to-side motion, protecting the ankle ligaments from sprain. The ankles are more the greater that movement deviates from the simple forward-and-back hinge actions of the joint.

This step-by-step procedure most effectively engages the foot's arches:

1. Press into the big toe ball mound (knuckle)
2. Press into the inner heel
3. Cross top of foot and press diagonally into outer toe knuckle
4. Draw back from the outer toe knuckle and press into outer heel

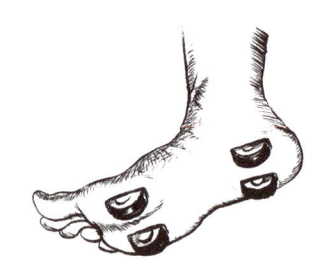

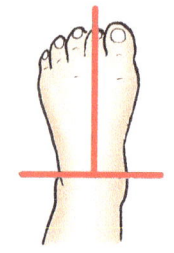

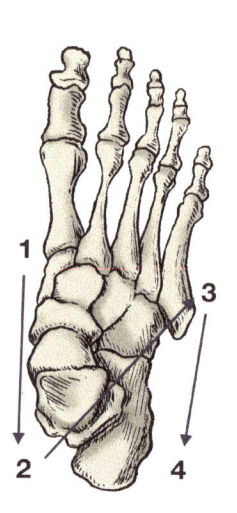

In all standing poses, such as Triangle, Warrior One, and Warrior Two, the four corners of both feet, and especially the rear foot, securely maintain equal pressure on the floor.

Muscles of the arches

The strength of the arches comes primarily from the ligaments and small muscles within the feet. The larger muscles that also create the foot arches are the *peroneus*, also called the *fibularis* and located on the outer shins, and the *tibialis* muscles, which are located on the inner shins.

The peroneus longus tendon crosses under the foot from behind the lateral malleolus and inserts into the first metatarsal (the most medial of the five long bones of the mid-foot) Due to its long-reaching course and insertion, it has significant leverage to strongly influence all three arches.

Balancing on one foot firmly engages the outer shin's peroneus muscles. One-legged standing poses lift and tone all three arches and are excellent therapy for flat or pronated feet. The peroneus muscles are the primary muscles used to establish "neutral" feet with all four corners pressing evenly.

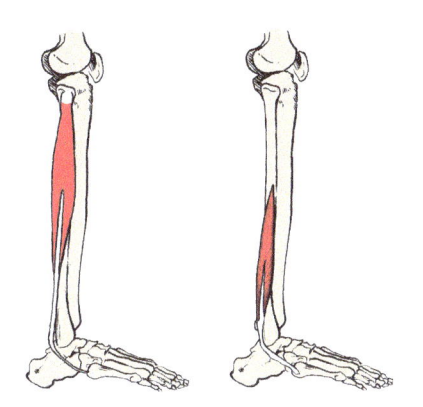

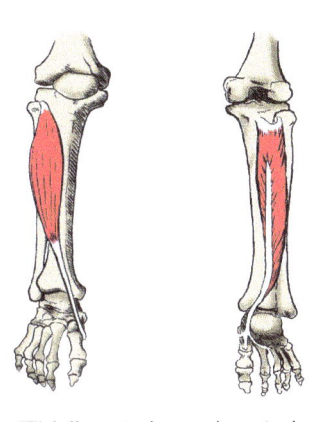

Peroneus longus and brevis

Tibialis anterior and posterior

How the feet remotely control posture

Some yoga teachers amaze their classes by looking at their students' feet and then advising them on seemingly unrelated, whole-body alignment issues. These teachers are aware of the alignment clues found in the distribution of pressure through the feet. When the foot is not balanced on its four corners, body alignment consistently shifts and is expressed in an anticipated fashion.

Pressing into the foot at each of its corners can specifically change alignment in the leg. As an example, pressing inferior through the inner heel increases ranges of motion at the hip. To determine how each individual corner of the foot can affect other joints of the leg, combine pressing front-and-back with increasing inner and outer pressure:

- Front-of-foot
  - Affects the ankle and knee
- Rear-of-foot
  - Affects the hip and pelvis
- Inner foot
  - Increases internal rotation and adduction
  - Increases *inward hip release* in the hip and pelvis
- Outer foot
  - Increases *forward tailbone scoop*
  - Externally rotates and abducts

These concepts may seem difficult to picture at first. One yoga posture that helps explore the effects of foot pressure is **Virabhadrasana Two** (Warrior Two Pose):

- Rear leg: Press into the floor with the inner heel of the rear foot to engage *inward hip release*. The goal is to increase hip flexibility. Pressing down through the inner heel internally rotates and draws the hip joint back. This releases the hip ligaments for safer flexibility. It also aligns the hamstring muscle fibers more vertically, which increases both strength and flexibility. In most standing postures, the rear leg will engage *inward hip release*.

- Front foot: Press into the outer, front portion of the front foot to externally rotate and abduct the knee and ankle of the front leg. This action aligns the knee with the outer hip and the outer border of the foot. This action and the alignment is creates stabilizes the hip and pelvis.

Pressing inferior through the inner heel is a common action taken in many asana to increase hip mobility. It is especially important for the hip of the rear leg in standing poses.

Additional details on foot mechanics

It is difficult to differentiate the mechanical actions of the ankle from those occurring in the twenty-six bones of the foot below. Movements within the feet are subtle with each range of motion small. However, control over the vaulting and flattening of the feet is extensive.

Alignment of the numerous small joints of the foot can be engaged as a whole with the following actions:

- Spread the *metatarsal* (transverse) arch from side-to-side across the front of the foot.
- Toes stretch straight out from the arch. They do not spread apart but instead extend from the knuckles. The toes act as outriggers that improve balance by providing additional surface contact.
- Press the toe pads down with equal pressure on each side of each nailbed.
- Lift the *inter-phalangeal joints* (the outer two joints of the toes) with a soft, claw-like gripping action.

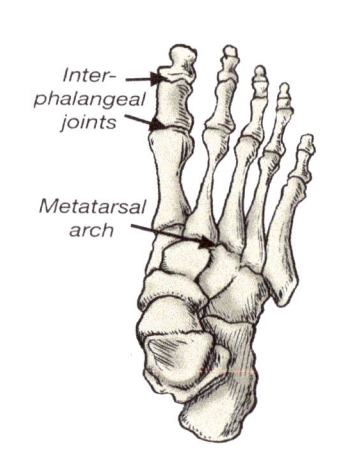

"Don't stand so close to me"   The Police

As discussed in previous chapters, the hips and knees play an important role in foot placement. The backward swing of the heel follows an arc that is in line with the ischial tuberosity of the pelvis. This mechanical relationship offers the rationale that the natural distance between the heels in standing poses is the distance between the ischial tuberosities, approximately 4-6 inches, or fist-width to head-width distance between the inner heels.

Some styles of yoga teach Mountain Pose using a stance that has the feet completely touching. Although this may seem like a minor teaching preference, it has a cascade of consequences, especially for yogis with naturally wide hips. Standing with the feet together increases the Q-angle, the angle between the hip and knee (see Chapter 19 for full discussion of the Q-angle). Additionally, because the front of the foot is usually wider than the back, bringing the heels together can externally rotate the legs and force the thighs anterior. This reduces *inward hip-release*, compresses the sacroiliac joints, and tenses the lower back musculature.

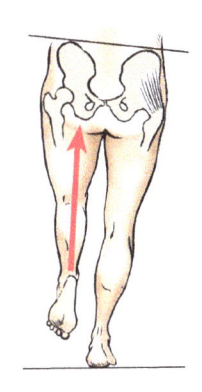

Eversion and inversion

Eversion is the dropping of the inner foot medially (inward) and lifting the sole laterally (outward). When eversion is greater than 25°, it is called *pronation*. When standing or in any weight-bearing postures, the feet will naturally pronate. The arches flatten and talus bones drop to the midline. The feet also pronate during the heel strike phase of walking. This may explain why the inner heel is fleshier than the outer heel so as to maximize shock absorption.

Inversion lifts the inner foot, turning the sole inward to face medially. The foot is considered to be in *supination* when it is inverted to approximately 50°. When the feet are non-weight bearing, as in **Dandasana** (Staff Pose), they naturally invert because of the asymmetrical mass and tone of the calf muscles (see Chapter 22 for details).

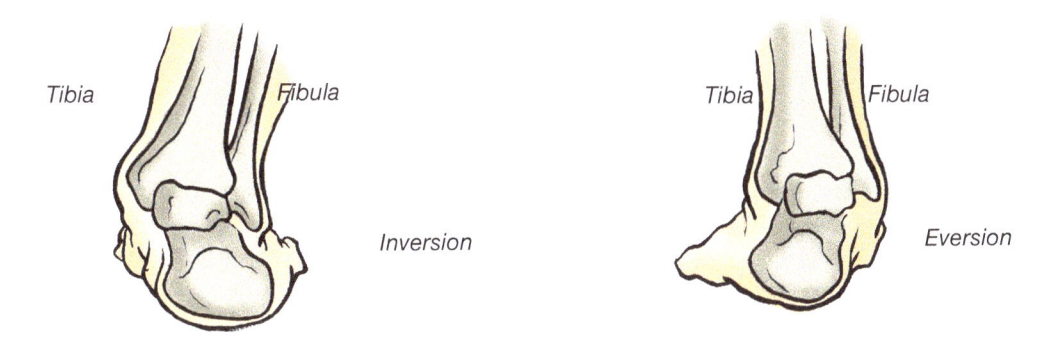

Plantar fasciitis

Plantar fascia, or the *plantar aponeurosis*, is a thick band of fascia that runs lengthwise along the bottom of the foot from the front of the calcaneus to the metatarsal arch. The plantar fascia consists mostly of collagen fibers. It supports the arches of the foot by enhancing muscular tension in the arch, acting like the string of an archer's bow. The arches of the feet otherwise cannot support the full static weight of the body and this arrangement compensates through a diaphragmatic, springy and pulse-like action.

Plantar fasciitis is a painful, inflammatory condition that affects the bottom of the foot. It is caused by excessive stretch or strain of the fascia, often the result of prolonged weight bearing. Pain is a primary indicator of plantar fasciitis. Diagnostically, a reliable method to determine the presence of plantar fasciitis is to lift the toes (dorsi flexion) when standing. If this increases the pain, some degree of plantar fasciitis is present. If the strain on the fascia becomes repetitive and chronic, the plantar fascial fibers tear and the protective fat tissue that cushions the fascia and bone from heel strike forces deteriorates. Plantar fasciitis is a common cause of heel pain and can lead eventually to heel spurs.

Plantar fasciitis rehabilitation suggestions

1. Reduce inflammation with ice massage, which can reach deep between the torn fascial fibers
2. Align the feet to develop well-toned arches to carry body weight and absorb heel strike forces
3. Engage *Shins-in-thighs-apart* to engage the peroneus muscles to tone the arches
4. Keep the calf muscles, the Achilles' tendon, and the plantar fascia flexible. A simple approach is to stand, elevating the front of the foot with a yoga block. This increases blood circulation and reduces inflammation. If the plantar fasciitis is initially too painful to stretch, first stretch the calf and Achilles' tendon alone. This can be accomplished by leaning forward over a flat rear foot in variations of Lunge or Warrior One poses
5. Yoga therapy: Hug all shin muscles inward and onto the bone; then going into a deep squat. This can improve chronic plantar fasciitis and heel spurs

Bunions

Then large toe naturally curves lateral toward the outer toes, a position that better enables foot flexion, an important action for bi-pedal locomotion.[1] *Hallux valgus* is the anatomical term for bunion. Bunions occur when the two phalanges of the big toe divert toward the small-toe side of the foot (valgus). Bunions are often accompanied by shortening of the ligaments that encapsulate the joints between the metatarsals and phalanges and by weakness and imbalance in the foot's interosseous muscles - the small muscles between the bones of the feet.

Bunions can occur in conjunction with flat feet, the condition of severe pronation called *pes planus*. Pes planus causes the heel to collapse medially, which increases pressure on the big toe where the bunion occurs. This abnormal pressure causes the tibialis anterior and peroneus muscles to weaken. As a result, posture often distorts, causing external rotation of the hip, weak gluteus muscles, and medial rotation of the knee. Over time, degenerative damage will develop in the soft tissues and joints.

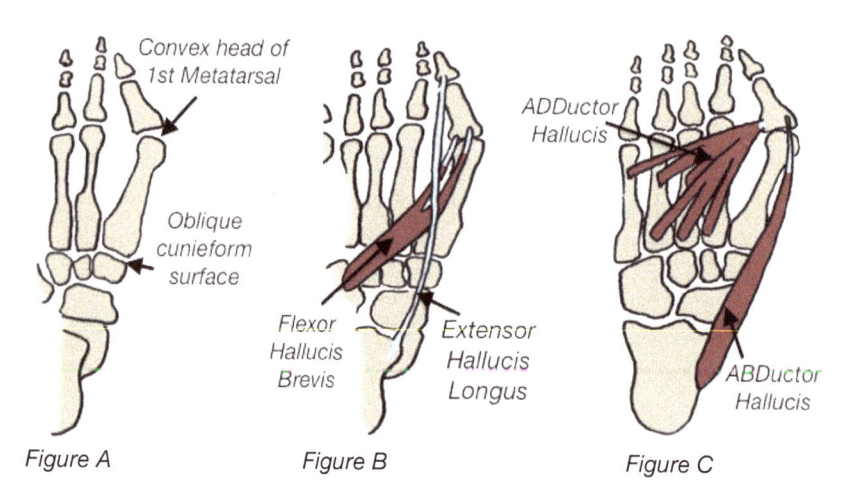

Figure A — Convex head of 1st Metatarsal; Oblique cunieform surface

Figure B — Flexor Hallucis Brevis; Extensor Hallucis Longus

Figure C — ADDuctor Hallucis; ABDuctor Hallucis

> **Technically Speaking - Structural causes of bunions**
>
> Although the blame for bunions is often placed on shoes that are pointy or have high heels and small toe-boxes, there are structural conditions that are known to be involved.
>
> Misshapen bones. The distal head of the first metatarsal can be overly convex. Alternatively, the distal surface of the 1st cuneiform (medial tarsal bone) can be abnormally oblique. Both conditions can cause the proximal metatarsal to slide varus. (Figure A)
>
> Muscular balance. The flexor hallucis longus and extensor hallucis longus lift and bow the foot laterally (Figure B). If the adductor hallucis (AdH) is hyper-contracted, it can overpower the abductor hallucis pulling the metatarsals towards the midline of the foot. (Figure C)
>
> Adductor hallucis. The adductor hallucis plays an important role in creating the transverse arch. If hyper-contracted, it can cause malformation of the arch. Adductor hallucis contracture can also increase pronation and increase pressure on the 1st metatarsal. An indication of pressure is callous formation on the 1st metatarsal mound.

Yoga and yoga therapy are most effective for rehabilitation when the cause of a bunion is muscular in nature, particularly when there is imbalance between the adductor hallucis and abductor hallucis (see Figure C). To address muscle-based bunion issues, use the following procedures in all standing poses:

1. Spread the large toe pad medially and engage all four corners of the foot. This assists a general release of tension in all of the hyper-contracted muscles of the foot
2. Stretch the metatarsal (transverse) arch, lengthening from the big toe to the little toe. This reduces hyper-contraction of the adductor hallucis muscle. Be sure that the big toe pad is lengthened
3. Contract the muscles of the medial longitudinal arch that runs from the big toe to the inner heel. This strengthens the weakened abductor hallucis muscle. Be sure that the big toe is spread medial
4. Balancing poses are valuable therapy when they utilize all of the above procedures.

> **A weighty subject**
>
> Individuals with a large body mass and those whose occupations require constant weight bearing, such as standing for long periods, place high levels of force on the arches of the feet. This causes the arches to collapse and flatten.
>
> When the arches collapse, the talus bone drops toward the midline, distorting leg alignment with the knee and hip rolling inward. Long-term weight bearing is a common cause of knee cartilage damage and joint degeneration of the knee and hip.

Twist walking

Twist walking is an excellent technique for addressing many conditions of the foot. It strengthens the arches, reduces plantar fasciitis, and can prevent and be rehabilitative for bunions.

How to twist walk:

1. Stand at the back of a sticky mat
2. Turn one the foot inward and step forward, pressing the pad of the big toe into the mat as if to extinguish a cigarette butt
3. Hold the big toe pad down and resist it from moving
4. Twist the rest of the foot inward until the ball of the big toe aligns medially with the big toe and can also press into the floor. This action will visually straighten the angled big toe
5. Continue twisting inward until the inner heel can be brought to the floor and in line with the big toe. This will square the foot to the front of the mat
6. Engage the same steps and procedure with the opposite foot
7. Continue to repeat Twist Walking with each foot, walking forward on the yoga mat its full distance

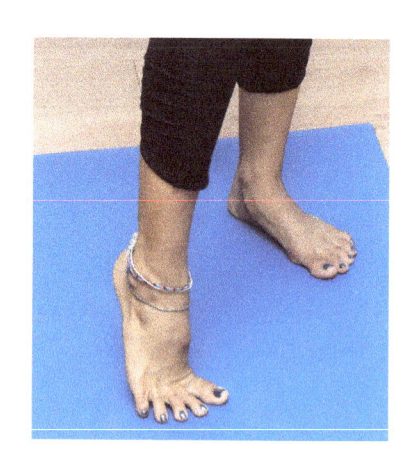

The high arch

Pes cavus (high arch or hyper-supination) is another challenge for the feet. One cause of this condition is weakness of the small, intrinsic muscles of the feet that attach from bone to bone. The small muscle weakness enables the larger antagonist muscles, the *tibialis posterior* and *peroneus longus*, to over-contract and lift the arches. In more chronic cases of high arches, the tibialis posterior or the peroneus brevis may reach a state of fixed contraction (contracture). Similar stretching to that utilized for plantar fasciitis can be an effective therapeutic.

> The feet form the essential foundation for many postures. Skillful engagement of these movable foundations allows us to engage in the dance of awareness that we call yoga!

24 Anatomy of the Spine

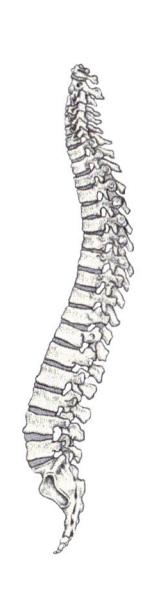

If the spine were able to move unencumbered by demands imposed by the shoulder and pelvic girdles and their respective extremities, it would achieve great freedom and fluidity. It would effortlessly adopt balance in posture and neutral spinal curves, both being the ultimate goal of **Tadasana**. In many ways, asana practice explores how the spine can maintain Tadasana while advancing through a series of increasingly complex positions and postures. Asana practice allows the spine to develop strength and flexibility while managing the demands of the girdles and extremities. Some postures, such as twists, are exclusive motions for the spine and not exploited by the readily movable shoulders, hips, and their extremities.

The alignment principles for the spine follow its anatomical design. Understanding the basic anatomy and mechanics of the spine will make alignment much easier to apply.

The adult spinal column consists of twenty-four free and movable vertebrae. Also considered part of the spine are the sacrum, comprised of five fused vertebrae, and the coccyx, which is formed from four fused vertebrae (three or five segments in rare instances). Although they consist of spinal vertebrae, they are considered pelvic structures. There can be additional variations in the number of spinal segments fusing, particularly with the sacrum, as well as the possibility of the upper sacral segments not fusing to the overall sacral body, at all.

The spine is the torso's structural center and is referred to as the "keel" of the body. Core musculature attaches to the many processes that protrude from the posterior portion of the spine. The anterior surfaces of the vertebral bodies are smooth, important for avoiding puncture of the internal organs. Attaching to the thoracic spine are ribs that protect the internal organs. The spine's mechanical duties provide locomotion and weight bearing. It also protects the delicate nervous system housed within its central canal before the spinal nerves exit from between the vertebrae.

The spine is a curved column consisting of three functional curves: the cervical, thoracic and lumbar curves. This curved design provides shock absorption and allows the spine's central vertical axis to resist forces (axial compression) up to ten times more than could be provided by a straight column.[1]

Architecture of a vertebra

Vertebrae have two sections. The anterior portion is called the *vertebral body*. The posterior portion is referred to as the *vertebral ring*.

The vertebral body is a thick, kidney-shaped block of bone, designed to carry the central vector of the force of body weight. Many of the internal organs abut and suspend from the smooth and rounded anterior surfaces of the vertebral bodies.

The vertebral ring is the "business end" of the vertebra. It is referred to as the *motor unit* because most mechanical movements of the spine occur here. Seven boney processes on the vertebral ring serve as attachment sites for the ligaments that stabilize the spinal segments and the muscles that move entire regions of the spine and torso. There are four spinal *facets* or hinges that are positioned at the top and bottom of each side of each vertebra (except for the top of the first cervical vertebrae). The facets are the primary sites of spinal movement. Their surfaces are flat and smooth to optimize a shearing, gliding motion.

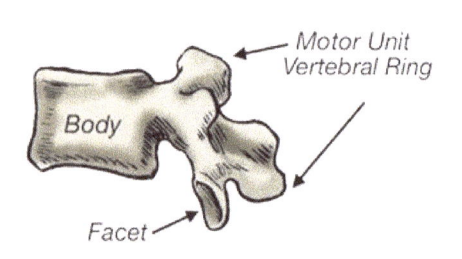

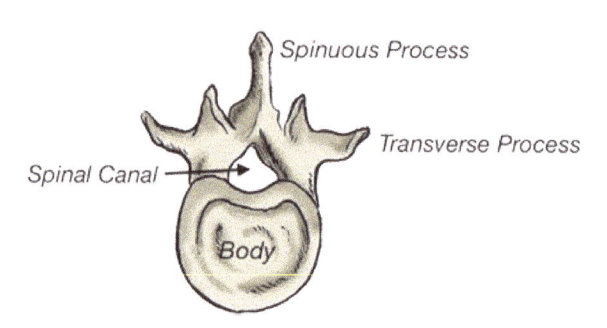

Spinal disc

There are 23 spinal discs located between vertebrae, beginning with the second and third cervical vertebrae and ending at the fifth lumbar and the sacrum. Spinal discs absorb shock and create space for the spinal nerves to exit the spinal column. The discs are equal in height, front and back. They do not naturally wedge or angle, even at sections where the spinal curves deepen. A disc has two distinct parts. The outer portion is the *annulus fibrosis*, comprised of concentric rings of fibro-cartilage. This portion absorbs 75% of the forces applied to the disc. The disc's center, the *nucleus pulposus*, is a round, gel-filled sac of protein that absorbs the remaining 25% of the forces.

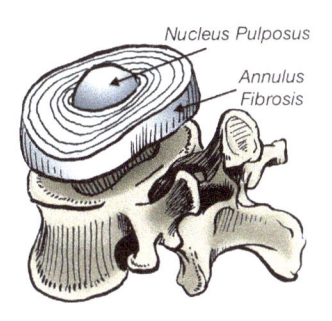

Rupture, herniation, bulge, protrusion, prolapse, slipped disc, and tear

There are many choices of words to describe vertebral disc injuries, which can become confusing. Following are the two terms that best and most simply cover the subject:

- A *disc protrusion* stretches and causes minor tears in the fibro-cartilage rings of the outer disc. It causes the disc to bulge like a bubble in a bicycle tire. There is, however, no leakage of protein from the nucleus. A disc can protrude laterally or medially into to the opening where the nerve exits the spine. A *bulge* is considered a protruded disc.

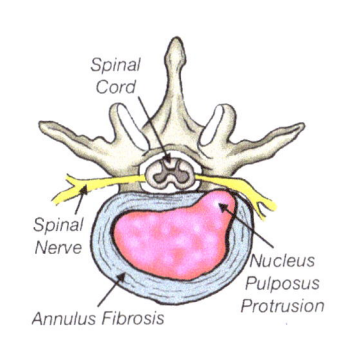

- In a *prolapsed disc*, the "bubble" breaks and the protein-filled gel of the nucleus pulposus leaks though the outer rings of the disc and comes into contact with the outer, surrounding tissues. A disc prolapse is the term best used to represent a *ruptured* or *herniated* disc.

The American Academy of Orthopedic Surgeons estimates that between 60 and 80 percent of North Americans will experience low back pain at some point in their lives. A prolapsed disc will be present in a high percentage of these cases.

Shear madness

In side-bending poses, teachers often instruct students to shift their hips toward the side opposite to where they are bending. Horizontally displacing the hips can cause the lumbar vertebrae to shear across the discs, increasing their likelihood of tearing. The risk increases if twisting, extension, or weight bearing is added to the form of the posture.

To safely side bend, first lift and lengthen the body on the same side that is being bent toward (the concavity). This elongates the concave side of the pose and initiates a smooth, curved movement instead of a sharp and angular one that causes the discs to wedge and compress.

Coming up from a forward fold posture, one-vertebra-at-a-time is a common teaching instruction. However, it is not recommended. It places too much weight on the disc without giving it adequate time to re-center the nucleus, which must shift as the positions reverse. It is best to come up and re-set the natural curves of the spine before fully lifting the torso.

Compressed concavity   *Lengthened concavity*

> **Technically Speaking - Tolerances before disc prolapse**
>
> Spinal discs are designed to manage high amounts of axial loading (vertical force). Discs in the young adult spine can support 800 kilograms per square inch before prolapse. Discs in the aged adult spine can only handle 450 kilograms/square inch. Carrying or lifting weight exponentially increases pressure on spinal discs. A 10-kilogram weight lifted with the knees bent and the spine upright produces approximately 100 kilograms of pressure on the discs. When 10 kilograms is lifted while bending forward and the arms and legs are straight, a force of up to 700 kilograms per square inch is generated. This is very close to the threshold that will damage a disc.
>
> Twists and torques produce shearing forces on the disc. Torsional forces on a disc that exceed 15° of rotation can tear the annular fibers in the disc's outer rings, leading to significant damage.

Back pain- mechanical or chemical?

The human body has developed a sophisticated strategy for distinguishing "I" from "other" via the immune system. It responds to invaders, whether they are viruses, bacteria, or biological venoms. The immune system can amass detailed memories of materials to which it has been exposed throughout a lifetime with a scope comparable to the brain. It is particularly responsive to proteins.

A prolapsed disc leaks proteins from the disc's nucleus into the nearby tissues, exposing them to proteins that until the moment of rupture had been sequestered, or walled off from the rest of the body. If the immune system fails to recognize the newly exposed disc material as "self", it may attack the protein. This initiates an inflammatory, auto-immune response in a narrow space that is already constricted by the swollen disc. Inflammation reduces the small opening between the vertebrae (foramen) through which the nerve root glides. It is the inflammation and not the disc itself that is often the cause of nerve compression and pain.[2]

The immune system's effect on disc injuries

Magnetic Resonance Imaging, or MRI is an imaging tool that can diagnose potential causes of back pain. It provides a detailed view of the spine, the discs, and surrounding soft tissues. An MRI might reveal that a disc prolapse has been identified as the probable cause of the current pain and also that older disc injuries are visible, ones that occurred many years prior. Bewildered, the patient wonders, "but I never had back pain before this happened!"

How could someone be aware of one injury and not others that are so similar? The answer may be found in how the individual's immune system responded to stress at the time of injury. If disc prolapse occurs during a state of panic or protracted stress, the immune system, unable to pinpoint the origin of the "foreign" protein can overreact. The system goes on alert and creates massive inflammation. If a person is managing the stress more effectively during the incident, the immune system may react only minimally, causing a more subdued inflammatory process, or possibly none at all.

The Valsalva effect: Breathless in Yogasana

Respiration has a direct effect on spinal disc pressure. When used effectively, respiration can provide great stability and protect the discs from damage. Improper breathing, however, increases disc pressure to dangerous levels. Many yoga traditions incorporate *Kumbhaka* (breath retention) and *Bandha* (breath lock). These two procedures are practices of *Kriya* (cleansing) yoga and *Pranayama* (energy flow), but they also play an important role in asana. More details on bandhas and breath practices are presented in Chapter 29.

Disc pressure increases when in a forward bending position, especially when the spine is loaded with weight or when breath is retained. Breath retention causes the pressure inside the discs rise significantly, reaching levels where adding a 20-pound weight can prolapse a disc.

Two physiological phenomena protect discs from what may seem like inevitable damage. Because the majority of vertical forces compressing the discs naturally shift from the gel-like nucleus to the outer rings of cartilage, aligned spinal curves and vertebrae are vital in order to access this mechanism.

The other protective action is called the *Valsalva effect*. The Valsalva effect occurs when air is compressed in the thoracic and abdominal cavities. The air that is trapped in the abdominal cavity creates a rigid, inflated beam. Gravity and heel strike forces that would otherwise be absorbed by the discs are transferred forward to the abdominal cavity, reducing pressure on the discs. The effect is short acting and designed primarily to protect the discs particularly when a heavy object is first lifted.

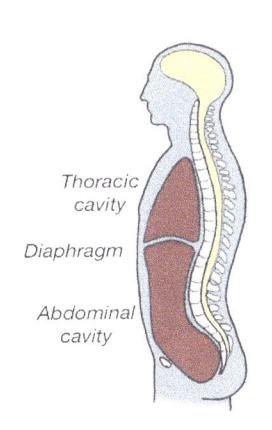

Thoracic cavity

Diaphragm

Abdominal cavity

The Valsalva effect, however, is significant. Axial compression on the discs is reduced throughout the lumbar spine. It is measured as a 50% reduction of pressure on the T_{12}-L_1 disc and 30% on the L_5-S_1 disc.[3,4]

The way to utilize the Valsalva effect in asana is to apply the bandhas. The bandhas are engaged but the breath is not held. When practicing asana, always breathe fully and evenly. A definite, critical difference occurs between holding the breath, which increases disc pressure, and engaging the bandhas, which traps air in the anterior body cavities and reduces disc pressure. Bandha application can be a purely mechanical action engaged by scooping the tailbone forward for Mula bandha and scooping the breastbone inward for Uddiyana bandha. These actions keep disc pressure normalized. Refer to Chapter 29 for details on full-cycle breathing and bandha engagement. Additionally, drawing the navel inward, accesses the *Maṇipūra*, or third chakra. This not only assists in abdominal pressure absorption but also helps protect the spine from compression that can result in lumbar hyperextension.

> Apply the Valsalva effect by engaging the bandhas
> Bandhas can reduce pressure from weight-bearing forces on the spinal discs by as much as 50%

How many movements can the spine make?

The spine is fluid and mobile yet detecting the small, individual movements it utilizes is not something easily apparent. Understanding what individual movements are possible allows the whole spinal system structure to achieve greater overall flexibility and how to advance a yoga practice more safely.

Each vertebra moves in seven different directions. Each vertebra is capable of:
- Flexion
- Extension
- Lateral Flexion (side bending) - left and right
- Rotation - left and right
- *Long axis extension* or distraction. This is a subtle movement where each vertebra can elongate through the spine's vertical axis, lengthening like a stretching spring. Traction physical therapies employ distraction and many yoga postures can engage long axis extension by their own actions.[5]

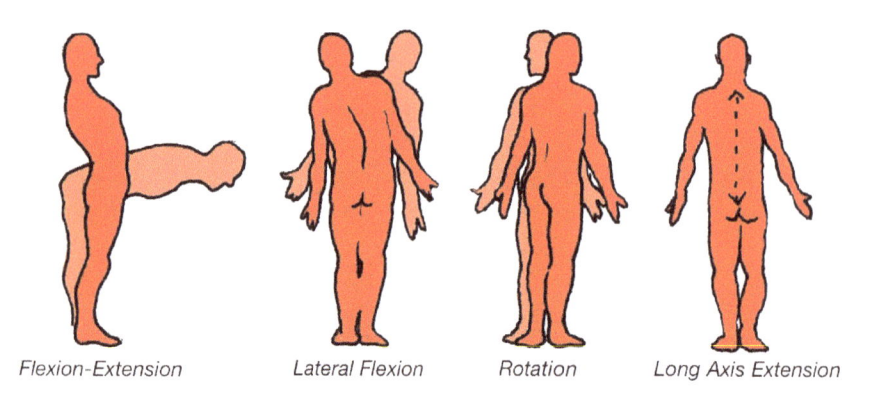

Flexion-Extension *Lateral Flexion* *Rotation* *Long Axis Extension*

At each facet joint, a vertebra is capable of moving independently in all seven directions. The angle of design of some facets may orientate them in favor of one direction of movement over another but all are possible at each joint. From the top of the sacrum to the highest vertebra, the Atlas, there are 48 facet joints.

With 24 moveable vertebrae in the adult spine, each moving in 7 directions, there is a sub-total of 168 joint movements within the spine. But this is not the final count. Each vertebra has four facets, each pairing with facets from the vertebra directly above or below. Facetal movement occurs between the pairs, which means each vertebra can move in the seven directions from separate two hinges. This brings the total to 336 independent, spinal movements. This is more enough potential mobility to create the snake-like, undulations we associate with a healthy spine.[6]

Most yogis have never learned to engage the spine's full capabilities. Poor postural habits and injuries limit the ability to utilize all of the spine's subtle movements. For instance, the thoracic spine, in conjunction with the rib cage, has 216 movements; yet often, out of habit, it is moved as one solid piece. Perhaps belly dancers fare better than yogis in accessing this extensive potential!

Independent facet movement

Facet joints often glide independently of each other. Even on the same vertebra, some joints can move freely while others become *fixated* (locked in place). When any movement is diminished, other joints compensate and take up the slack. Hypermobile facets are ones that move beyond their normal range. They frequently create subtle compensations that rarely reach the yogi's awareness until the joints have already become weakened, inflamed, and unstable. Fixated joints, due to their lack of motion, eventually develop adhesions. Over time, degenerative changes can develop in either hypermobile or fixated vertebrae. Degeneration includes calcium deposits, spur formation and, eventually, spinal fusion. Although these changes may be attributed to the "typical" aging process, restoring normal mobility and precise spinal alignment can curtail the occurrence. Yoga practice is an ideal system to develop alignment, mobility, and stability.

Hypermobility, clicks and cracks

Some students are familiar with the creaking and cracking that can occur in the spine when twisting or turning. When the back elicits a crack, substances are released called *endorphins*. Endorphins are neurotransmitters that are chemically similar to morphine. They produce sensations of pleasure and relief that cause some students become almost addicted to constantly cracking their backs.

Most cracking that recurs from the same movements have little therapeutic value. These generalized cracks and pops do not release fixated joints as chiropractic procedures may do but instead, generally move already unstable ones. Repetitive movement of unstable joints stretches the ligaments and leads to chronic hypermobility in those joints. Moving hypermobile joints can feel empowering at first as it produces a sense of freer mobility but this "benefit' rapidly diminishes. The hypermobile regions eventually become inflamed and painful. If hypermobility continues to be exploited, the slow process of degeneration begins in the spine and its discs. Over time, hypermobility is replaced by immobility.

Mobility – the curse or the cure?

Certain regions of the body may seem more flexible and more freely moveable in comparison to other, less mobile regions. The objective in asana practice is to *move from the least mobile regions first*. This lessens the exploitation of the mobile regions. Precise postural alignment reduces potential damage from these discrepancies in mobility and stops the hypermobile joints from exceeding normal limits.

Rag Doll Forward Bend is a deceptively complex pose that offers more than a simple stretch of the back body. In Rag Doll Pose, the body can experience the energetic contrast between *ease* and *tension*. Areas of tension are often less mobile and stuck, unable to find ease.

In Rag Doll, identify the tense regions of the spine. Release deeply and allow the tissues to soften around the spine by bringing the focus of the breath to those specific areas. With each exhalation, release the tension. Be aware of the freer, adjacent areas of the spine that may try to compensate by overstretching.

Rag Doll Forward Fold

In the **Cat-Cow** series, add undulating spinal movements to engage all seven directions of movement at each vertebra. Move the spine in an effortless, wave-like motion. Try to identify any segments that are either not fully participating or conversely, are moving excessively. Once a pattern is detected, use each successive wave of Cat-Cow to increase motion in the limited regions and reduce motion where the spine over-engages. Eventually, more fluidity and balance is attained.

Author's perspective: Rounding the upper thoracic spine as instructed in the Cat Pose is not beneficial unless the yogi has a markedly flat thoracic curve. As an option, lift only the lower rib cage and drop the chest, allowing the upper thoracic spine to flatten toward the mat. While maintaining this orientation, attempt undulate the spine. Although the undulations will be more challenging in this position, they mobilize the vertebrae with the spine in alignment, a better approach.

Thoracic extension in Cow Pose

Rounded thoracic spine in Cat Pose

Changes in spinal curves throughout life

If we observed a group of people at a distance and could barely make out their general form, we would instinctively sense their individual ages and levels of vitality. The uprightness of their spines and the rhythm of their gaits send visual cues that are hard-wired into our basic consciousness and innate drive for species survival.

At birth, the spinal curve retains its fetal form, rounded in a "C" shape. The concavity of the fetal curve faces anterior, what is referred to as *kyphosis*. As the infant begins to lift its head in search of food from the breast or bottle, the neck muscles strengthen, and over a few months, the cervical spine reverses its curve to exhibit a posterior concavity, or *lordosis*.

When the baby begins to crawl, the hipbones recess into the acetabulum, stimulating the formation of deep-seated hip sockets. Crawling, and eventually walking forces the lumbar spine, like the cervical spine, to reverse into a *lordotic* curve. Because ample space is required for the lungs and heart, the thoracic spine retains its original *kyphotic* curve.

The final form of the normal adult spine is "S" shaped, with three major curves: cervical, thoracic, and lumbar. Similar to the loading mechanics of a spring, the S-shape design best manages the shocks and forces transferred through the body's central axis.

As humans advance in years, the overall curve of the spine often returns whence it came. A lifetime of spinal wear and tear, a front-body lifestyle, and general loss of muscular strength result in degenerative changes that cause the spine to collapse and return to a C-shaped curve.

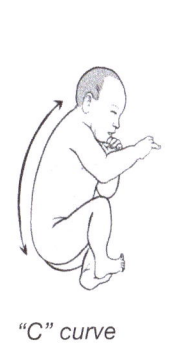

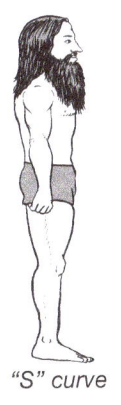

"C" curve   "S" curve   "C" curve

The "Dead Zone"

Twisting poses, such as Revolved Triangle or Revolved Lunge, can be used to detect and compare relative mobility in regions of the spine. This procedure is best performed by an assistant. The student assumes either pose, depending upon their skill. The assistant palpates the spine, looking for tight or immobile regions by pressing into each vertebra at the lower side of the spinous process.

Sections where the spine feels stiff and hard, as opposed to springy, are "affectionately" called *dead zones*. With practice, the subtle variations in rigidity can be identified. If immobility is caused by muscular tension, it will have a springy quality when the spine is deeply pressed.

Joint immobility, in contrast, presents with a hard end resistance. The assistant may also be able to detect the hyper-mobile segments, often interspersed between immobile ones. If no evaluator or assistant is available, the student can perform gentle undulations while in one of these revolved postures. If carried out with focus and attention, the student can evaluate their movements. To test further, attempt to lengthen and move more fully from immobile areas.

Scoliosis   Dangerous curves ahead

A lateral (side-to-side) spinal curvature is called a *scoliosis*. Although it is not a normal curve and many of its causes are unknown, scoliosis is a common occurrence. In fact, the spine is pre-disposed to scoliosis. The heart in the initial stages of human embryonic development is large and positioned at the center of the growing fetus. The developing spine will deviate to the right to accommodate for the heart. This deviation may persist after birth and increase as the spine continues to grow through childhood.

The greater the angle of curvature, the more problematic the scoliosis becomes. At extreme degrees of curvature, scoliosis can compromise the function of the heart, lungs, and diaphragm, as well as other internal organs.

Rotation of the spine can occur in the region of scoliosis and is called rotatory scoliosis. It is commonly seen with larger curves, although not in every case. To attempt to balance the scoliosis, smaller, secondary, compensatory curves often develop.

The cause of scoliosis is often unknown. The term given in these cases is *idiopathic*, from the Greek, "one's own suffering". However, besides the potential cause scoliosis being heart-development interference, a few causes are identified:
- Uneven hips or uneven leg lengths
- A wedge-shaped vertebra (in place of a square one)
- Muscular asymmetry or underdevelopment
- A flat, hyper-flexible spine that attempts to stabilize by deepening its curves laterally to form a shock absorbing spring

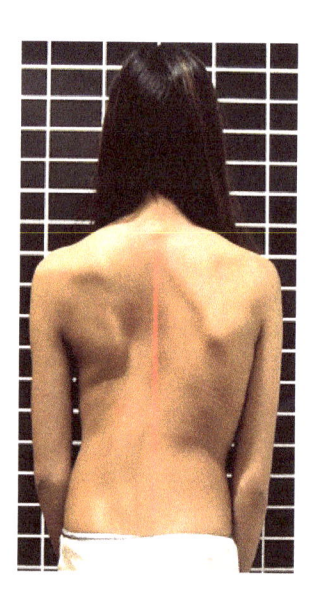

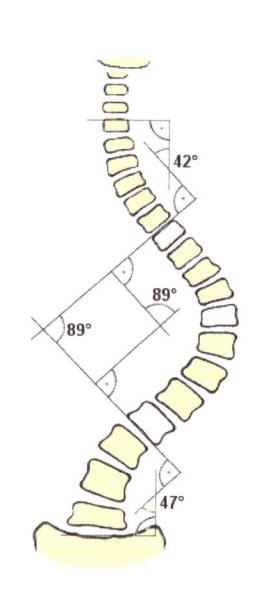

Scoliosis and mobility

Mobility is essential for spinal health, especially when scoliosis is present. Mobility must be balanced and evenly dispersed throughout the spine to be safe. Mechanically, mobility is reduced in the sections of the spine with scoliosis. To compensate, hypermobile regions can develop that overtime become degenerative. The better able the spine can access its 336 subtle, inter-segmental movements within their normal ranges of motion, the fewer the compensations and degenerative damage. It may seem like a daunting task to maintain full and fluid mobility in the spine but with practice and awareness, it is possible to come close. Professionals, such as chiropractors can assist in discovering imbalances and restore mobility. Belly dancers do pretty well at accessing the spine's subtle movements, too!

Scoliosis formation may be a compensatory response to a flat, "wobbly", hypermobile spine with its anterior and posterior spinal curves straighter and flatter than average. Lateral curvature becomes the body's attempt to increase stability, which is necessary for scoliosis management.

Essential yogic principles for scoliosis

- *Samasthiti* is the first and most essential principle to apply in asana when scoliosis is present. Samasthiti can be considered the quality in the body where every muscle and tissue is physically and energetically balanced - in every direction, along every surface, and along each joint. Being conscious of and keeping the intention to maintain Samasthiti can significantly reduce the compressive effects of scoliosis.

- *Move from the least mobile regions first.* Initiating movement from the least mobile joints will produce fluid and graceful action, even when imbalances are present. Moving from the least mobile regions of the spine first can re-pattern its habitual mechanics. These changes transform the asana practice and become an excellent therapeutic help reduce the acceleration of scoliosis and its resultant spinal degeneration.

- *Lengthen along the vertical axis of the spine (long axis extension).* With scoliosis, the spine tends to compress and shorten. Over many years, the spine in effected regions may collapse, resulting in an observable loss of height. Extending the spine through the deepest portion of a scoliotic curve lengthens spinal joint spaces and the spinal muscles, which, in time, become more balanced and their muscle memory will help keep the spine straighter.

Scoliosis and yoga therapy

Yoga asana and therapy offer many approaches for reducing the effects of scoliosis. It is an extensive topic that a number of masterful teachers and their books address comprehensively. This book does not attempt to cover the topics of yoga therapy and scoliosis fully and it is recommended to seek out additional sources to more fully explore the subject. Offered here are two examples of how restorative asana and props can effectively address scoliosis.

Example 1: With the student lying in side posture, place a bolster at the convexity of the lateral curve (point where the spine rounds toward the side). On the inhalation, lengthen the spine; on exhalation, relax the muscles and release deeper into the bolster.

Example 2: With scoliosis, the muscles along the convex (rounded) side of the spine overstretch and become less efficient and weaker. The muscles on the concave-curved side (recessed) of the spine are over-contracted and shortened, enabling them to overpower the convex-side's musculature. Yoga teacher and researcher Doug Keller teaches an effective approach to lengthen the musculature along the concavity of a lateral curve and traction and lengthen the muscles of the concavity.[7]

Procedure:

1. Determine which side of the spine has the lateral concave curve (recessed) and which side is convex (rounded).
2. Stand with concave side closer and perpendicular to the wall.
3. On a yoga block, stand and balance on the convex-side's foot, which is the leg farther from the wall. Balancing strengthens the musculature along the convexity that are presumed weaker. It also strengthens the outer leg muscles.
4. Allow the inside leg to suspend. This creates downward traction on the spine along the concavity of the scoliosis.
5. The inner arm raises overhead and lengthes up the wall. This elongates the concavity of the scoliosis while stretching the tighter musculature of the concavity using upward traction. The raised arm focuses on the thoracic spinal curve, which is a more common location for scoliosis. The suspended leg focuses on lengthening the lumbar curvature on that side.

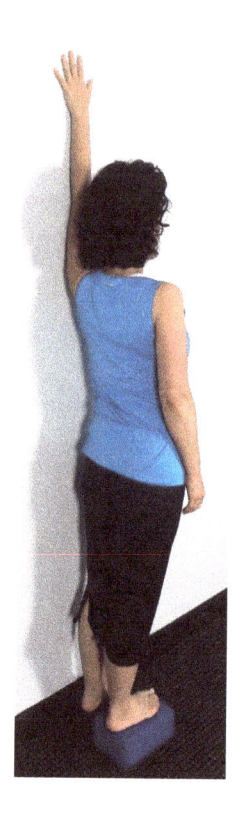

> Regardless of whether or not scoliosis is present, yoga students can learn about their spines and identify where movement may be excessive, absent or normal. This information is essential for a safe and therapeutic asana practice.

Mapping out the spine

Whether or not scoliosis is present, our postural habits determine what issues require attention in our asana practice. Could you answer with certainty, "Which shoulder do you habitually hold higher?" or "Which hip is forward?" This information is important and unique to each person and best learned by direct observation.

A useful approach is to "map out" our bodies. When our baseline posture and personal habits are identified and their causes are understood, improvement is more likely, affecting our vitality both on and off the yoga mat.

Assess the appearance of these anatomical markers, looking for levelness and balance:

- Are the collarbones square and level?
- From the anterior, is the nose centered over the breastbone?
- Does the back of hand face forward? (Not desired, indicating rounded shoulders)
- Does middle finger align with the seam of pants?
- Are the bottom tips of the shoulder blades level?
- Is one shoulder blade farther back? Does one rotate forward and lift away from the spine?
- Are the lower ribs symmetrical and level with each other?
- Does the rib cage jut forward on either side?
- Do the crests of the hips appear level?
- Is one hip forward of the other?
- Is the soft tissue and muscle over the hips evenly toned or does one side bound up?
- Are the knees in line horizontally with each other?
- Can scoliosis be visually observed when looking at the back?
- When bending forward to a half forward fold, does the spine straighten or are there areas that remain curved or that bound up? Straightening indicates that the scoliosis is *functional,* a consequence of movement patterns. A remaining curved indicates a *structural* curve, one resulting from bone and muscle development.

Once we understand our spinal movements and are able to locate our imbalances, we can utilize this knowledge and bring these vital statistics to yoga practice. Subtle adjustments can be made to our unhelpful postural tendencies that are discovered through this personal evaluation.

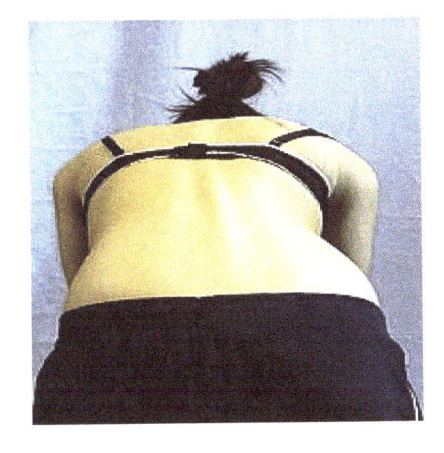

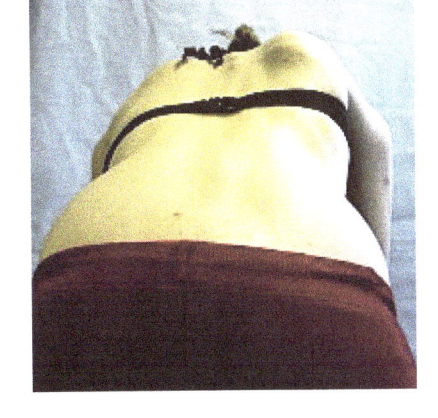

Mild Scoliosis

Severe scoliosis

Imbalanced spinal mobility

Whether scoliosis is present or not, each vertebra of the spine is designed to move individually and without strain or excessive wear-and-tear. Every vertebra's movement is also interdependent with all others, especially the ones in the same region. If one vertebra becomes "stuck", nearby vertebrae will often compensate by becoming hypermobile, increasing their ranges of motion. Over many years of compensation, the spine will show signs of degenerative damage. Spinal discs will wear thin and the surfaces of hypermobile vertebral bodies become rough and irregular. Boney extensions of the vertebrae, called *spurs*, may form as an attempt by the spine to stabilize. This degenerative, wear-and tear scenario is called *osteoarthritis*. In advanced stages, two or more vertebrae may fuse together, a condition called *boney ankylosis*.

Students with less a flexible body type must be careful not to overly force their asana practice. Immobile joints simply may not release and participate movement and freely moveable segments may compensate, dominate, and move beyond their normal ranges of motion. It is best to move immobile regions slowly, with precision, and with the awareness of breath to create space between the vertebrae to increase mobility.

Over time, hypermobile regions of the spine that have shown excessive flexibility throughout life may begin to lose mobility as degenerative changes, such as arthritis eventually reduce the ability of the spinal joints to move. To avoid this outcome, especially for highly flexible students, resist exploiting the regions of instability and hypermobility. Instead, take a true, focused inventory of the spine and learn to initiate all movements from the regions of least mobility.

Applying all the alignment principles shared in this book, including for regions seemingly unrelated to the spine, can effectively change time-hardened movement patterns. Manipulative therapy, such as those from chiropractic and some forms of physical and yoga therapies, can help identify and restore lost movement to regions difficult to reach through asana alone. All help in reversing the destructive outcome of imbalanced spinal mobility.

Keep the lumbar curve intact

The lumbar curve needs to be preserved in poses such as Happy Baby, which can flatten the spine. Keep the legs lengthened or suspended or press the knees toward the hips to maintain the curve.

25 The Lumbar Spine

The incidence of lower back pain is approaching near epidemic levels. It has become the nemesis of our western civilization. Our sedentary lifestyle is epitomized by riding in automobiles, sinking into soft couches, never learning the basics principles of posture or sitting, and not being required to sit in a deep, comfortable squat - makes the lumbar spine vulnerable to injury.

In a recent study of lower back pain trends, it was reported that 25% of Americans, more than 76.2 million people, experienced lower back pain during a recent three-month period. Following a one-year period, that number increased to nearly 50%.[1] Over a lifetime, an estimated 80% of Americans will experience some back pain.[2] Most lower back injuries and pain result from mechanical dysfunctions that affect the spine and pelvis. Fewer back complaints will originate from organic conditions, such as inflammatory arthritis, infection or cancer.

There are no "magical" cure-alls for lower back pain, nor is yoga the exception. Yet the experience of many yoga practitioners corroborates asana as an excellent approach to managing lower back issues. Yoga can potentially deliver a personalized physical therapy session, using sophisticated tools for improving posture, uncovering muscular weaknesses, and addressing unhealthy mechanical patterns - all causal factors for the development of spinal arthritis and degenerative joint disease, as well.

Motion of the lumbar spine

The lumbar facets are capable of seventy potential movements. Their facets orientate front-to-back, or along the sagittal plane, which more easily allows for flexion and extension. More limited in movement are the directions of lateral flexion and vertebral rotation. The subtleties of these movements can be challenging to evaluate and moving from one facet in one direction is not a reasonable expectation. Traditional range-of-motion tests provide only a limited assessment of gross thoraco-lumbar movement. The more specific facetal movements can be analyzed by methods such as motion palpation, a chiropractic procedure.

Lumbar Ranges of Motion	
Flexion	40°
Extension	30°
Lateral flexion	20 -30°
Rotation	10° (2° at each segment or 1° each)

The ilio-lumbar ligaments

The pelvis functions as the foundation for the spine. Anchoring the lower lumbar spine to the pelvis are large bands of dense, highly collagenous tissue called the *ilio-lumbar ligaments*. They connect the lowest two lumbar vertebrae, L4, L5, with the iliac portions of the pelvis. By their location and design, they stabilize the entire lumbar spine but in the process, reduce the mobility of L4 and L5 when compared to the rest of the lumbar vertebrae. When necessary, the ilio-lumbar ligaments can stabilize the sacroiliac joints against forceful or improper movements.

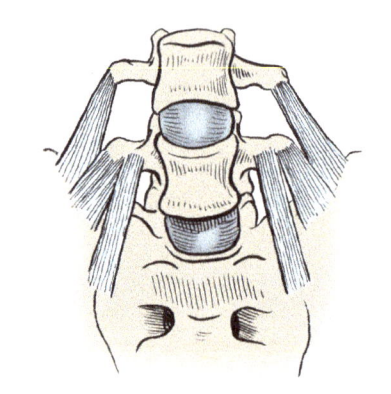

Ilio-lumbar Ligaments

Happy baby, indeed!

It may be a topic of curiosity how infants can effortlessly round their backs and able to easily chew on their toes. Tauntingly, they look up at their mothers with an expression that seems to say, "Go ahead and try this, you Big Strange Creature!"

At childbirth, the ilio-lumbar ligaments are only rudimentary. They begin their actual development during childhood but do not fully form until the adolescence years. Poor posture or low back injury during these critical years can result in the ligaments developing improperly elongated and unable to provide the stability they are meant to offer.

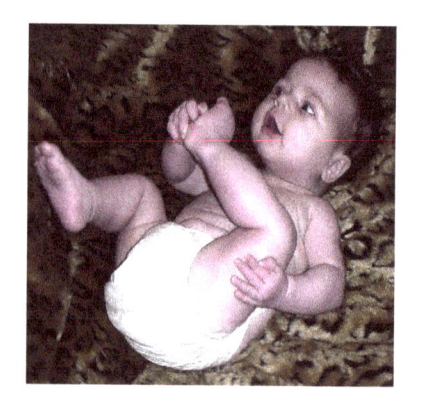

Once a ligament is elongated, its length is permanent. Elongated, loose Iliolumbar ligaments cannot provide adequate support to the lumbar spine. If poor postural habits persist into adulthood, structural weakness in the lower spine takes its toll. The result of prolonged instability at the L4 and L5 vertebrae is disc damage and degenerative osteoarthritis. Spinal instability is a key factor in the high incidence of lower back pain.

Stiff, or not stiff? That is the question!

Low back pain and stiffness are often the reasons students find their way to their first yoga class. New students often believe that lack of flexibility is the primary cause of their troubles. Supporting this assumption, while in a standing forward fold, they display an excessively rounded lower back. Their finger-reach to the floor is a discouragingly great distance. It is fair to assume that these students may never make the final cut for Cirque du Soleil®. Yet, when on all fours, these students often have no difficulty deepening the arch of the lumbar spine toward the floor. If the cause of their "stiffness" were the result of an inflexible spine, it would have continued to remain rounded and unable to reverse its configuration into a concavity, regardless of their posture or position.

Contrary to the lower back's rounded appearance in poses of flexion, the apparent inflexibility often has at its core ligament instability. Weak, overstretched ligaments are unable to stabilize the spine while maintaining its natural, anatomical curve. In response, spinal muscles become contracted and tensed and the surrounding tissues become inflamed. It is tense muscles and inflamed tissues that cause stiffness and the misperception of inflexibility. Also of note, hamstring muscle tightness and limited hip flexibility can be significant culprits in this scenario. Increasing both of their flexibilities can significantly aid in reducing lower back pain and stiffness while stabilizing the lower back. Yoga asana overall is an invaluable tool for addressing instability and/or lower back stiffness. Correctly performed postures can stabilize weakened ligaments and allow mobilization from the appropriate places.

Spinal stenosis

When observing seniors, you may recognize common characteristics shared by these wonderful sages. Beyond their obvious slow shuffle and rounded upper backs, another feature is often present: straight, flat lumbar curves and flattened buttocks. These observations are often the result of *spinal stenosis*, a common, usually age-related, degenerative condition. Weakened ilio-lumbar ligaments, spinal instability, or spinal disc injuries lead to the development of spinal stenosis.

Spinal stenosis is the narrowing of the spinal canal. It results from a thickening of the posterior vertebral rings. Anteriorly, the large, block-like vertebral bodies are designed to carry the weight of the human frame. As one potential cause for stenosis, should the body's center of gravity shift posterior, weight is transferred to the slender posterior rings. This can cause the rings to thicken to handle the extra structural loading and eventually narrows the canal.

The space within the canal is intended for the spinal cord and its multi-fibriled, tapered tail to travel unencumbered. If narrowed, it compresses the nerve tissue, causing severe discomfort in the lower back and lower extremities. Stenosis is most common in the lumbar spine but can develop in other spinal regions, most frequently in the cervical spine.

Half Boat Pose, **Ardha Navasana**, is an excellent therapeutic pose to address the flattened curve and the effects of lumbar stenosis.

Narrowed Spinal Canal

When to surrender

The lumbar spine is most stable and safest from injury when its lordotic curve maintains its size and shape throughout every change in posture and position. Always maintain an egg-shaped lumbar spinal curve!

In Chapter 20, focused on the hamstring muscles, a two-step procedure was introduced that is to be applied to all forward bending postures and directly effects the lumbar spine. In **Paschimottanasana**, step one establishes the lumbar curve, which is held for as long as possible throughout the pose. In step two, the lumbar spine may flatten, as the pose becomes an expression of complete surrender that releases all muscular tension. This usually does not produce negative consequences for someone with a healthy, flexible spine. However, the full surrender position may not be beneficial for students who have lumbar instability or stenosis. For these students, it is better to maintain an egg-sized lumbar curve throughout both steps of the forward fold, regardless of how deep the posture goes. This approach keeps the pose fully therapeutic. It strengthens the muscles of the lumbar spine around a proper curve. It also reduces the effects of spinal stenosis. Modifications to the posture using props, such as sitting on blankets or a bolster, can keep the lumbar curve intact while in engaging both steps of the seated forward fold.

Step-one: forward fold with lumbar curve intact

Step-two: release back, allowing it to round if the lumbar spine is healthy

Rounded Postures

Most people have a lifestyle that requires the back to round in countless situations. Since the front of the body is generally 30% more flexible and 30% stronger than the back, it is important to develop the strength and flexibility of the upper back to bring better structural balance. There is little value in overdeveloping the already overpowering chest and anterior shoulders while ignoring the upper back. If a yoga practice omits the rounding of the spine altogether, students would be better for it!

Tip of the pelvis, wag of the spine

The upper aspect of the sacrum is called the *sacral base*. It provides a moveable foundation for the spine. As the entire pelvis tilts, the sacral base shifts, angling forward or back relative to the horizontal plane. Changing the angle of the sacral base directly changes the shape of the spinal curve.

An *anterior* pelvis tilts the sacral base forward, causing the lumbar curve to increase and the thighs move posterior. A *posterior* pelvis tilts the sacral base back toward the horizontal. A more horizontal sacral base flattens the lumbar curve and flattens the buttock while bringing the thighs forward.

The action of "scooping the tailbone forward" is the method to engage *Mula Bandha*. The Mula Bandha action does not excessively force the pelvis to tilt posterior, which would flatten the lumbar spine and press the thighs anterior despite all the actions being linked together. To prevent over-efforting, firmly keep the thighs back, an element of *Inward hip release*, while scooping the tailbone forward. (Review Chapter 14 for more details).

Anterior tilt

Posterior tilt

Lower torso muscles

Abdominal exercises such as crunches and sit-ups are often considered to be essential for strengthening the lower back. This is partially true. If the abdominal muscles are weak, an abdominal exercise regimen will have noticeable, positive results for lower back stability. To achieve total structural integrity, however, all three primary, lower core muscle groups need to be strong and balanced. These groups are the *abdominal muscles*, the *psoas major/minor*, and the *lumbar paraspinal muscles*.

At first, it may seem like a complicated task to balance the three muscle groups together. The good news is that the pelvic integrative alignment principles taught in this book will easily accomplish the job. The point where the three lower core muscle groups find their ideal balance and effort is when an egg-sized lumbar curve is established.

Lumbar Paraspinals
Abdominals
Psoas

Lower Crossed Syndrome

A deeply curved lumbar spine with an anterior pelvis creates a common, recognizable pattern. As illustrated, it produces an imbalance that creates weak and corresponding tight muscles. Correct pelvic alignment and engaged bandhas are fundamental to breaking the pattern.

Weak abdominals
Tight erector spinae
Tight iliopsoas
Weak gluteus

> A lumbar curve that is approximately the size of a small egg provides ideal spinal stability and muscular balance

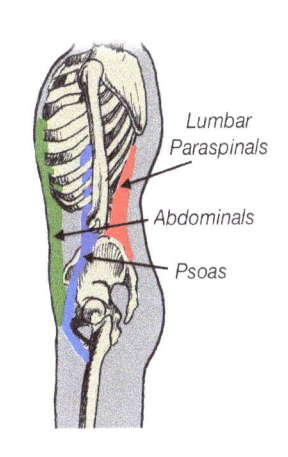

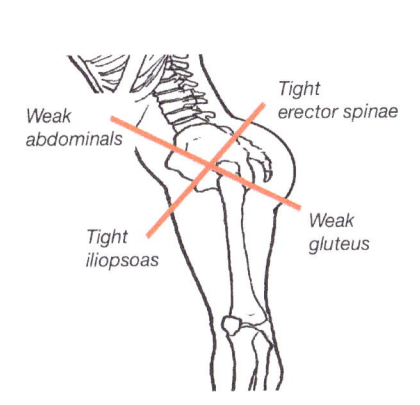

Don't crush the egg!

Which comes first: the egg-sized curve or balanced core muscles? As with riddle, the answer can be both, or perhaps, either.

To form the correct lumbar curve, deepen the space at L4-L5 to the size needed to nestle an imaginary egg in between the two long, thick bands of paraspinal muscles. When lying supine, the egg could theoretically fit under the lower back and would neither roll out nor get crushed. Of course, differently sized eggs are appropriate for different sized bodies!

In **Ananda Balasana** (Happy Baby Pose), the hips are flexed deeply, causing the pelvis to tilt posterior, often causing the pelvis to lift off the mat and the lumbar spine to flatten. Despite this common tendency, it is not desired or safe. Flattening the spine weakens the lumbar ligaments. Instead, keep the sacrum pressed to the mat and attempt to retain an egg-sized curve in the lumbar spine. This will be more attainable if instead of immediately pulling the knees toward the shoulders, the knees are first pressed away and in the direction of the hips. Maintain that pressure as knees are then brought toward the shoulders.

In contrast, when lowering straightened legs from a supine, leg's at 90° position, the upper lumbar curve tends to increase excessively. Resist this occurrence by firmly engaging the lower core musculature while applying *Scoop the tailbone, Scoop the breastbone, Draw in the navel*. This will help maintain the curve at one egg and not produce a half dozen!

The abdominal obliques

The abdominal obliques are a diamond-shaped muscle group that encircles the lower torso in a girdle-like fashion. The *internal, external,* and *transverse* abdominal obliques stabilize the lumbar spine. Superiorly, the abdominal obliques attach to the lower seven ribs. Inferiorly, they anchor to the pelvic brim, in part, by the *inguinal ligament*, the most common location for hernias. Besides providing lumbar strength and stability, the obliques help twist the lower torso in lumbar rotation. The abdominal obliques' contraction is *parabolic*, which forms the hollowed shape of the waist.

Abdominal Obliques

This parabolic action can be understood by picturing closely packed strings spanning between two rings. The upper ring represents the lower rib cage and the lower ring depicts the pelvic brim. For a twist to occur, at least one of the rings must remain stable while the other turns. Both rings can turn but in opposite directions. If both the lower rib cage and the pelvis turn toward the same direction, neither spinal twisting nor the waist-narrowing engagement of the muscles can significantly occur.

As a rule, twists occur in the spine, not in the pelvis or shoulder girdle.

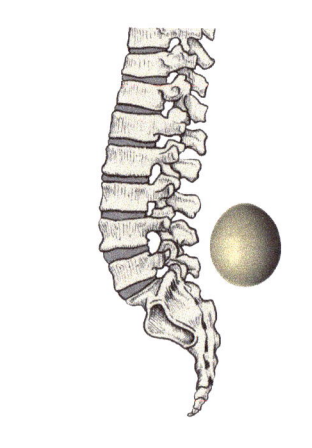

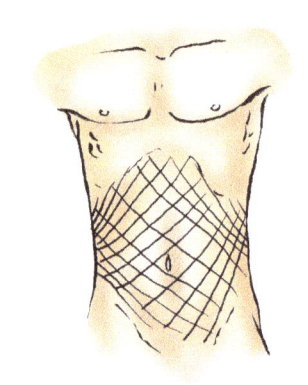

Twist not engaged   *Twist Engaged*

In the Revolved Lunge Pose, stabilize the pelvis (lower ring), keeping it square toward the front of the mat. The lower rib cage (upper ring), joined by the chest, and shoulders twist in the opposite direction. This action strongly engages the abdominal obliques and provides a therapeutic "squeezing" of the internal organs. Since the lower side of the spine tends to become concave in twists, engage the abdominal obliques with equal length and tension along both sides of the torso.

Conversely, in the Supine Twist Pose, the lower rib cage (upper ring) remains stable or rotates in the opposite the direction that the pelvis rotates. The pelvis (lower ring) rotates to stack the hips vertically.

Spinal twists have great therapeutic value when performed correctly.

Revolved Lunge

Keep the central axis aligned when twisting

In Supine Twist Pose, the central axis should remain in a straight line. To allow for this to happen, the hips must first be re-positioned before the knees drop to one side. The incorrect position introduces unwanted torque and shearing across the spinal discs while deviating the bandhas off of their central axis.

To perform correctly: after setting both shoulders flat, slightly lift the hips and shift them the distance of half the width of the pelvis toward the side opposite that the knees will be dropping. This is approximately a 10-14 inch shift to the side of the upper leg.

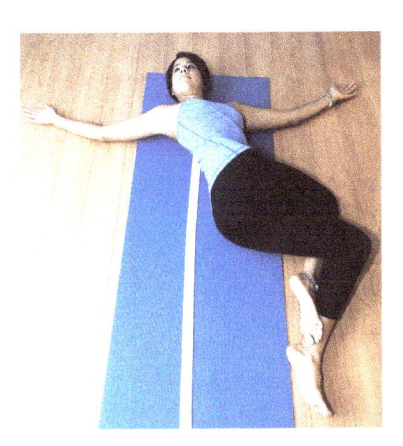

Supine Twist- incorrect

Abdominal six-pack

If a six-pack abdomen were desired, yoga would not be the first choice as the physical method. Most yoga postures do not isolate the abdominal muscles or engage them with the firmness required to develop a six-pack. There is certainly nothing incorrect in developing "washboard abs" as long as doing so does not overdevelop the abdominals to the point where they overpower the psoas and lumbar paraspinal muscles.

The "black hole" of the belly

A universal *black hole* draws everything from all surrounding directions into its center. Engaging the navel is analogous to the behavior of the black hole. Abdominal musculature draws inward from every direction. This action is subtle, yet a definite effort. The goal is to maintain a subtle abdominal muscle contraction in yoga practice and beyond. This action engages the *Manipuri chakra* and works together (synergistically) with the *Uddiyana bandha*.

The action of *Drawing in the Navel* prevents lumbar spinal curve from overarching. This valuable action protects spinal discs and nerves from compression, especially when attempting backbending poses.

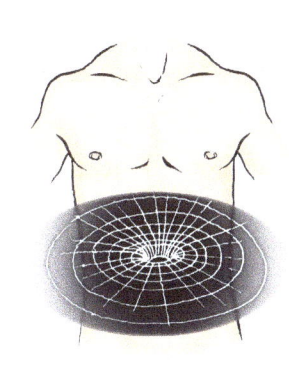

Spondylolisthesis

Approximately 5% of the general population has a structural defect in one or more vertebrae called a *spondylolisthesis*. A spondylolisthesis is a break in, or a non-fusion of a vertebra's posterior ring. It causes a permanent instability of the vertebra and a tendency for the impaired vertebra to slip anterior. With spondylolisthesis, managing the force of gravity becomes problematic. Weight-bearing activities and back-bending postures must be performed with caution to not aggravate the condition by forcing the vertebrae to drift forward.

Spondylo, as it is often called, appears as an inward depression in the lower back. One discerning characteristic of the presence of a spondylo is that the pelvis does not excessively tilt forward as might be expected where there is a deep curve but instead, retains a neutral position. This differentiates it from a deep lumbar curve formed by an anterior pelvis.

There are no special yoga postures, nor medical interventions that can repair actually repair a spondylolisthesis. However, it is possible to limit the pain and limitations, as well as degenerative changes it creates on the spine. Yoga asana and the integrative alignment principles being discussed throughout this book address the forces of gravity that cause spondylo vertebrae to continue to drift forward. Precise alignment can allow most asana to be practiced safely, despite the presence of spondylolisthesis.

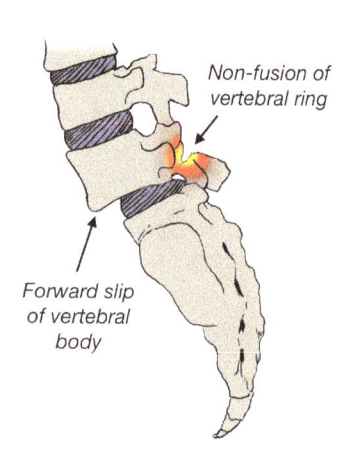

These previously presented alignment principles help to safely manage spondylolisthesis:

- *Move from the Mula:* This prevents the center of gravity from shifting forward and placing stress on the spondylolisthesis.
- *Scoop the tailbone; Scoop the breastbone; Draw in the Navel* (Chapter 11). These actions are the most effective way to reduce anterior-driving forces on the lumbar vertebrae.
- Draw in the "black hole" of the navel. This limits the drifting forward of the spine.

Supported bridge for spondylolisthesis

Supported Bridge Pose is an excellent restorative posture for spondylolisthesis management. The pelvis is lifted and supported on a block that crosses the sacroiliac joints. The lumbar spine suspends and lengthens, receiving axial traction while the vertebra with spondylolisthesis drops posterior.

Place the block at any height. Be sure the block is resting on the sacrum, not the lower lumbar spine. The force of gravity allows the lumbar spine to traction (lengthen mechanically in long axis extension). The spondylolisthesis drops posterior, reducing the anterior stress on the spine. Supported Bridge can often provide rapid relief of pain in acute conditions.

Supported bridge

Sciatica

The sciatic nerve is the largest and longest nerve in the human body. It travels along the posterior leg and into the foot. It is comprised of a combined network of nerve root fibers that exit the spine from the L-3 to S-2 vertebral segments (some variation is common). *Sciatic neuritis*, the proper term for sciatica, results from inflammation of one or more of the sciatic nerve's roots. It can produce deep buttock pain that can radiate down the leg to as far as the foot.

Sciatic neuritis is most often caused by disc protrusion. Individual roots of the sciatic nerve exit through foramen, small openings between the vertebrae. Normally, there is approximately 12 millimeters of slack with which the nerve root can slide. The 12 millimeter limit is usually reached when the straight leg is raised to approximately 60° of flexion. For a healthy disc, the roots of a free-moving sciatic nerve can stretch beyond 60° without producing any tension or irritation. However, if a disc is protruded or swollen, the nerve root becomes trapped or must travel an additional distance beyond 12 millimeters to pass around the bulge.

The Lasegue's test, an orthopedic test for sciatica, lifts one straight leg at a time. If pain is elicited before reaching 60° of leg lift, it is indicative of sciatic neuritis. Leg pain beginning after 60° is more likely not caused by sciatica but resulting from other conditions, such as hip problems, sacroiliac dysfunction, or issues originating in musculature, particularly in the piriformis or the hamstring muscles.[3]

The straight leg raise test is not conclusive for a disc protrusion, however, it does provide a yoga teacher with information that can be used to responsibly advise students about how to modify their practice and when to see a healthcare professional.

The **piriformis muscle** externally rotates the hip. Its path courses directly over the sciatic nerve. For some individuals, the sciatic nerve travels through the center of the muscle or between two separate muscle bellies. Should the piriformis muscle becomes weak and flaccid, it can collapse onto the nerve and produce pressure. Should the muscle spasm, it can create a vise-like grip around the nerve. Either of these piriformis conditions can produce sciatic pain. With the piriformis, pain is usually confined to the buttock region; it less often travels down the leg as is more common with a disc related condition.

A tear in the **hamstring muscle** at its attachment to the ischial tuberosity can be equally uncomfortable and mimic the pain of sciatic nerve inflammation. It is a common cause of pain often observed with "yoga butt". A torn or strained hamstring muscle attachment at the ischial tuberosity can also produce leg pain that follows the course of the muscle as it runs down the back of the thigh and behind the knee.

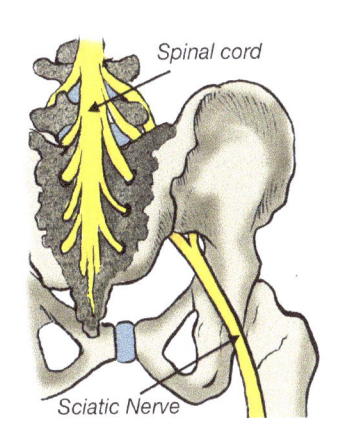

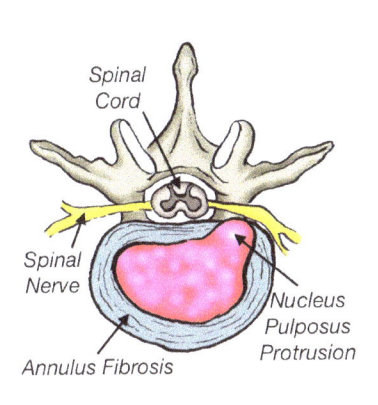

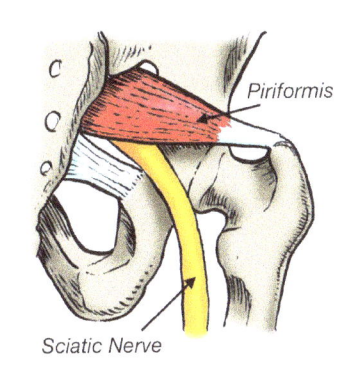

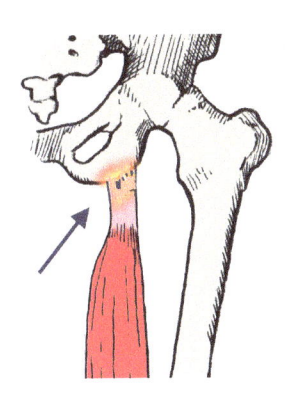

Hamstring pain tends to be localized within the leg muscle itself; pain or sensitivity is not produced when the lumbar vertebrae are pressured, as may occur with a disc-related condition.

Muscle-related back pain

Injury to spinal discs occurs frequently. With the advent of MRI, herniated discs are seen often, even if pain is only slight or non-existent. However, the most common cause of back pain is localized muscle strain. An overly diligent yoga student may strain a muscle and cause it to become irritated or slightly torn. Pain from muscle strain is relatively short lived, lasting two to five days. In most cases, short-term muscle strain results from over-exuberance and pushing a day's practice beyond the 10% safe limit. Muscle pain that occurs repetitively and becomes chronic is often a sign of underlying mechanical dysfunction. Chronic muscular pain indicates that muscles are attempting to compensate for structural weakness elsewhere and straining beyond their limits. As an example, muscles will become tense and shortened if a vertebra is unstable and its ligaments cannot provide adequate support. Conversely, when joints lose mobility, muscles weaken as a result of a lack of mechanical stimulation. In either case, when the cause of the pain/strain is mechanical or structural, utilizing the alignment principles is essential to stop repetitive injury and to rehabilitate the injured area. Pain reduction becomes an almost immediate feedback indicator when correct alignment is engaged in areas of strained muscles.

With any type of pain, it is always important to consider other potential causes that are related to non-structural health considerations. Pain that persists and does not improve through rest, rehabilitative exercise, or greater precision of alignment may have organic or metabolic origins. The conscious yogi never ignores unresolved pain but seeks to uncover and understand its true cause, promptly and fully.

Move gently out of backbends

Immediate or rapid reversal of the lumbar curve is not safe practice. Deep back-bending poses such as **Urdvha Dhanurasana** (Reversed Bow Pose) are best followed with a neutral lumbar curve posture, such as Downward Facing Dog than immediately hugging the knees to the chest or attempting a full forward bend such as Happy Baby Pose or **Uttanasana**. This allows the inner disc material to re-center itself before the counter posture shifts it to the opposite direction.

Of course, the safest way to perform any backbend posture is to consider it a chest opener and not to excessively deepen the already concave lumbar spine. This approach encourages the less mobile, thoracic region to initiate the posture and participate more fully in the backbend before the lower back engages and takes over most of the work. It is good practice to first engage the alignment principle of *Scoop the tailbone, Scoop the breastbone, Draw-in-the-navel* with all back-bending postures to prevent hyperextension and disc compression in the lumbar spine.

Most injuries in asana can occur when moving into or out of a posture. Often the transitions between postures are rapid and leave less time allotted to properly establish the asana's foundation and engage good postural alignment. These are also times when emotionally, the final destination outweighs the journey. Careful, conscious action and precision in alignment during the initial and final stages of asana are essential for a safe practice.

26 Yoga Butt

In any crowded yoga studio, it is not uncommon to observe what is sometimes affectionately referred to as the *yoga butt*. More often, this phenomenon is the natural physique of the yogi, possibly genetic, and not resulting from a specific type of yoga practice. Some students and teachers consider the yoga butt a desirable trait and encourage it! Natural or not, yoga butt can predispose some who possess it to injury, shifting the claim from compliment to curse.

The cause of yoga butt is not always obvious. Besides genetics, sports that require powerful hip extension, such as sprinting, climbing, or gymnastics, can overdevelop the gluteal muscles. The tensor fascia lata muscles on the lateral hips counteract the pull of the gluteus maximus. They may become irritated along the iliotibial band's tract and possibly produce a snapping sound with movement.

Yoga butt is recognizable by a number of features. Most obvious are an exaggerated anterior pelvic tilt and lifted and prominent buttocks. The lumbar spine will have a deep anterior curve. The quadriceps and iliopsoas (hip flexors) will shorten and become large and often tight. Knees are often hyper-extended, a compensation to balance the muscle tension in the legs, although it may cause the knee's posterior collateral ligaments to weaken. The hamstring muscle tendons where they attach on the ischium can become chronically overstretched as a result of being lengthened by the flexed, anterior pelvis. It is common for yogis with yoga butt to have a flat thoracic spinal curve and the lower rib cage jutting forward.

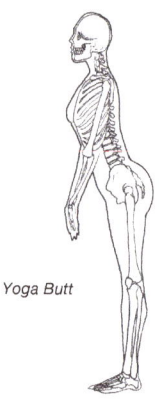

Yoga Butt

A flat lumbar curve can also initiate the yoga butt

If the lumbar spine is very mobile, it is less stable and causes the natural lumbar curve to flatten. In order to find stability, the curve must deepen. This requires the pelvis to tilt anterior and angle the base of the sacrum forward. As a result, the body's center of gravity shifts forward and the buttocks lift and protrude. To resist this shift in posture, all the muscles that anchor the pelvis to the legs increase in size and strength. The thighs become tight and more massive, which increases hyperextension of the knees.

With a flat lumbar spine, when seated, the large buttock muscles may restrict the pelvis from rocking forward and keep the lumbar curve too flat. This can result in lower back pain. Yogis with a large, protruding buttocks should engage more internal rotation to compensate and to help form the egg-shaped lumbar curve. The yogi should avoid any IT band snapping or pain that is repetitive.

Lower Crossed Syndrome

A deep lumbar curve and anterior pelvis creates an imbalance in the lower body musculature. This relationship is recognized as a pattern where muscular weakness criss-crosses the torso with tightness in the opposing muscles. Alignment and bandha engagement as discussed throughout the book can effectively break the pattern.

Weak abdominals · Tight erector spinae · Tight iliopsoas · Weak gluteus

Springing into action

The spinal curves and a coiled spring function in a similar fashion. Both increase their mobility and flexibility through elongation; by widening the space between the vertebrae or by spreading the coils within the spring. A straight spine has flat, shallow curves, giving it greater flexibility but resulting in much less stability.

Conversely, when spinal curves or a spring's coils deepen, compress, or round, they become more stable and better able to absorb shock. Curves are formed as a compensatory response to the forces of gravity that are placed on the spine as it adapts to the demands of weight bearing.

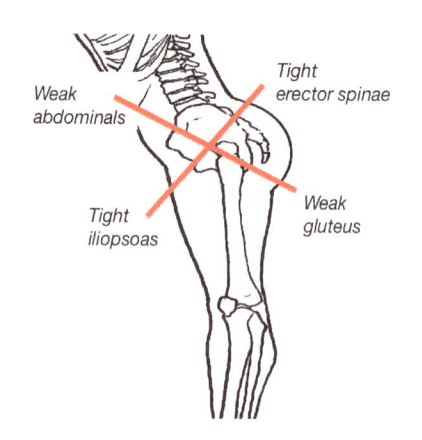

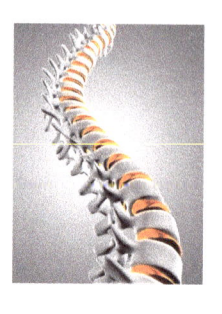

Pain in the yoga butt

Pain from yoga butt usually emanates from the hamstring attachment at the ischial tuberosity. The anterior tilt of the pelvis that results from yoga butt lengthens the hamstring muscles. Elongated hamstring muscles place greater tension on their tendons, making them highly susceptible to injury. The hamstring muscle attachment at the ischial tuberosity can easily tear, producing deep buttock pain. If the tendon tears and any portion might separate from the bone, it is called an *avulsion*.

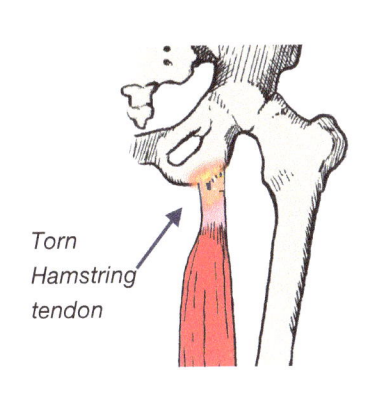

Torn Hamstring tendon

Pain from a hamstring tendon tear is easily confused with pain that might originate from a lumbar disc causing a sciatic nerve injury or from the piriformis muscle becoming strained or entrapping the sciatic nerve.

Unfavorable postural habits can develop as a result of yoga butt. Due to the often accompanying, flattened thoracic or cervical spine, the shoulders tend to round forward and the head thrust forward. Upper back muscles overstretch and weaken. It is common for the knees to hyper-extend to create counter-resistance to gravity.

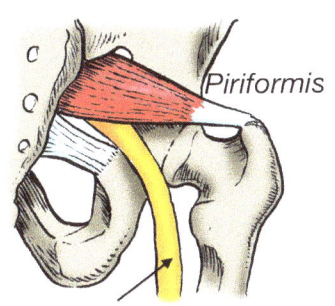

Piriformis

Sciatic nerve

Yoga butt can injure the spine. Adopting this posture increases the lumbar curve, compressing and intensifying pressure within the spinal discs. The vertebrae shift forward and produce forces that shear across the spinal discs that cause damage. Degenerative wear and tear develops over time, causing the vertebral discs to narrow and the vertebral joints to become rough and irregular.

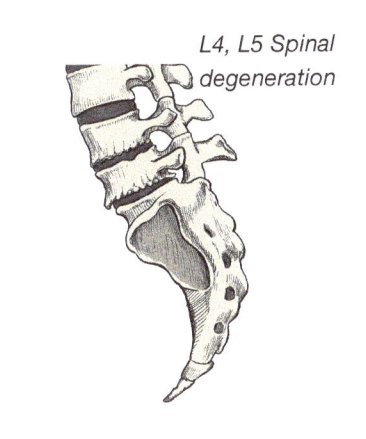

L4, L5 Spinal degeneration

Four to five percent of the population has a vertebral defect known as *spondylolisthesis* may produce a postural appearance similar to yoga butt. As a differentiating sign, with spondylolisthesis, anterior pelvic tilt is often minimal or absent. (See Chapter 24 for additional details about spondylolisthesis)

The righting reflex

Our nervous system possesses a *righting reflex* - an instinctive impulse that perpetually adjusts our center of gravity. The purpose of the righting reflex is to keep the head balanced in relation to the horizon so that the eyes and ears receive stimulus symmetrically. To maintain this balance, the body may adopt complex postural distortions. For example, if an injury or postural defect causes the neck to slant toward one side, the shoulders might tilt down on the opposite side in order to re-establish the horizontal orientation of the head. Another example is when a visual imbalance such as astigmatism of the eyes causes the head or the entire body to contort in order to bring about visual symmetry. The righting reflex is an important adaptive mechanism but can result in undesirable effects.

You may be asking yourself, how does this relate to yoga butt? If the pelvis tilts forward and the buttocks protrude, the righting reflex may be triggered to restore structural balance by jutting the lower ribs forward or flattening upper spinal curves - two undesirable responses. Although these postural shifts can be re-aligned, it usually requires considerable alignment effort- from the bottom up!

On a positive note, the righting reflex plays an important role in rehabilitation. The effort it takes to develop better postural habits become instinctual more quickly when the body uses sensory cues from the righting reflex to establish its innate balance. By improving one region's alignment and orientation to gravity, the corresponding regions more easily become better aligned.

Steps to reduce yoga butt posture

- Establish a stable foundation by balancing on each foot along its four corners.
- Engage *shins-forward* and *thighs-back,* even if only an isometric contraction. This action prevents the knees from hyper-extending. Squeeze the outer sides of each knee. This action contracts the muscle fibers embedded in the collateral ligaments and assists stabilization of the knees.
- Align the center of the hips vertically over the ankles.
- Engage *inward hip release* and *forward tailbone scoop* to balance the pelvis in a neutral position. The key indicator of a neutral pelvis is having an egg-sized curve in the lumbar spine. In most cases, *forward tailbone scoop* needs to be more aggressively engaged than used if there is not a pronounced yoga butt.
- Draw the lower rib cage back and scoop the lower tip of the breastbone back.
- Increase the volume of the upper thoracic region by engaging *chest integration*. This is achieved by breathing 360° around the upper chest at the level immediately below the collarbones. Be sure to expand the rib cage in all directions including the inner armpits and across the upper back. This action may at first be challenging but this chest integration creates a fuller, rounder thoracic spine and a broader foundation for the neck that allows the cervical curve to deepen. More details regarding chest integration are provided in Chapter 29.
- If a forward fold is attempted, firmly root down from the center of the hip sockets to the ankle bones to help reduce hamstring tendon strain.

Caveat: In Chapter 11, the concept of the three "S"'s was introduced. When standing against the wall with the sacrum in contact with it, students who have yoga butt or enlarged and lifted gluteal muscles, should not force either the shoulders or skull to align over the sacrum if it causes the lower thoracic rib cage to protrude forward. It may also be appropriate for the egg-shaped lumbar curve to be larger than a single egg size, although some reduction of the curve will probably be more integrative. The yogini should be careful not to hyper-extend her knees or round the shoulders forward when attempting to bring the ribs posterior.

Following the above steps is effective in reducing damaging stress on the body resulting from excessive anterior tilt, or yoga butt. These procedures change the very nature of all postures, not just physically, but mentally and emotionally, as well. Some yogis might emotionally resist or challenge the need to correct the yoga butt posture since it has gained some cultural desirability. However, if pain is present, the student will eventually be motivated!

27 The Psoas Muscle

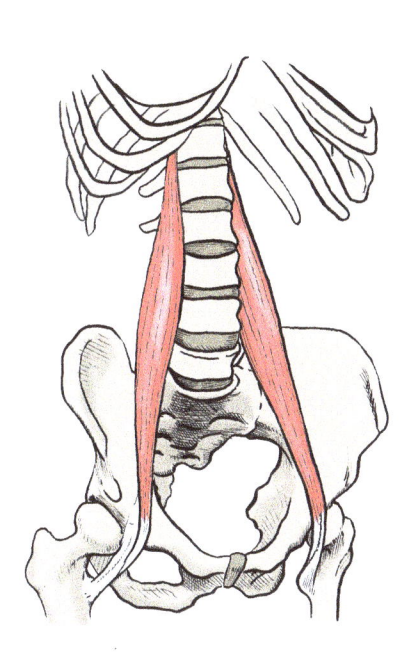

The psoas muscle, pronounced "so-as," has emerged from relative anatomical obscurity to compelling curiosity for its role in body mechanics. The function of the psoas poses differences of opinion about how it works in every situation. Despite the ongoing debate, the psoas plays an important role in yoga alignment.

The psoas is comprised of either one or two portions - the psoas major and minor. Only 40-50% of the human population has a psoas minor with its only task being contribution to the efficiency of the psoas major. Otherwise the psoas minor has no separate role. Both muscles together are referred to as simply the psoas muscle.

The psoas major is long and tapered (fusiform) at both its ends. It has an extensive blood supply and a rich red color. For those who find obscure facts interesting, the psoas muscle of cattle stock is highly coveted, being the filet mignon and tenderloin cuts of steak.

The psoas muscle is located deep within the lower torso and pelvis. It runs an oblique course, anterior to the large quadratus lumborum and helps to form a supportive, muscular wall behind the internal organs of the abdomen. The psoas originates from the transverse processes and lateral vertebral bodies of the twelfth thoracic vertebra and each of the five vertebrae of the lumbar spine. Its extensive attachments send fibers directly into the spinal (annular) ligaments and the vertebral discs. This enables the psoas to directly influence the curve of the lumbar spine. The psoas also interweaves its fibers into the quadratus lumborum and thoracic diaphragm, allowing it to contribute to respiration.

As a core muscle, the psoas helps to integrate the lower spine, pelvis, and hip joints. It inserts on the lesser trochanter, a small thumb-like protrusion on the upper, inner neck of the femur. The psoas does not attach directly to the pelvis, which increases its mechanical advantage by allowing it to execute long, unobstructed, and efficient pulling of the spine. Working in sync with the abdominal and lumbar paraspinal muscles, the psoas helps balance and stabilizes the lower back.

Psoas function is better understood if its methods of contraction and stretching are viewed with the perspective of concentric contraction (muscle shortens as it loads and contracts) verses eccentric contraction (muscle lengthens as it loads and contracts). The primary action of the psoas muscle is hip flexion, bringing the thigh to the torso. It also provides limited external rotation of the hip. Both actions occur using concentric contraction. Some anatomists however, believe that the psoas muscle does not shorten during hip flexion (concentric contraction) and cannot be considered a flexor muscle.[1] This controversy and others make the mechanics of the psoas muscle a complex topic. In one-legged standing poses, the weight-bearing psoas eccentrically contracts (sometimes considered to be isometric contraction) and is a demonstration of the psoas' fundamental role in every posture. When actively stretching the psoas, eccentric contraction may routinely occur with the action. Fortunately, in its role in yoga practice, engaging the psoas is relatively easy to understand.

Iliopsoas

The psoas couples with the iliacus muscle to form the iliopsoas muscle. The iliopsoas is the primary hip flexor and also contributes to external rotation of the hip. The psoas and iliacus share a common tendon that attaches to the lesser trochanter. Each muscle, however, has a separate and distinct nerve supply.

The iliacus is a short muscle, originating on the inner bowl of the pelvis. Unlike the psoas, the iliacus has no contact with or influence on the spine. The iliacus flexes the pelvis anterior. It also flexes the thigh but is unable to contribute beyond 90° by itself as it requires leverage from the long-length psoas. The extra work of the psoas in two directions causes loss of its efficiency and is one reason why in Navasana, Boat Pose, it is harder to maintain the lumbar curve when the knees are straight.

Indirectly and conversely, along with the gluteal muscles and long head of the biceps femoris, the iliacus assists in forward tailbone scoop.

Hip flexion by iliacus as part of iliopsoas

The elusive psoas

Function of the psoas is difficult to observe because of the muscle's deep location. The psoas cannot be directly palpated or massaged from the body's outer surface nor can its action be isolated from that of surrounding muscles. Psoas contraction and the degree that it deepens the lumbar curve will adjust in response to the position of the spine, hips, or changes in the angle of the leg.

Each of the following conditions alters the normal action of the psoas muscle:

- Tight hamstrings
- Misalignment of the pelvis
- Limited hip mobility
- Limited lumbar spine mobility
- Conditions that restrict the shape of the lumbar curve, such as stenosis

The dual nature of the psoas

Yoga asana usually focuses on either strength or flexibility. For example, a forward fold may concentrate more on the flexibility of the hamstring muscles and less on its strength. For the psoas, however, focus on strength and flexibility seems to coincide. The following lists present actions caused by the psoas, either during stretching or contraction. As mentioned earlier, although eccentric contraction can be involved with stretching, the focus in asana can remain simply on flexibility.

Psoas Stretch

- Femur heads draw back
- Engages *inward hip release*
- Anterior pelvic tilt
- Opens the sacroiliac joints
- Increases the lumbar curve

Psoas Contraction

- Femur heads press forward
- Engages *forward tailbone scoop*
- Posterior pelvic tilt
- Closes the sacroiliac joints
- Lumbar curve flattens

Technically Speaking

Stretching
Psoas stretch pulls the spine from segments T-12 to L-5 from a vector posterior of the central axis. This causes the lumbar spine to protrude forward and its curve to deepen.
Psoas stretch tilts the pelvis anterior, increasing the sacral base angle and with that, the lumbar curve.

Contraction
Psoas contraction pulls the lumbar spine from a vector anterior of the central axis, causing the curve to flatten.
Psoas contraction tilts the pelvis posterior, causing the sacral base to become more horizontal and flatten the lumbar curve.

Stretching lengthens a muscle, increasing flexibility and mobility
Contraction shortens a muscle, producing force that provides strength and power

Psoas strength in asana

In **Paripurna Navasana** (Full Boat Pose), the abdominal muscles keep the lower torso straight and stable. The iliopsoas provides strength in the Boat Pose, acting as the primary muscle to flex the hip to hold the legs close to the trunk. The psoas muscle engages as part of the iliopsoas and alone if the legs lift beyond 90°. The psoas balances between stretch and contraction (or concentric and eccentric contraction) to support the lumbar curve to prevent the lower back from rounding. Setting up the pose with forward pelvic tilt helps establish the lumbar curve by enabling psoas contraction to pull posterior to the central axis and deepen the curve. If the hamstring muscles are short and tight, the ability to tilt the pelvic forward is challenged, causing the psoas to contract anterior of the central axis, pulling the lumbar spine flatter and making the natural lumbar curve more difficult to establish and maintain.

With straight legs extended on the ground as in Pilates Roll Up, the psoas focuses on stabilization, providing minor assistance to the rectus femoris and abdominals in pulling the trunk toward the thighs, Note: Performing this action is not recommended if the lumbar spine is hypo-lordotic or unstable.

Single-sided (unilateral) psoas engagement flexes the lower trunk laterally. This occurs in standing side-bending poses such as the Standing Side Bend, **Urdhva Hastasana Ardha Chandrasana** (not shown).

In **Parivrtta Hasta Padangusthasana One** (Revolved Hand-to-Big Toe Pose) the standing leg's psoas helps to rotate the trunk in the opposite direction (contra-lateral rotation).

In one-legged, standing balancing poses such as Tree Pose, the psoas eccentrically or isometrically contracts firmly on the side of the standing leg, stabilizing the torso in a straight and vertical position. On the raised leg side, the psoas provides hip flexion and helps lift the leg.

> The hallmark of a well-toned psoas is a lumbar curve the size of a single egg

Paripurna Navasana

Pilates Roll-up

Parivrtta Hasta Padangusthasana

Anterior pelvic tilt, *inward hip release,* and psoas stretching

When the pelvis tilts forward, *inward hip release* activates, causing the thighs to draw back and internally rotate. The lesser trochanters located on the upper, inner femur bones, which are the psoas muscle attachments, are rotated back with this action. When the musculature of the lumbar spine is balanced and an egg-size spinal curve is established, the psoas is stretched to the point where it achieves optimal efficiency. If the pelvis does not tilt forward, the sacral base does not drop anterior. In this position, the psoas is unable to stretch to its ideal length and the correct lumbar curve is unable to form.

Anterior Pelvis

Hamstring muscles and the psoas

The hamstring muscles attach to the bottom of the pelvis at the ischial tuberosities. If the hamstrings are tight, the pelvis anchored and pulled posterior, limiting how much forward pelvic tilt is obtainable. Posterior pelvic tilt also presses the hip joints forward, which prevents *inward hip release* and its component, internal hip rotation that is needed for correct psoas stretching.

A short, tight psoas muscle

Muscle that is stretched to more than 10% beyond its baseline length is at risk of being overstretched, inflamed, or torn. The psoas muscle flexes the hip, bringing the thigh toward the front torso. To stretch, it must move in the opposite direction, hip extension. Overstretching the psoas rarely occurs because the hips have a limited range of motion in extension. Limited hip extension does safeguard the psoas by preventing it from exceeding dangerous lengths while stretching. But, a short, tight psoas is more common and predisposes yogis to other injuries, even with relatively conservative asana stretches.

Posterior tilt

Poor structural alignment is the most common pre-cursor to a non-traumatic, musculo-skeletal injury. Since the psoas plays a central role in postural alignment of the lower spine and pelvis, its tone is vital. The psoas depends on precise balance between *inward hip release* and *forward tailbone scoop*. The egg-sized lumbar curve that results from correct alignment of the lumbar spine is the hallmark of a well-toned psoas.

Sitting with the knees higher than the hips or the legs splayed apart cause the psoas to become short and tight. Some causes of short, tight psoas muscles are:

- Sitting in car seats or soft sofas that causes the hips to drop lower than the knees
- Practicing old-style sit-ups that position the pelvis in a posterior tilt
- Bicycling with seat that is too low, causing the knees to over-bend and splay outward
- Over-engaging Mula Bandha (*forward tailbone scoop*)
- Bridge or wheel Pose with the knees wider than head-width distance apart

A tighter psoas on one side can cause the trunk to laterally distort to that side. Unilateral psoas imbalance shortens the length of the torso and spine on the affected side. Psoas imbalances are associated with scoliosis. One-legged standing balance poses strengthen the psoas on the weight-bearing leg along with the tensor fascia lata and peroneus muscles of the same leg. Single leg balance poses are valuable therapy to assist in the rehabilitation of unstable sacroiliac joints.

Urdhva Hastasana (Upward Hands Pose) offers a general but useful evaluation of side-to-side psoas length and tone. If, when raising both arms overhead, one arm appears shorter than the other, this may be an indication that the psoas muscle is short and over-contracted on that side. Of course, many other causative factors, unrelated to the psoas, may be at the cause and should also be considered. As a general evaluative tool, this test has shown to be reliable.

Urdhva Hastasana

Psoas stretching increases sacroiliac joint mobility

The psoas has no direct muscular interaction with the sacroiliac joints; however, they have a significant effect on each other. When the psoas lengthens during *inward hip release*, the sacroiliac joints open and become mobile. If the sacroiliac joints are unstable, the psoas muscle may be forced to overexert in an attempt to provide the missing support. Kinesiological muscle testing often discovers that the psoas becomes inefficient and tests weak from the extra demand of an unstable sacroiliac joint.

Asana for psoas stretching

Setu Bandha Sarvangasana   Bridge Pose

Bridge Pose is an excellent posture for evaluating the psoas muscles. If the knees are unable to remain in line with the hips but instead splay outward in Bridge or **Urdhva Dhanurasana** (Wheel Pose), it often indicates that the psoas muscles are short and tight. If this is the case, hold a block between the knees to prevent the outward splaying of the knees and to encourage *inward hip release*. This will lengthen the psoas muscles. *Forward tailbone scoop* is engaged at the completion of every posture and is especially important in Bridge. "Back-bending" pose, such as Bridge and Wheel, require *forward tailbone scoop* to stabilize the sacroiliac joints and protect the lumbar spine from an overarching, compression injury.

Bridge Pose

Anjaneyasana (Low Lunge Pose), High Lunge, and **Hanumanasana** (Forward Split) are excellent postures to stretch the psoas. In these and similar poses, the psoas of the rear leg gets stretched. To prevent injury in these poses, *Inward hip release* is engaged in the rear leg to loosen the hips ligaments and contract the hamstring muscles. The inner heel is pressed inferior to additionally release the hip. The effectiveness of the overall pose depends on the rear thigh remaining straight and firmly integrated into the posture.

Lunge Pose

Hanumanasana

Supta Padangusthasana One

In Supine Single-Leg Lift, the back of the leg on the mat remains flush with the floor. If it lifts away from the mat, buckles, or turns outward (external rotation), those are common indicators of a psoas muscle that is short and tight.

In **Savasana**, the Corpse Pose, the legs stretch out straight. If an egg-sized lumbar curve naturally forms, it indicates that the psoas muscles are properly toned. When the knees are bent and the feet placed on the floor, the well-toned psoas allows the lower back to flatten onto the floor.

Supta Padangusthasana One

The psoas and the abdominal organs

Adverse health conditions have been associated with chronically shortened and tight psoas muscles. The lower abdominal organs rest against the psoas muscles as a posterior supportive wall. When the psoas is tight, the organs can roll forward and trigger dysfunction of the digestive track, bladder, or sexual organs. It is not unusual to see athletes who have well-developed abdominal muscles but short tight psoas muscles, resulting from lack of stretching.

When the psoas muscles are habitually tight, the thighs protrude forward and the spine flattens. This also results in limited space in the abdominal cavity. Each kidney is nestled into a pocket made of connective tissue that attaches directly to the psoas muscle. Tension in the psoas muscle may have an impact on kidney function. Improving psoas function and drawing the lower rib cage posterior as when engaging Uddiyana bandha may provide positive changes in some of these organic issues.

Of course, psoas dysfunction is not to be considered the root cause of all organic ailments and responsible attention is always given to organic symptoms.

What is the sound of one hip snapping?

In **Supta Padangusthasana One**, the Supine Single-Leg Lift, lowering the straight leg may elicit an audible "pluck", perhaps heard across the entire yoga studio. The iliopsoas tendon usually causes the snapping sound as it shifts lateral to medial, getting trapped against either the iliopectineal eminence on the front of the hipbone or the labral collar that surrounds the femur head. Most often the snap is produced when bringing the hip from a flexed position through extension or abduction. This indicates a tight psoas. If pain accompanies the sound, the psoas tendon may be inflamed. To improve this condition, increase *inward hip release* and press through the inner heel as the leg is being lowered. These actions will often eliminate the snap of the tendon.

Another unrelated hip snap can possibly occur at the tensa fascia latae muscle on the lateral hip.

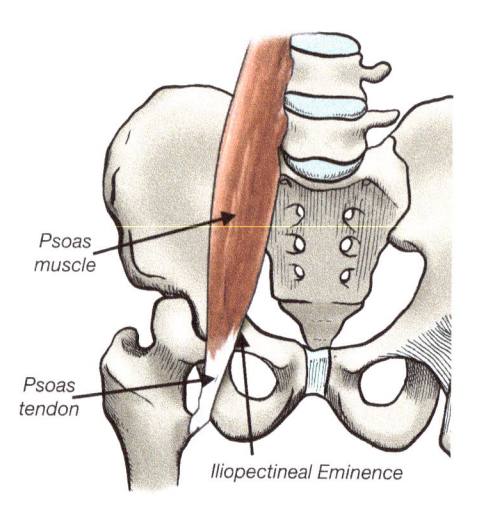

Psoas muscle

Psoas tendon

Iliopectineal Eminence

Iliopsoas muscle stretch

This supported restorative stretch was also presented in Chapter 18 where the iliopsoas muscle was presented as the antagonist to the gluteal muscles. Similar to the quadriceps, with hip extension being so limited, the iliopsoas has few opportunities to be fully stretched.

In a supine position, drop one leg over the edge of a massage table or a set of raised yoga blocks. The dropped leg extends the hip, stretching the iliopsoas. The torso must firmly maintain **Tadasana** alignment. To prevent lumbar hyperextension and sacroiliac strain in this position, engage *Scoop the tailbone, Scoop the breastbone, Draw in the navel*. The leg that is dropped is considered the rear leg of the posture. It actively engages *inward hip release* by rolling in and drawing back the thigh.[2] This stretch can also be performed more simply by placing a block under the sacrum and letting both legs stretch toward the floor.

28 The Thoracic Spine

Imagine an early morning yoga class with the rising sun's soft light initiating the start of practice. The teacher begins by instructing the students in **Urdhva Hastasana** (Upward Hands Pose) to draw the lower ribs back, extend the upper chest forward, and lift the arms overhead. For many students, being instructed to move the lower thoracic cage in one direction while moving the upper thoracic region in the opposite direction is perplexing. The very idea that individual regions of the spine and rib cage can move independently, not as one fixed unit, is a new consideration for some students.

The thoracic spine is the posterior pillar that supports the rib cage. Its mobility is restricted, obviously, because of its attachments to the ribs. Although the thoracic spine is not a region with broad, sweeping mobility, the number of subtle, intra-joint movements is extensive. The twelve thoracic vertebrae alone move in seven different directions at each of their four facets, producing a surprising total of 336 subtle, individual inter-segmental movements! The essence of yoga practice is to increase awareness of this potential for mobility and the mechanical skills to access more of it.

Spinal movement is best understood as having two components: broad *ranges of motion* that move an entire region or section of the spine and *inter-segmental movements*, the small, subtle movements that occur at each individual facet joint. The thoracic spine has significantly less range of motion than the cervical and lumbar regions in spite of its nearly double number of joints as the other regions.

The thoracic spine facets open along the frontal plane, moving with a shearing, side-to-side glide. This orientation makes lateral flexion more easily available while flexion and extension are significantly more limited. Extension is further restricted by the long, central spinous processes on the posterior of each vertebra that spike downward. The processes are already in close contact with each other with the spine in its neutral spine position before any extension is added to the tight spacing.[1]

Range of motion in the thoracic spine

As mentioned in Chapter 25, it is difficult to isolate and measure the thoracic ranges of motion with reliability. Orthopedic testing resigns itself to using a single, combined, *thoraco-lumbar* measurement of the thoracic and lumbar regions. Most authoritative texts do not provide specific measurements, as the reliability of findings has not been substantiated.

From a tally of opinions on thoracic spine ranges of motion, the following values can be offered: [2]

- Lateral flexion   20- 40° bilaterally
- Rotation   35° bilaterally
- Flexion   20-45°
- Extension   25-45°

The rib cage

The rib cage encloses and protects the vital organs of the upper body. The rib cage must also move freely to expand and contract with the complex mechanisms of breathing. Similar to the keel of a wooden boat, the thoracic spine is the central support for the framework of ribs that creates the rib cage.

There are twelve pairs of ribs that form the rib cage. Not all ribs are the same. The top seven ribs are called *true ribs* because they attach directly to the breastbone (sternum) at the center of the chest. Because of their boney attachment to both spine and sternum, the upper thoracic spine has the least mobility within the thoracic region. The thoracic curve is rounded posterior. Its apex is at the seventh thoracic vertebra (T-7) and is mechanically the least mobile vertebra of the region.

The five lower sets of ribs that follow are called *false ribs*. The upper three sets do not attach directly to the sternum but to cartilage that extends from the sternum to the ribs. The bottom two sets of ribs are given the additional name, *floating ribs* because they are free of any type of anterior attachments.

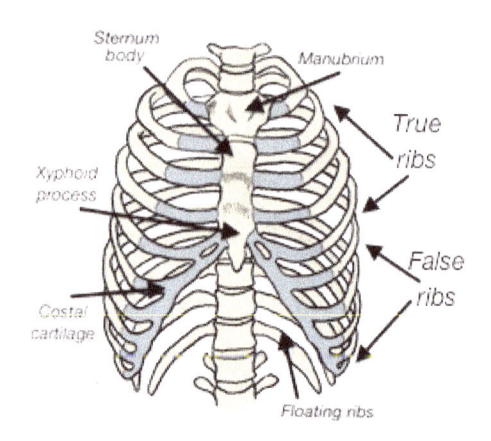

In aging individuals, increased limited thoracic mobility is common. Over time, the upper back becomes rigid, immobile, and rounded and can fairly be considered a sign of diminished health and vitality.

The upper back is highly subject to bone loss, or osteoporosis. Bone loss often produces compression fractures in the upper back. Hormonal and dietary components play a significant role in osteoporosis; however, poor posture and loss of mobility are perhaps the strongest causative factors. The limited ability to move and mechanically stimulate the upper thoracic vertebrae makes them particularly susceptible to bone density loss and eventually, the structural fracture and collapse.

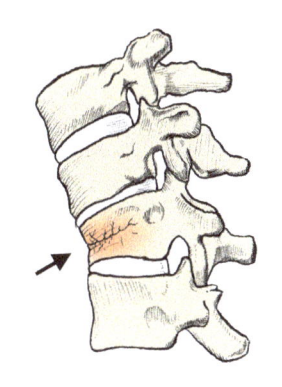

Compression Fracture

The 12th thoracic vertebra

Located at the transition point between the thoracic and lumbar spinal curves, the twelfth thoracic vertebra is mechanically unique. It functions as the primary swivel (rotation) point within the entire vertebral axis. To facilitate rotation, the vertebra's anterior body is uncharacteristically larger than its posterior ring. The deep muscles of the spine do not attach to the posterior ring of T-12, presenting a distinctive design that reduces muscular hindrance to rotation. While having extensive rotation, the 12th thoracic vertebra, like the rest of the thoracic spine, is limited in flexion and extension.

T-12 Vertebra

Additionally, the diaphragm, psoas, and trapezius muscles all attach to the 12th thoracic vertebra. With a common point of connection at the T-12 vertebra, these important muscles are able to influence each other's action, albeit subtlety. Since respiration involves the diaphragm and the abdominals, which attach to the T-12 ribs, it is important to correctly align the lower thoracic vertebrae for efficient breathing.

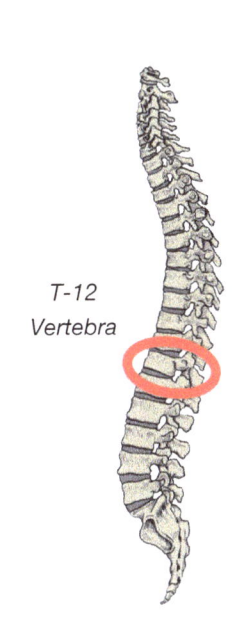

Moving the thoracic spine

For yogis and non-yogis alike, moving each vertebrae of the thoracic spine independently is quite challenging. Injuries, poor postural habits, or the presence of scoliosis add to the challenge.

The best strategy for moving the thoracic spine is to initiate movement from the least mobile vertebral segments. After that, continue to access lesser mobile areas until regions of full mobility are reached. Use the breath to expand the thoracic region to help discover the most restricted areas. If there are no obvious overlying restrictions, initiate movement from the level of the heart, which is anatomically the least mobile portion of the thoracic spine. Movement of the lower thoracic spine has additional considerations that are presented in Chapter 15.

Thoracic spinal mobility is essential for enabling the shoulder girdle to move safely and efficiently. The ability to extend (flatten) provides the space necessary for the shoulders to glide onto the back body; otherwise they will tend to round forward over the chest.

As will be presented in Chapter 31, the principles of shoulder integrative alignment require the thoracic spine to first be mobilized before full shoulder alignment is engaged.

Two helpful procedures that were presented in previous chapters enhance learning how to engage the subtle movements of the thoracic spine:

- Spinal undulation (Chapter 24): develops the skill to isolate independent movement of the spine, vertebra to vertebra.
- The *lock and load* principle (Chapter 11): draw the navel inward; then draw the lower rib cage back, locking it into a posterior position. Once the lower thoracic spine is stabilized in this fashion, the upper thoracic vertebrae are able to move independently forward.

Kyphosis – the rounded back

Kyphosis

The thoracic spinal curve is convex, coming to a rounded apex on the back body. This type of curve is called *kyphosis*, kyphos from the Greek word for hump. Kyphosis refers to any posterior curve, whether normal depth or exaggerated. The rounded, thoracic curve provides the necessary space into which the heart and lungs can expand. As mentioned in Chapter 24, the spine of the fetus and newborn is kyphotic (C-curve). Although the term kyphosis is often used to reference an excessively rounded, posterior curve, the correct way to describe an excessive thoracic curve is *hyper-kyphosis*.

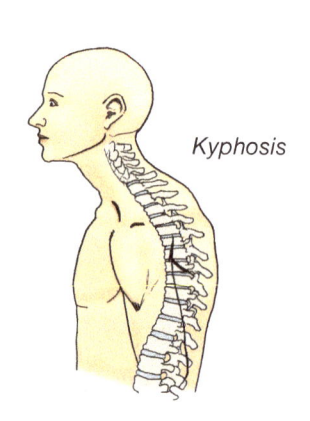

Deeper curves have less mobility than flatter ones; therefore, the greater the degree of kyphosis, the more difficult it is for the curved region to move. The larger the thoracic kyphosis, the less mobility that is available in the spine and rib cage. Deep kyphosis can compromise the function of the cardio-pulmonary system due to stress on the chest musculature and circulation.

Viewed from the back body, a spine with an excessive kyphosis protrudes beyond the posterior edges of the shoulder blades. The upper back and paraspinal muscles will mound up, blocking the natural gliding of the shoulder blades toward the center of the back, a necessary action that occurs with arm movement. Yogis with thoracic hyper-kyphosis often compensate by rolling their shoulders and rib cage forward when raising their arms, a position where the shoulders are highly vulnerable to traumatic injury. When the thoracic spine is rounded, it is more difficult to create the broad foundation across the shoulder girdle that is necessary for the neck and head to be properly supported.

To address kyphosis therapeutically, lie over a rolled yoga mat or blanket, placed vertically between the shoulder blades. The roll presses the spine anterior and allows the shoulder blades to release posterior. The roll should also support the head for neck, aligning the ear with the shoulder sockets.

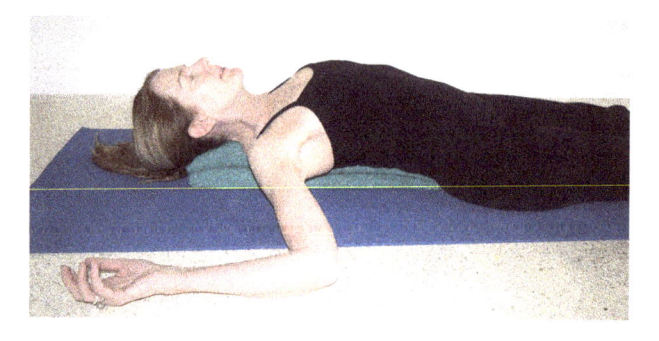

Flat thoracic curve

The thoracic spine can also be too flat. A flattened curve creates an elongated spine that is generally hyper-flexible. The term assigned to a flattened thoracic spine is *hypo-kyphosis*. With a normally curved spine, the skin and muscle tone across the back from shoulder to shoulder appears as a relatively smooth, continuous arch. With hypo-kyphosis, a pronounced, depression is formed between the scapulae and their medial borders are clearly visible.

A useful technique to improve the flat spine is using Cow Pose, which normally rounds the upper back and shoulders. Keep the heads of the arm bones firmly back and only round the upper thoracic spine without the changing the position of the shoulders. Breath techniques can also address a "flat back" to expand the thoracic cage and the spine. On the inhalation, increase the volume of the lungs across the upper back (use the Chest expansion in following section to assist this action).

Flat thoracic curve

Chest expansion

Inhale fully, inflating the upper chest 360° around at the level just below the collarbones. Inflate across the upper back and side-body in the region located along the inner armpits. Importantly, remain inflated, even during the exhalation phase.

Chest expansion is similar in idea to pumping a small bicycle tire. The tire remains full and inflated, even in between pumps while the handle is lifted and no air is injected into the tire.

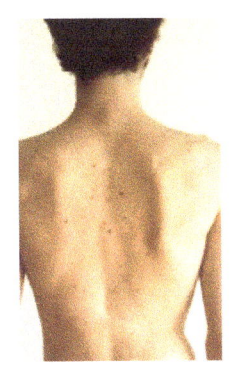

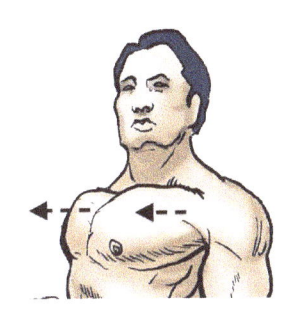

Expand the upper quadrants of the lungs in 360° of direction

As well as being an excellent tool for hypo-kyphosis, chest expansion across the collarbones helps prevent the shoulders from rolling forward. It creates the required broad-base foundation for the head and curve of the neck. Additional chest expansion details are presented in Chapter 31 along with a full exploration of shoulder integrative alignment.

Rounding the upper back? Not on my watch!

A common teaching instruction that is given is to lift up from a forward folded position, one vertebra at a time. This action is basically rounding the back. This instruction is commonly given in **Upavistha Konasana** (Wide-Angle Forward Fold).

This incorrect instruction loads the spine with significant weight before its stabilizing curves have been established. If the thoracic spine is kept flat, it is easier for the lumbar spine to maintain its egg-sized curve and the vertebrae to reverse their position one bone at a time.

Upavistha Konasana

Since extension in the thoracic spine is already challenging for many students, the instruction to "round the back" is not only injurious but perpetuates this unsafe habit. A typical yoga student need never purposely round their upper back in any pose. Asana practice will be safer if the student flattens the thoracic spine each time he hears the instruction to round it!

To be fair, initiating a posture may first require some rounding of the thoracic spine. Yoga postures that bind the arms, for example, may require an initial forward rounding of the spine and shoulders in order to clasp their hands. To minimize this counterproductive action, first lengthen the thoracic spine and integrate the shoulders by drawing the inner heads of the arm bones back and gliding the outer edges of the shoulder blades toward the spine. Maintain the elongated thoracic spine by pressing the bottom tips of the shoulder blades anterior.

When binding, it is also important to remember that the forearms flex and rotate independent of the shoulder girdle. It is often possible for the thoracic spine and shoulders to remain well aligned and the bind occurring by the action of rotating the forearms to clasp. The concept of integrating the shoulders into alignment first also applies not only to binding poses but to spinal twists, as well.

In the classic **Balasana** (Child's Pose) the open palms face up. The shoulders roll forward of the chest and the thoracic spine rounds and stretches. Instead of relaxing muscular tension as intended, the thoracic spinal curve increases, causing the upper back musculature lengthen to weaken. Simply turning the hands over, placing the open palms down significantly improves the pose. Many yoga instructors have replaced traditional Child's Pose with Extended Child's Pose with the arms overhead. When the front of the palms face anterior in this example and in all asana, the shoulders more easily remain on the back body and the thoracic spine extended.

Child's Pose  *Extended Child's Pose*

"What is round will roll!"

Anusara yoga teacher, Jaye Martin, offers this simple, but insightful mantra. It is evident in **Navasana** (Boat Pose) where rounding the spine causes the yogi to roll backward. A straight, extended thoracic spine helps keep the pose straight and erect.

Rounded Navasana  *Navasana*

Paschimottanasana, the Seated Forward Fold, is considered a pose of "surrender" in its final position. For many yogis, the thoracic spine will round and the shoulders roll forward. The thoracic spine can be protected from excessive rounding by fully integrating the shoulders into alignment before deepening into the pose. This action lengthens and extends the thoracic spine and expands the chest as it leads the pose forward. First initiating shoulder integrative alignment is essential especially if the style of yoga calls for the arms to be raised overhead when folding into the pose. Engagement of the upper back musculature, particularly the latissimus dorsi, is essential. Thoracic extension allows for greater flexibility in the upper back and deeper folding. Keeping the elbows point either next to the side body or lifted and in line with the shoulders as in "Cactus-arms" will help extend the thoracic spine and keep the shoulder blades on the back. Chapter 31 fully explores the action of shoulder integrative alignment.

Paschimottanasana – Full fold

Paschimottanasana – Extended thoracic spine

Seesaw rib cage

A common tendency with students is to move the thoracic cage as one unit, rocking it from top and bottom in a solid, seesaw fashion. Learning to move the upper chest forward while the lower rib cage moves posterior is certainly challenging for those not trained in belly dancing! With the numerous segmental movements available in the thoracic spine and rib cage, it is possible to break this habit pattern. Uddiyana bandha is the key action in preventing the lower rib jutting forward. In poses such as Warrior One, firmly engage Uddiyana bandha by drawing the lower ribs and tip of the breastbone back and inward. Use that action for stability and press the chest forward and break the seesaw pattern.

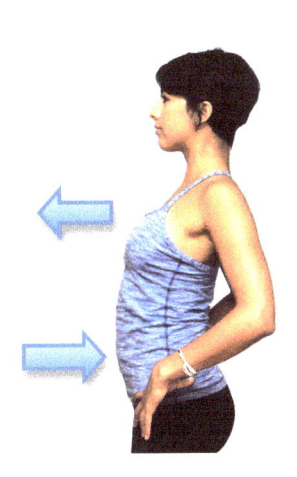

Shoulder shrugging

A similar reasoning for not rounding the spine is not to shrug the shoulders upward toward the ears, another unnecessary action to engage. In the activities of daily life, most people spend an inordinate amount of time with their shoulders tucked up under their ears. There is no need to further develop the "skill" of lifting the shoulders and tightening the neck muscles. That said, there could be a small degree of shrugging when initiating lifting the arms, as it requires the Rotator Cuff to engage and moderate shoulder lifting may occur. However, this shrugging action is quickly released from the pose once the arms the inner arm bones are drawn posterior.

Thoracic spine extension with blocks

An effective method for increasing thoracic spine mobility and reducing the effects of rounding of the upper back is to lie supine over two yoga blocks. This is a powerful, posture-changing therapy.

The correct placement of the blocks is specific to each student and the relationship between their shoulder blades and thoracic spine. Normally, the spine is anterior to the shoulder blades. It is also not uncommon for the spine to protrude beyond the shoulder blades on the back.

Getting the block positioning correct is fundamental to this pose. If the spine is anterior of the shoulder blades but the thoracic block is placed vertical, it incorrectly encourages the shoulder blades to wing off the back. It the spine protrudes beyond the shoulder blades, a vertical block is desired and will effectively integrate the upper torso and shoulders. Otherwise, and in most cases, the block is paced horizontal to assist the shoulders blades and serratus anterior to press forward into the rib cage and toward the chest.

Procedure:

1. With the help of an assistant, use a block to determine if the spine is positioned anterior or posterior to the shoulder blades. If anterior, there will be a small space created when the block is placed across the scapulae. If the spine is posterior, there will be no space and block may not rest evenly across the scapulae.

2. Once determined, set the blocks into the correct orientation.

3. The upper edge of the horizontal block is placed two finger distances from the upper armpit (axilla). This ensures that the block can support the scapula fully. It also allows the lower floating ribs to descend (Uddiyana bandha). A vertical block will be positioned at a similar height. Placing the narrowest side of the block upward allows the lower ribs to best descend but if it is necessary for comfort, using other heights or a blanket is fine.

4. If the spine is limited in extension, the lower back may become overarched and the pose will be uncomfortable. Place a blanket under the sacrum to limit lumbar overarching and help "unwind" the spine. This action effectively engages Mula Bandha.

5. The block placed under the head allows the throat to drop posterior and to lift base of skull (Jalandhara bandha). Ideally, the ear aligns with the shoulder sockets, although the height of the block can be at any level that is comfortable. It is not advised to allow the neck to hyperextend without any support.

6. To avoid muscle spasm when coming out of the pose, it is best to slowly roll off to one side and wait for a few breaths before coming up to allow any tension to subside.

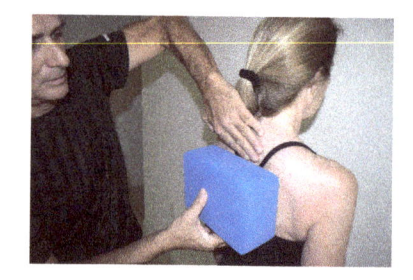

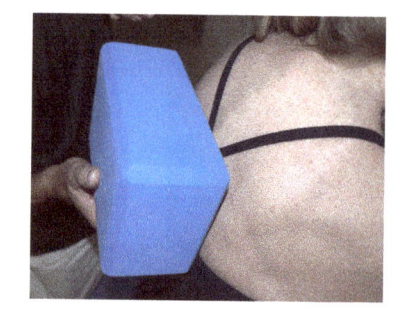

"It's not how far you go, it's how you go far!"

29 The Breath and the Bandhas

The coordination of movement and breath is a distinctive feature of yoga. In some yogic traditions, breath control is essential to asana practice, synchronizing the cycles of breath with specific positions and postures. Other systems of yoga take a less stringent approach. A few yogic approaches make breath a primary focus within the practice. *Pranayama* practices harness the power of breath to complex patterns and specific postures. Regardless of how formal the attention someone chooses to place on the breath, yoga practitioners all seem to agree that free flowing and unimpeded breath is necessary for an auspicious asana practice.

The technical term for breathing is *respiration*. It describes the lung's biological activity that exchanges oxygen with the discharge of carbon dioxide and water. It consists of two stages, *inspiration* or inhalation, and *expiration* or exhalation. The average adult carries out 12-20 cycles of respiration each minute. Highly trained yogis can reduce the respiration cycle to as low as 1-2 per minute. The ability to control the fullness and rhythm of breath is an indication of the health and vitality of the heart and lungs. Yoga practice provides a great opportunity to influence this vital body system.

The diaphragm – primary muscle of respiration

The *diaphragm* is a dome-shaped structure that separates the thoracic cavity from the abdominal cavity. It forms a barrier between the internal organs located in each cavity. The diaphragm is the primary muscle of respiration, composed of large flat sections of muscle, interwoven with fascia, uniting into a large central tendon. When inhaling, the diaphragm contracts, expanding downward. In exhalation, the contraction of the diaphragm releases, causing it to lift. The unique design of the central tendon allows both contraction and relaxation of the diaphragm to change the shape of the thoracic cavity in all of its three dimensions.[1]

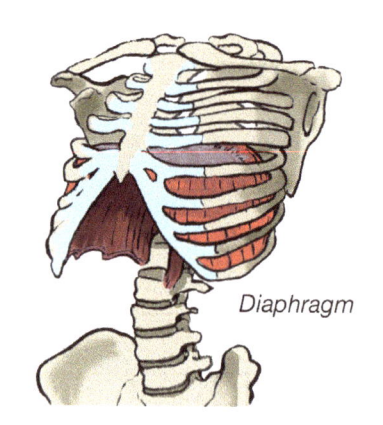

Diaphragm

The diaphragm's attachment to the skeleton is extensive. On the front of the body, the diaphragm embeds muscle fibers into the base of the two lowest ribs, the costal cartilage (cartilage that attaches the ribs to the sternum) and into the tip of the breastbone (xiphoid process). The posterior insertion of the diaphragm is on the vertebral bodies of T12 to L2 where it interweaves its muscle and tendon fibers into the psoas and quadratus lumborum. These interconnections establish a mechanical relationship between breathing, the lower back and the pelvis.[2] The diaphragm's nerve supply is the *phrenic* nerve, which emanates from the mid-cervical spine.

The diaphragm during respiration

At first glance, the diaphragm's movements may seem counter-intuitive. When a typical muscle contracts, it shortens, reducing the space between the bones to which it attaches. Contraction of the diaphragm, however, lengthens the muscle and increases space between the ribs.

On inhalation, the diaphragm is pulled down from its central tendon and contracts. The thoracic cavity expands upon inhalation, causing the belly and chest to broaden.[3] The cavity expands vertically down into the abdomen cavity and lifts up into the base of the neck. The thoracic cavity also expands transversely, from side-to-side and front-to-back. The diaphragm uses leverage from the diaphragm's attachment to the sternum and xiphoid process to lift and widen the lower rib cage.

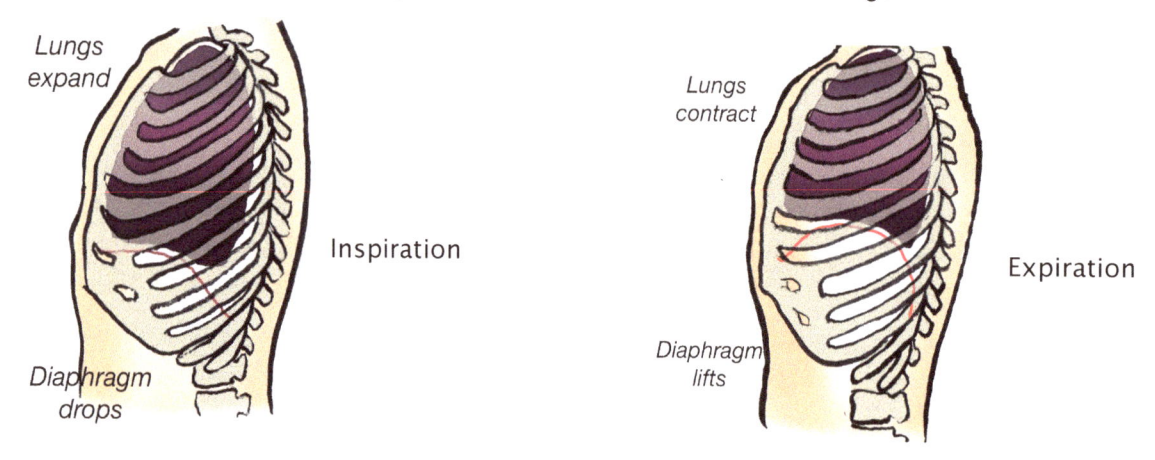

Lungs expand

Diaphragm drops

Inspiration

Lungs contract

Diaphragm lifts

Expiration

Diaphragm upon expiration

Expiration reverses the muscular action of the diaphragm. When the diaphragm relaxes, its central tendon is drawn up and into the thorax. The size of the chest size contracts and the belly flattens.

Protruding lower ribs

The diaphragm is stretched when the lower rib cage juts forward, increasing tension on the diaphragm and allowing the lungs and thorax to expand more effectively.[4] The efficiency of the diaphragm, as well as any muscle, increases when stretched, as long as its gross length does not exceed more than twenty percent. The efficiency of the diaphragm is maximized within the first 1-3 inches of the lower ribs jutting forward. Beyond that point, forward protrusion of the lower ribs reverses and decreases the diaphragm's overall efficiency. Increasing diaphragm efficiency is useful in backbending postures where flattening of the thoracic spine reduces the volume of the thoracic cavity. Excessive rib protrusion, however, can compress of the lower thoracic spinal curve and cause disc injury.

The decrease in diaphragmatic efficiency that accompanies excessively protruded lower ribs can be experienced during deep backbend poses. Initially, lifting the lower ribs in backbending postures may "feel good" as the diaphragm is stretched but once the ribs protrude too far, the pose becomes painful and breathing is restricted. The ability to breathe comfortably in a backbend pose is a reliable sign that the asana is being executed correctly and safely. Backbends are initiated during inhalation.

"Backbends" are best considered to be thoracic spine extensions, where the thoracic spine is lengthened and deep lumbar extension limited. One method is to engage Uddiyana bandha. As the ribs begin to jut forward, the tip of the breastbone draws posterior, which reduces strain on the diaphragm. In all backbend poses, Uddiyana bandha, Mula bandha, and Manipura chakra are routinely engaged.

Other muscles that play a minor, accessory role in inspiration are the sterno-cleido-mastoid, serratus anterior, pectoralis major and minor, the trapezius and latissimus dorsi. Avoid excessive use of these minor muscles in inspiration. Their overuse can tense the shoulder girdle and neck, causing them to lift and shrug and produce increased stress, strain, and potential injury.

Overuse of accessory muscles tightens the neck, shoulders and upper back

The abdominals

The abdominal muscles are the primary antagonist muscles to the diaphragm. The abdominals are the predominant muscles of expiration, although they have a role in both cycles of respiration. The abdominal musculature and the diaphragm both actively engage during both cycles of respiration. Contraction of the diaphragm vs. the abdominals varies inversely in a reciprocal relationship that is referred to as a *floating equilibrium*.[5]

During expiration, the abdominals draw the lower ribs and sternum inward. The diaphragm's lifts, pulled from its central tendon. The volume within all three dimensions of the thoracic cavity reduces. Coordinated action between the diaphragm and abdominal muscles also circulates air through the abdominal cavity, providing important stabilization to the spine.

Forward bending into yoga postures coordinates with the exhalation. Abdominal contraction engages the muscles that empty the lungs, flatten the abdominal cavity, and flex the torso to the hips.

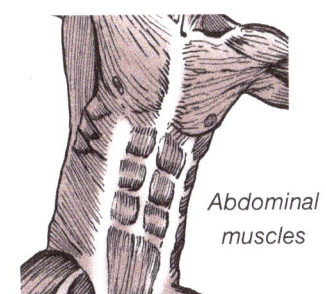

Abdominal muscles

Effective diaphragm engagement

Instinctively after a race, exhausted runners bend forward, drawing back the lower ribs. In this position, the overstressed diaphragm relaxes, allowing it to recover more quickly from fatigue. By pushing down on the knees, the chest opens anteriorly and shoulders draw onto the back, another instinctive exhaustion response that increases lung volume.

To increase lung efficiency, vertically lengthen the side body rib cage. This action will lengthen the spine, expand the ribs, and broaden the diaphragm while staying within its 20% range of efficiency.

Paradoxical respiration

Although breathing is a natural, unconscious action, some students have difficulty coordinating their cycle of breath with the correct phase of diaphragm or abdominal contraction. In these uncoordinated cases, the abdominal muscles are contracted on the inhalation instead of being relaxed. On exhalation, the abdominals relax when they are supposed to contract. This is called *paradoxical respiration*. This inefficient pattern of breathing fails to properly expand the abdominal cavity and leaves the spinal discs vulnerable to injury during weight bearing activities. Details on the relationship between the abdominal cavity and the lumbar spinal discs can be reviewed in Chapter 24's discussion of the *Valsalva effect*.

Should paradoxical breathing be the habit, yoga students can practice *Kabalabhati* breathing to learn to draw the abdomen inward on exhalation. Do do this, on the exhale, gently lift both the perineum and diaphragm and draw in the navel and engage the bandhas. Practing inhalations, place a hand on the abdomen and attempt to expand the region forward. This exercise and other yogic breath practices can help re-establish the natural breath cycle.

Nose breathing

There is an uncomplicated adage that is attributed to Sri B.K.S. Iyengar: "The mouth is for eating and the nose for breathing." When asked about breathing during asana, Mr. Iyengar simply said, "yes". With long-held postures and limited flow in his tradition, coordinating breath and asana is not a focus, quite different than the vinyasa flow of Astañga Yoga.

Nose breathing offers numerous advantages over breathing through the mouth. Nose breathing filters bacteria and pollutants, preventing them from entering the interior of the body. Circulating breath through the nasal cavities and sinuses warms and humidifies the air before it reaches the inner tissues of the body. Breath entering the sinuses and the porous, inner surfaces of the skull is necessary for oxygenating the outer surfaces of the brain cavity. Nose breathing retains moisture in the lungs and their linings, which helps hydrate the body.[6]

Other benefits of nose breathing are less obvious but quite important. Breathing exclusively through the nose has been shown to produce a positive shift in brain waves. Laboratory studies performed at Maharishi University (MUM) in Fairfield, Iowa explored the physiological effects of breathing. Test subjects used nose and mouth breathing combinations during various physical activities, such as running and weight lifting. When mouth breathing was used during either cycle of breath, test subjects produced mostly beta brain waves. Beta waves are cycles of electrical energy elicited when the mind is engaged in highly cognitive activities, such as thinking, talking, remaining alert, and concentrating. The studies further demonstrated that when nose breathing exclusively was used on both inhalation and exhalation, the brain produced an increase in alpha, delta, and theta waves. These are the electrical wave patterns associated with states of deep relaxation, meditation and the "higher" states of consciousness. These wavelengths have been associated with healing from illness, better-balanced hormone levels, and heightened states of spirituality.[7]

Nose breathing during exercise can help to maintain a heart rate below 60% of the maximum heart rate (MHR).[8] Yogic breathing techniques used by researchers at MUM kept the heart rates of athletes below 60% while performing sports that usually produce high spikes in the heart rate. Athletes were tested in sports ranging from long-distance running to power lifting. With only nose breathing, lower heart rates were sustained without loss in performance. The benefits from this are inexhaustible! Nose breathing delays sports fatigue because lactic acid production is reduced as a result of the blood and muscles remaining oxygenated for longer periods.[9]

Nose breathing techniques stimulate the metabolism to burn fats. Exercising with a heart rate at 60% of MHR or lower will stimulate the body to burn stored fat to fuel the muscles. Faster heart rates are interpreted as the body being in a state of stress or survival. Instead of fat burning, the body shifts its energy source to carbohydrates and proteins. These are fuels that require less energy for the muscles to metabolize and better suited for fight-or-flight, stress-type demands. Unfortunately, the body will utilize blood cells and muscle tissue to attain its fuel consumption needs.

Mouth breathing and stress

After a stressful situation, the first instinct is often to release a deep and breathy "Whew!"; that mouth exhalation being a typical expression of the stress response. Habitually exhaling through the mouth, however, may over-produce and release stress hormones and can lead to deleterious health effects. Meditation and breathing techniques that focus on nose breathing and discourage mouth breathing help develop the yoga's ability to generally reduce anxiety and stress.

The heart of the yogi

Breathing habits that may have developed to manage high physical demand or stress can create an abnormal overdevelopment of the heart muscle and its surrounding blood vessels. This is a condition know as cardiac hypertrophy. The heart and its large blood vessels function as other muscular tissues; they change in size to meet the demand placed upon them. One result in an increase in the amount of blood pumped by the left ventricle of the heart, known as high stroke volume. This causes the heart and aorta to enlarge. A heart that has over-developed and enlarged to meet the demands of explosive power sports and stress creates a life-threatening situation once the level of exercise lessens. Once tossing baseballs is replaced by tossing back beers, an ex-athlete's risk of heart attack increases.

Once the constant demands of intense exercise on a hypertrophied heart are no longer present, the heart can become fatty and inefficient, making it prone to cardiac incidents. An enlarged heart is a serious energetic drain on the body. By only engaging nose breathing, even when the cardio-vascular demands are high, keeps the heart rate low. Nose breathing helps reduce the volume of blood that is pumped per stroke, or beat, that passes through the heart's walls and chambers. With lower demand on the heart and aorta, there is less likelihood of enlargement. The character of the heart can remain yogic: appropriate in size and highly efficient.

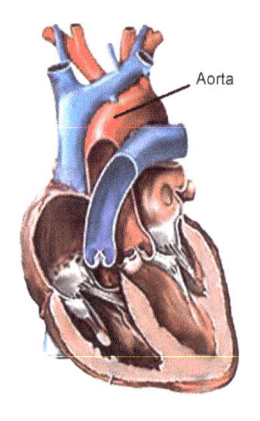

Normal heart muscle

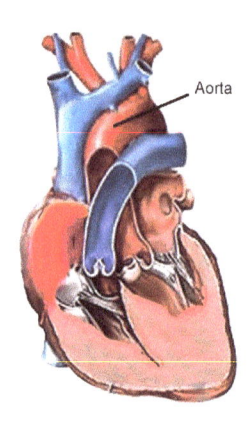

Thickened heart muscle

Blood pressure and yoga

This subject can produce strong and contradictory opinions. Medical advice is generally cautious about cardiovascular issues and what is a safe blood pressure for practicing vigorous exercise, including some forms of yoga. Concerns regarding the risk of performing **Sirsasana** (Headstand Pose) and its effect on blood circulation in the brain are well established. Yet, some authors with backgrounds in both medicine and yoga conversely regard a vigorous practice and inversions as not only preventative but also restorative and therapeutic, even in cases of advanced cardiovascular disease.[9, 10]

There are no absolute answers to the many questions and considerations that arise regarding high blood pressure. A fundamental tenant of yoga is Ahimsa, *do no harm,* and caution and consideration are the wisest approach to take for all students. The benefits derived from advanced postures can often be achieved in modified and remedial poses. This approach is best not only for cardiovascular issues, but orthopedic ones as well.

Bringing wisdom and concern about illness or disease is the very essence of yoga. Resisting the drive of the ego to attempt risky postures is the true practice of yoga. Managing any major health condition requires great personal responsibility. A student with heart issues must do diligent research and seek unbiased medical consultation. Then, with all the wisdom the student can muster, a skillful yoga practice can be designed.

> Serious yoga students eventually realize that wisdom is not necessarily in the answers, but in the deepest exploration of the questions themselves.

Advanced yogic breathing techniques

Yoga incorporates a sophisticated breath practice that is not considered asana but is still part of Hatha yoga. Techniques are derived from the practices of *Kriya* yoga and *Pranayama*.[11] Kriya yoga is a system of cleansing practices designed to remove obstructions in the relationship between the body and mind. Kriyas include practices for both psychological and physiological cleansing. These include hygienic bodily cleanings as well as *mudras* (body and hand positions that increase prana), *mantras* and *meditations*.

Pranayama is primarily the yogic practice of breath control. Derived from Sanskrit, Pranayama refers to the practice of extending and drawing out the breath and life force. Yoga teacher Richard Freeman describes Pranayama as "the release of life energy from its bounds."[12]

Pranayama is considered an advanced, esoteric practice of yoga. As with most spiritual and religious practices, esoteric studies are the pursuit of the more advanced practitioner. The sometimes, simple-appearing breathing techniques require a level of awareness that often requires years of Hatha yoga practice to fully prepare the student. Breathing techniques, nonetheless, are presented by teachers to students at even the early levels of yoga practice. A few basic Pranayama techniques are explored in this chapter that can be appropriate for students of all levels of asana practice.

Basic three-part breathing

1. Lie supine
2. Inhale and fill the abdomen; retain the breath in the abdomen
3. Inhale and fill the chest; retain the breath in the first two cavities
4. Inhale into the lower neck; retain the breath in all three regions
5. Hold the breath for a count between four and sixteen; then exhale, releasing the breath in the reversed order taken during inhalation

This exercise can be practiced with both hands placed over the belly or one hand over the belly and the other on the chest to bring better focus. A rolled mat or blanket can be placed lengthwise under the spine and the head slightly lifted to facilitate the practice.

Alternate nostril breathing

Alternate nostril breathing is a slow and controlled, rhythmic series of inhalation, breath retention, and exhalation. Known as **Nadi Shodhana** and **Anulom Viloma Pranayam**, these popular practices are found in many yoga classes. They are excellent methods for developing breath control and considered a valuable technique for balancing neurological activity between the two brain hemispheres.

1. Gently close one nostril, with or without finger pressure.
2. Inhale for a count of four; retain the breath for a count of four; reverse whichever nostril is closed and slowly exhale out for a count of eight through the side that was retained.
3. Inhale through the now-unobstructed nostril; repeat the procedure starting from the other side.
4. With each successive round of breathing, the count for each step, especially retention, can be lengthened proportionally.

Throughout the Kriya, the shoulders remain fully on the back body, square to each other, and aligned. Since the forearms can independently rotate 180°, the humerus can remain externally rotated and the shoulders do not have to be compromised.

There are many breath-counting methods and variations of the finger configurations (mudras) used for alternate nostril breathing. A traditional approach can be followed or one's own method can be applied.

Of course, alignment of the body is maintained throughout the practice, particularly in the shoulders, upper back, head and neck. If fatigued, students might begin to round forward. This impedes breath flow by constricting the upper quadrants of the lungs. Nose clasping by the fingers will be more stable if the shoulder adopt a **Chaturanga-like** position with the collarbones lengthened wide and the inner elbow held physically or energetically in line with the side body.

Kumbhaka   Breath retention

Kumbhaka, or breath retention is one part of the alternate nostril breathing sequence. Breath retention strengthens the muscles of respiration, however, holding the breath also increases cardio-pulmonary pressure. If practiced in a safe, conscious manner, Kumbhaka not only avoids cardiac risk but also provides heart-health benefits similar to those received from cardiovascular exercises such as running.

The term Kumbhaka derives from the Sanskrit word for a "pot". It is a container that is sometimes full or sometimes empty. In breath retention, the chest is the pot that alternates between being full, chest filled with air, and being empty, air emptied from the chest. Kumbhaka occurs naturally in every breath cycle: at the end of each inhalation (*anatara*- breath inside) and the end of each exhalation (*bahya*- breath outside). Kumbhaka is skillfully harnessed in Pranayama and used as preparatory practice for meditation. Kumbhaka retention is applied when engaging the *bandhas*, or breath locks. Kumbhaka also re-vitalizes the soft tissue linings that form a friction-reducing buffer between the ribs and lungs.

Kumbhaka can increase pressure in the spinal discs so it must be practiced skillfully. Breath retention should not be performed in extreme forward bending postures, heavy weight bearing, or during active asana. **Padmasana**, the Lotus Pose, and **Siddhasana**, the Easy Sitting Pose, are suitable postures for the practice of Kumbhaka.

The Bandhas

The translation of the Sanskrit term *bandha* is a lock, knot, or bond. It is associated with the processes of binding, restraining, or capturing. The "lock" compares to the lock of a canal, not of the lock-and-key variety.

Bandhas accompany breath retention. They help to regulate the flow of prana. Bandhas are used in the cleansing practices of Kriya where the bandhas are used to stimulate the internal organs.
Bandha locking can be practiced as an asana unto itself or an energetic action that underlies all other asana. As asana, bandhas are performed most often in a sitting or standing position. Muscular action is firm.

Uddiyana bandha

Bandhas create core a tension that is both muscular and energetic. In the *Astañga* yoga tradition of *K. Pattabhi Jois*, teachers often instruct their students to engage a subtle form of Mula bandha - "24/7"!

Bandha engagement is a valuable and nearly essential component in asana practice. Often the action is barely physical, being only subtle and energetic, allowing them to be a constant backdrop to asana practice. For example, engaging Mula bandha forcefully by firmly squeezing the anus or performing a *Kegel-type* contraction is generally not recommended.[13] In contrast, the Pranayama practice, Uddiyana bandha is performed with a firm and intense contraction, although it is not considered to be an asana practice.

The three most common bandhas and methods to engage them:

- Muladhara (Mula) bandha: a gentle upward lift from the center of the perineum [14]
- Uddiyana bandha: drawing in the xiphoid process inward and lower ribs back
- Jalandhara bandha: leveling the soft palate with ear canal and base of occiput, causing a soft closing of the glottis

From an esoteric perspective, Mula bandha is the "root" lock that helps propel the upward flow of prana. Prana rises molten-like in a spiraling fashion through the *Sushumna*, the central channel of the core. Prana exits the body through the posterior fontanelle of the skull. The pranic channels also form bandhas as they travel into the palms of the hands (hasta bandhas) and the soles of the feet (pada bandhas).

Bandhas, from an anatomical point of view

As previously discussed throughout the book, bandha engagement is a simple, but valuable method of protecting the spinal discs and establishing postural alignment through the central axis of the body.

To lock the bandhas physically, the diaphragms are contracted. Diaphragms vault up in a dome-like fashion as the bandhas are engaged. Bandha/diaphragm engagement initiates at the perineum, rises through the thoracic diaphragm, continuing through the posterior roof of the mouth at the soft palate, and exits energetically through the posterior fontanelle of the skull.

Forward tailbone scoop engages the Mula bandha. This action stimulates a plexus of parasympathetic nerves nestled above the tailbone's inner surface (coccyx). The parasympathetic nervous system produces calm within the body that may be experienced when Mula bandha is engaged. Bandha contractions, in general, provide a deep sense of integration, calmness, and safety.

Ujjayi Pranayama   Victorious breath

If you walk past an Astañga yoga class, you might hear *Ujjayi*, a deep, synchronized, hissing sound that causes you to quickly turn around with your light saber in full force, looking for Darth Vader. The sound is created when both the inhalation and exhalation are directed across the back of a partially closed glottis, the area of the throat that closes when swallowing. Air passing over the partially closed glottis resonates, creating a deep sound from the throat and vocal chords.

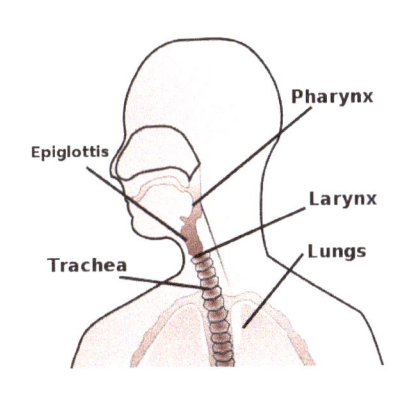

212 YOGA ALIGNMENT PRINCIPLES AND PRACTICE

Ujjayi is a valuable tool that calms the nervous system and maximizes the exchange of airflow. It is performed with a closed mouth but not as a nasal "sniff". It produces a deep, guttural sound that reverberates from the back of the throat. When performed correctly, more air enters the lungs than is possible from sniffing or simple mouth breathing. Ujjayi breathing can be challenging for the beginner student to learn but can quickly be mastered by most.

How loud is the Ujjayi call to victory?

The question of how audible Ujjayi breathing should be has differences of opinion by yoga traditions. Some schools of yoga insist that the Ujjayi breath be distinctly heard from all students, in unison. Other traditions believe that constant, deep vibration of the vocal chords is irritating and dries out the delicate mucous membranes of the vocal chords, leading to sustained damage.

Physiologically, it can be easily performed in silence, avoiding any potential risk of injuring the vocal chords. This seems to win out as the safest method.

The Valsalva Effect and breath

Introduced in Chapter 24, the Valsalva effect is a hydraulic-like action of the diaphragm and abdominal muscles that transports air into the two anterior cavities of the body. It creates an inflated, structural support for the lumbar spine.

Some students hold their breath as a method to engage the bandhas. Continual breath retention is not recommended since it increases spinal disc pressure. Instead, learn to apply the bandhas in a gentle, deliberate fashion during *both* inhalation and exhalation.

Thoracic cavity

Diaphragm

Abdominal Cavity

Back off the bandha!

The instruction to firmly squeeze the buttock in a *Kegel-type* fashion to engage Mula bandha is an instruction passed down by many teachers who perhaps have not fully exploring the subject. A hard, forceful action causes students to miss the subtle effects of the action that are calming to the nervous system.

Anatomists often consider the pelvis to be a bowl. Imagine the pelvis bowl is filled with thick, soapy water. The amount of force that is used to engage Mula bandha has been suggested to be equivalent to the effort used to lift a small bubble up from the pelvic bowl's center.

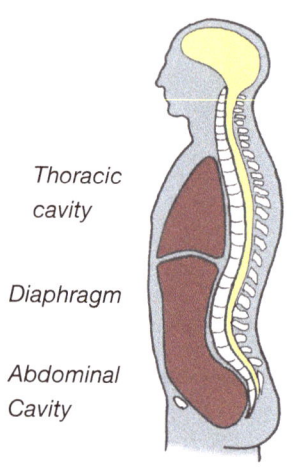

> Cycling inspiration and expiration with a continuous, gentle application of the bandhas aligns the core and keeps the spine safe

30 Shoulder Anatomy

You may need to put your shoulder to the wheel in order to master the principles in this chapter but it will certainly be worth the effort. Developing strong, flexible, and well-aligned shoulders is necessary for asana practice to advance safely. Learning the shoulder's anatomical and mechanical concepts make it easier to understand and apply the alignment principles presented in the next chapter.

The basic class instruction often given is to bring the shoulders back; this is too simple and insufficient. The shoulders are not merely "ball and socket" joints. Each shoulder has various internal components and is overall part of a functionally integrated system involving the thoracic spine, upper extremities, and head and neck. Often, injuries in these regions, especially repetitive injuries, can trace their origins to shoulder misalignment.

With its extensive fluidity and range of motion, it is easy to take the shoulder's far-reaching movements for granted, evidenced by the high incidence of shoulder injuries attributed to yoga.[1] The greater a joint's mobility, the greater the need for mechanical precision for maintaining graceful and safe action.

Shoulder mobility

The shoulder complex provides both powerful muscular strength and the broadest ranges of motion found in the body. Shoulder movement has evolved to follow a forward arc, intended to allow the eyes remain transfixed with the varying position of the hand.

The mechanical term used to describe the general movement of the shoulder joint is *circumduction*. Circumduction combines all directions of the shoulder's movements as it courses through a single, cone-shaped arc, radiating outward from its one central axis. Its movement is seamless yet its graceful fluidity comes from precision in its engagement.

The shoulder's individual ranges of motion, called degrees of freedom, are as follows:

- Flexion   180°
- Extension   50°
- External rotation   80°
- Internal rotation   110°
- Abduction   180°
- Adduction   0°

Full flexion and abduction bring the arm to the same final position that ends at 180° overhead.

Adduction must be combined with flexion or extension because the arm's neutral position is next to the torso, which will obstruct pure adduction.

Circumduction requires a significant degree of external rotation to avoid joint damage between the outer collarbones and the outer shoulder blades as the arm courses overhead beyond 90°.

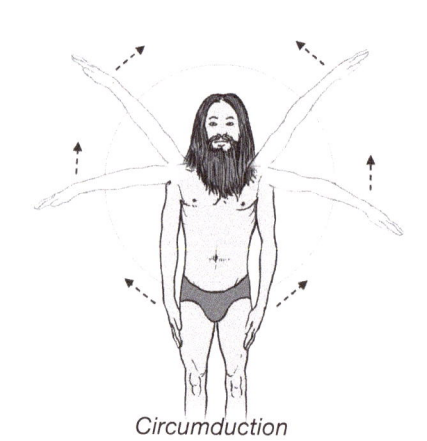

Circumduction

The three mechanical components of the shoulder

- Gleno-humeral joint
- Clavicular components: Acromio-clavicular joint   Sterno-clavicular joint
- Scapulo-thoracic "joint"

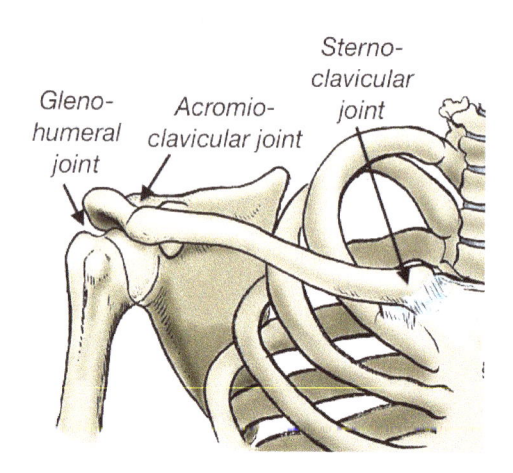

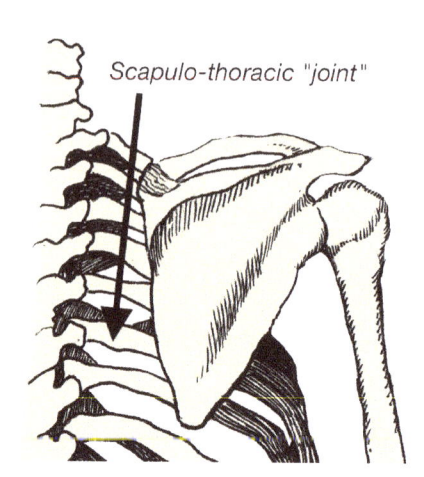

Gleno-humeral joint   *Acromio-clavicular joint*   *Sterno-clavicular joint*

Scapulo-thoracic "joint"

The gleno-humeral joint

The *gleno-humeral joint* forms the upper border of the armpit (axilla). It is the primary joint of the shoulder girdle and most mobile. The *humerus* (upper arm bone) and the *scapula* (shoulder blade) fasten into a ball-and-socket system. The *glenoid fossa,* from the Greek *glene* for socket, is a shallow, pear-shaped impression located on the lateral aspect of the scapula. Technically, the glenoid fossa is too flat to be categorized as a socket. The head of the humerus serves as the "ball" of the joint. The humeral head connects to its shaft by only a rudimentary neck, a design significantly different than the femur bone where its neck is substantial and enables the femur bone to angle inward.

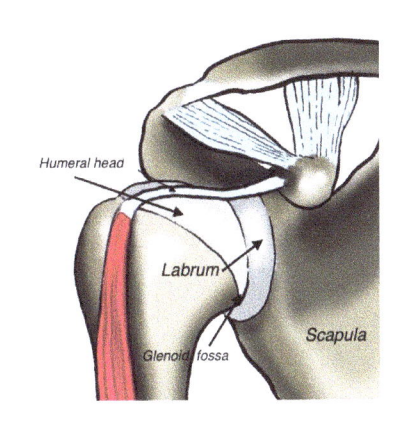

As with the hip joint, a cartilage gasket called the *labrum* surrounds the glenoid fossa. This collar forms a snug fit with the large, spherical head of the humerus.

To allow for extensive mobility, the gleno-humeral joint is loosely bound by capsular ligaments that envelop the entire joint to fully contain the joint's lubricating synovial fluid. For mobility, the capsule is looser inferiorly but more taut along its top border to aid with stability.

Unlike the hip joint, which has a sharply defined central axis of rotation and limited *joint play*, the gleno-humeral joint can shift its central axis by as much as one to two inches. This floating central axis design allows the shoulder to move smoothly through large arcs and to rapidly change the direction or angle of the joint.

The first 90° of circumduction occurs primarily in the gleno-humeral joint. The *deltoid* and *supraspinatus* muscles provide the majority of strength in this phase of circumduction that requires that the arm lift away from the side of the torso.[2]

Comparing the hip and shoulder "sockets"

The hip: the *acetabulum* is deep and round. Its large rim holds the femur head snug within the cup of its socket, preventing the hip from deviating off its central axis.

The shoulder: the *glenoid fossa* is a shallow depression on the scapula that forms the joint with the head of the humerus. It is a loose, lax joint with a central axis that shifts between one and two inches during circumduction. This provides the shoulder joint with the widest range of motion of any joint in the body.

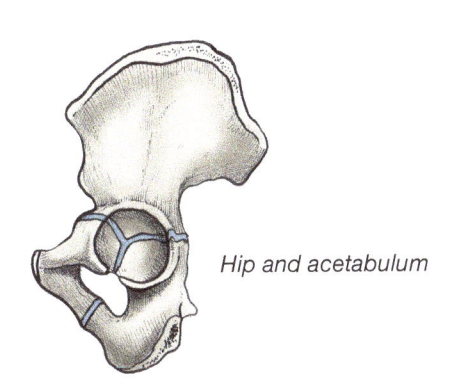

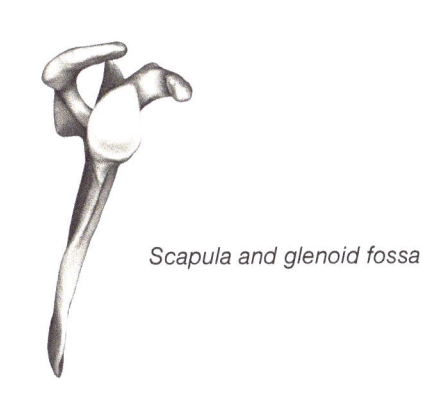

Hip and acetabulum

Scapula and glenoid fossa

Biceps brachii tendon

The *biceps brachii* is the primary flexor of the elbow. It also flexes the shoulder to a lesser degree. It provides external rotation, or supination to the forearm when the elbow is flexed; this being the action used to turn a screwdriver. The biceps is bi-articular, meaning that it crosses and mobilizes two joints. As the term *biceps* implies, it has two heads (muscle bundles), a long head and a short head. In approximately 10% of the population, there can be multiple short heads with three to seven most reported. The short head provides the power needed to initiate the long head and assists the pulling action of the humerus to the chest.

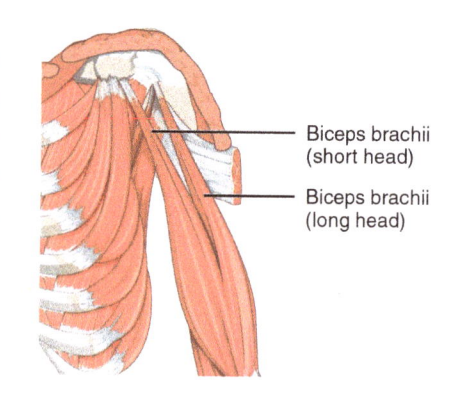

The long head's tendon enters the joint and attaches to the top of the humeral head deep inside the capsule. It helps stabilize the gleno-humeral joint, preventing dislocation of the humerus when the arm is pulled forcefully from any direction but mostly from superior and lateral. Extreme forces on the joint can tear the tendon. The long head of the biceps tendon is vulnerable to injury due its tenuous path through the constricted and mechanically demanding upper shoulder joint region.

To keep the biceps tendon safe when it is engaged:
- Draw the inner head of the humerus posteriorly into the joint. Use equal force on both its inner and outer aspects
- Lengthen the collarbones horizontally across the upper chest

---

Specific joint and muscle involvement in circumduction

Because shoulder mobility does not conform to standard terminology, the term *circumduction* is used to describe the shoulder's full range and the multiple directions it travels through its wide arc. For example: The initial 90° of arm raising from the side body is the direction of *abduction*. Continuing past 90°, the arms begin moving "toward" the midline of the body, or *adduction*. When the arms are completely overhead, 180°, they are essentially in the same position after this abduction/adduction as they would be if that had been lifted overhead from forward flexion. Using the term circumduction reduces this confusion.

From **0-90°** of either flexion or abduction, shoulder movement occurs almost exclusively in the gleno-humeral joint. The supraspinatus and deltoid muscles primarily provide the action. At 90°, the upper ligaments of the joint capsule become taut, helping to support the arm.

Continuing from **90-150°**, the clavicle rotates at each end and the scapula begins its glide away from the spine. Additional muscles begin to contribute at this stage: the middle and upper trapezius and the serratus anterior. The rhomboids do not play a significant role in circumduction.

In the final **150-180°** of flexion or abduction, all previously mentioned muscles are engaged with additional aid from the paraspinal muscles. When one arm is lifted, that scapula and the spine shift toward the lifted arm side as the spinal muscles on the opposite side (contra-lateral) contract. If both arms are raised together, muscles on both sides of the spine engage and the lumbar spine deepens its lordotic curve.

Joint clearance requires external rotation

Similar to the greater trochanter of the hip and its proximity to the ilium, the outermost process of the shoulder, called the *greater tuberosity*, is located close to the *acromion* process of the scapula. The space under the acromion (see illustration) can become compressed, pinching ligaments and other soft tissue when the arm is abducted overhead. To avoid injury, the shoulder must externally rotate once the arm is abducted 90° from the torso. Failure to externally rotate can result in compression damage to the biceps tendon, the ligaments of the shoulder joint, the supraspinatus muscle, or directly to the acromio-clavicular joint.

Although the forearms can rotate separately from the shoulders, when the palm faces downward, the shoulder externally rotates. When the palm faces down, the shoulder internally rotates.

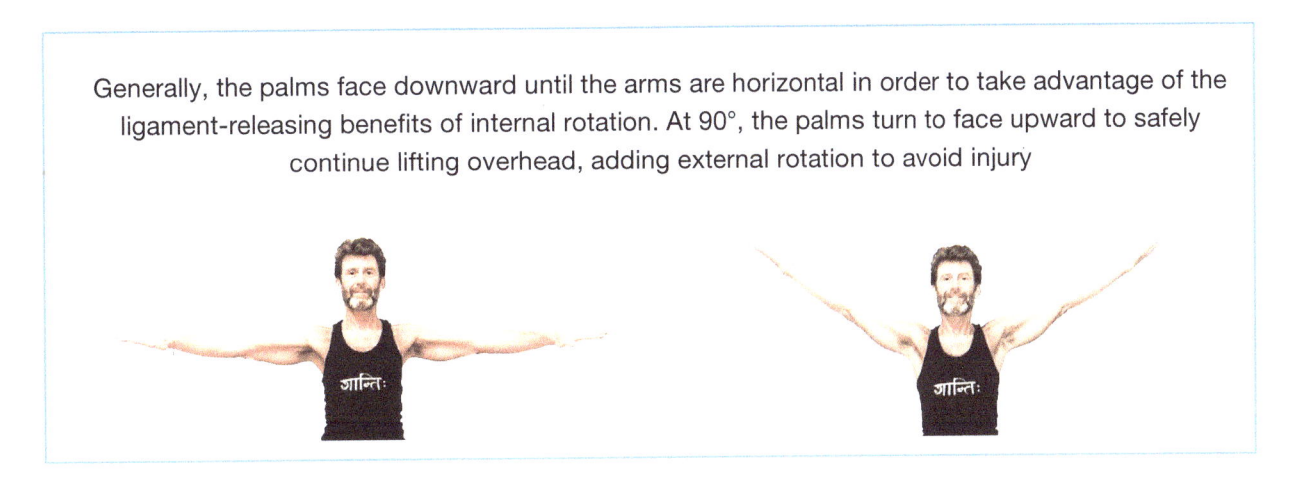

Generally, the palms face downward until the arms are horizontal in order to take advantage of the ligament-releasing benefits of internal rotation. At 90°, the palms turn to face upward to safely continue lifting overhead, adding external rotation to avoid injury

Shoulder dislocation

The most vulnerable direction that the gleno-humeral joint can move is the combination of anterior and inferior, which is movement downward and toward the breast. The anterior-inferior portion of the joint is its least stable. It receives limited support from the shoulder's ligaments, rotator cuff muscles, and biceps tendon, which results in structural weakness and limited stability. The gleno-humeral joint is most vulnerable to dislocation when in the anterior and inferior directions. This usually occurs when the shoulder rolls forward and the head of the humerus moves anterior.

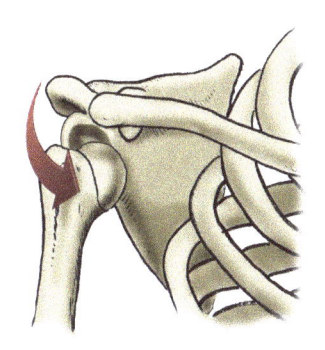

The clavicle and its two joints

Commonly known as collarbones, the two clavicles bridge across the two scapulae (shoulder blades) and are supported centrally by the *sternum* (breastbone). They provide anterior bracing of the shoulders and join with the other structures to form the shoulder girdle. The two clavicles are the only long bones of the body that are horizontally positioned. From the outer shoulder, they angle 60° forward to where they medially attach to the sternum. The clavicles rotate 30° at each of its ends along a horizontal axis (imagine hot dogs spinning on a commercial griller). The clavicle's participation in shoulder movement occurs primarily when circumduction is between 90° and 150°.[2]

The human clavicles have evolved to firmly brace the front of the shoulder girdle. Other mammals have only remnant-sized clavicles, and some, none at all. Animals that are primarily runners benefit less from collarbones than do climbers. Cats, being both runners and climbers, have small clavicles embedded in their muscles, but they do not form a functional joint. This unique design gives cats the ability to squeeze through small spaces. Humans, however, have long, broad collarbones that keep the shoulders from collapsing forward. This might be why we humans get into tight spots from which we seem unable to get out!

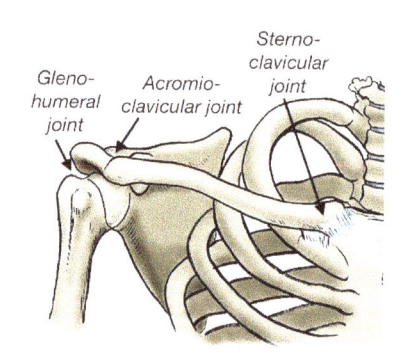

Not a fashion statement!

Touted perhaps as stylish, prominent collarbones indicate poor shoulder alignment. It displays a loss of postural integrity, which causes the shoulder girdle to round forward. Rounded shoulders predispose the biceps tendons to tearing and the rotator cuff muscles more vulnerable to injury (details reviewed later in this chapter). Prominent collarbones will often shift the body's central axis forward, tighten the jaw, and tense the upper back muscles. This is not so chic a posture!

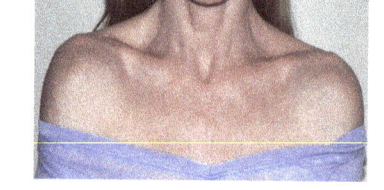

Prominent collarbones

The acromio-clavicular joint

If you hang around a group of physically active people, it is not too long before someone mentions their A-C joint injury. The A-C, or *acromio-clavicular* joint is formed where the clavicle joins the acromion portion of the scapula at the upper, lateral shoulder. The A-C joint pivots on its long (horizontal) axis when the arm raises overhead. Rapid movements and frequent positional changes of shoulder motion put significant strain on the A-C joints, making them a common site for sports-related injuries. Ligament tears, joint separations, and fractures are the most common injuries to the A-C joint. While the joints are just forming in a young person, rigorous activities such as rowing or archery can cause a non-fusion of the acromium to the rest of the scapula. A large bump is sometimes noticed at this joint; a permanent enlargement that often indicates the history of an acromio-clavicular stress or injury. From my personal experience of having suffered bilateral A-C separations, these bumps are useful for keeping backpack straps from sliding off the shoulders!

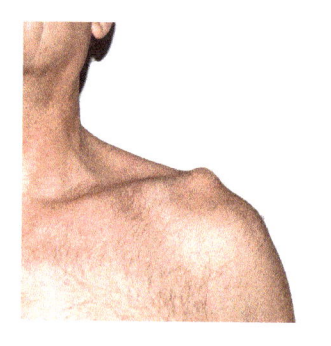

Enlarged acromio-clavicular joint

The sterno-clavicular joint

The medial end of each clavicle affixes to the breastbone to form the *sterno-clavicular* joint. It also attaches to the first rib. The joint is wrapped by thick ligaments and cushioned by an articular disc. This juncture is the only skeletal connection between the shoulder girdle and the torso.

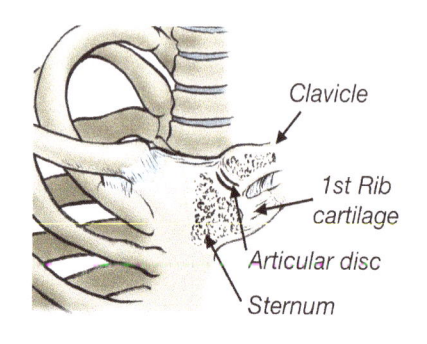

The entire weight and muscular power of the shoulders and upper extremities are structurally supported, bone-to-bone, only at this location on each side of the breastbone.

Poor alignment of the shoulders can produce excessive strain on the sterno-clavicular joints and create joint hypermobility. Although the ligaments that surround the joint normally form a sturdy capsule, they can become overstretched and result in the joints becoming unstable. Unstable A-C joints are unable to brace the front of the shoulders properly against forces such as those produced when jumping back into **Chaturanga Dandasana** (Four Limbed Staff Pose).

Mechanical stress on a hypermobile sterno-clavicular joint can cause the joint to become enlarged and swollen. This site is susceptible to arthritis and bone infection.

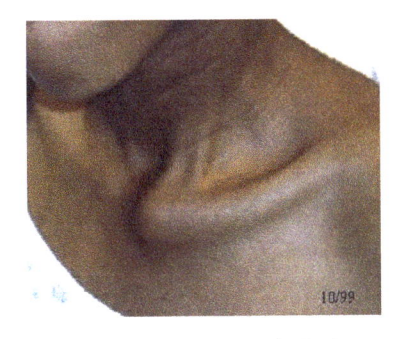

Dislocated sterno-clavicular joint

The scapulo-thoracic "joint"

A joint is defined as two or more bones directly united or connected through an intervening substance, such as cartilage.[3] Since none of the spinal vertebrae attach directly to the scapulae, the scapulo-thoracic "joint" is technically not an anatomical joint. Physiologists, however, consider the scapulo-thoracic articulation a "pseudo" joint because mechanically it functions similar to a joint between the scapula and the thoracic spine except being interconnected by musculature and not cartilage.[4]

Regardless of how it is categorized, the scapulo-thoracic articulation is extremely important to the mechanics of the shoulder. The principles of shoulder integrative alignment presented in Chapter 31 explore the scapulo-thoracic joints, essential to balanced shoulder functionality.

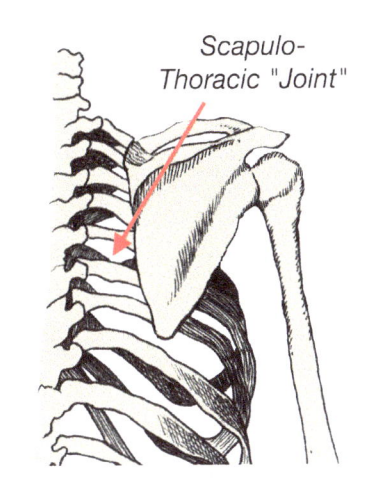

The scapula   Fun facts

The shoulder blades, or scapulae, are located on the upper back. Vertically, their length stretches from the second to seventh thoracic ribs. The upper medial edge of each scapula is located approximately 3 inches (5-6 cm) from the spinous processes. The distance from their lower medial border to the spine increases to nearly seven to eight inches. The *acromion* and the *coracoid* are two boney projections, or processes, that arise from each scapula. They extend superior over the gleno-humeral joint. The coracoid provides ligament and muscular attachment for the biceps brachii, the coracobrachialis, and the pectoralis minor. The acromion attaches to the clavicle to form the A-C and provides tendon attachment for the deltoid and trapezius muscles.

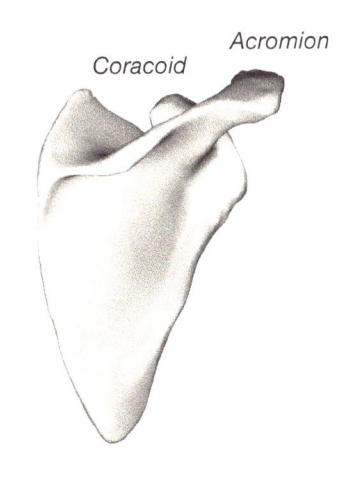

Scapulo-Thoracic "Joint"

Posterior view of scapula

The shape of the scapula is contoured with the radius of its inner surface curved at 30°. They align with the upper back, not flat their neutral position, but angled 30° toward the front of the torso. Their curved shape and angled orientation it what allow small movements of the scapulae to amplify into large arm swings. Scapulae movement, drifting laterally, is most active at 90-180° of circumduction. [5]

The glenoid fossa, its contoured surface, acts as a movable base that stabilizes the humeral head while it shifts into all its positions. It acts like a radio-tracking dish following the humeral head as the arm moves in space. Because the shoulder joint is relatively close to the body's core, small movements at the gleno-humeral joint become large and rapid movements by the time they reach the hand. For that reason, hand and arm movements can injure the shoulders. Engaging the core before the extremities participate in upper body movement is essential. Shoulder integrative alignment engaged prior to moving the extremities is vital to avoid injury.

The outer borders of the shoulder blades can be palpated high within the inner armpits. This is a useful teaching instruction when learning how to "slide" the shoulder blades onto the back by drawing their outer borders inward toward the spine. This will be further discussed in the following chapter.

Where go the palms, so go the shoulder blades

An instruction that yoga teachers sometimes use is to "imagine putting your shoulder blades into your back pants pockets." This visualization can be useful to students as it reminds them to draw the shoulders down and away from the neck, reducing the unproductive habit of shoulder shrugging. The action is an indirect use of the latissimus dorsi muscle.

Another valuable principle: the shoulder blades move in the same direction as the palms of the hands. The heels of the hands correspond with the bottom tips of the shoulder blades. In asana, especially with arm balances, to keep the shoulder blades on the back and closer to the spine, torque the heels of the palms and thumbs toward the midline of the body, as if turning two faucets inward. As with many actions taken to align the body in asana, the muscles utilize an isometric or eccentric contraction. Actual movement in the position of the bones and joints is either subtle or there is no movement, at all.

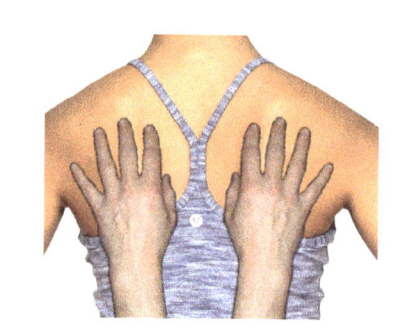

Keep your angel wings folded in!

Despite the fact that the scapulae will broaden and wing off the upper back when the arms are raised overhead, effort is made to keep them on the back and close to the spine. Do not physically force them not to spread but keep the muscles that hug the scapulae to the upper rib cage engaged. As long as we remain in human form, our angel wings stay furled.

Postural adaptation to shoulder limitations

Poor posture limits shoulder mobility and make them vulnerable to injury. If the thoracic spine is overly rounded, the shoulder blades are unable to fully retract onto the back. This causes the shoulder blades to habitually remain wide on the back and for the front of the shoulders to round and drop forward. The upper spine and neck cannot align properly when the shoulder girdle is rounded and will develop dysfunction in their own movements. The lumbar spine also compensates to shoulder dysfunction, usually by increasing its curve. After years of this poor postural pattern, the lumbar spine eventually flattens as conditions such as spinal stenosis develop.

Upper-crossed syndrome

If the shoulders and upper back round forward and the head and neck drift forward of the midline, the *Upper-Crossed Syndrome* is likely present. The muscles that become lengthened become weak and the muscles that shorten become tight.

Correcting this imbalance requires the ability to fully extend the upper thoracic spine and learning how to press the scapulae onto the upper back. Simply drawing the head back will not be effective. The Three "S" alignment exercise is a valuable therapeutic tool to rehabilitate the Upper-crossed syndrome (Chapter 11).

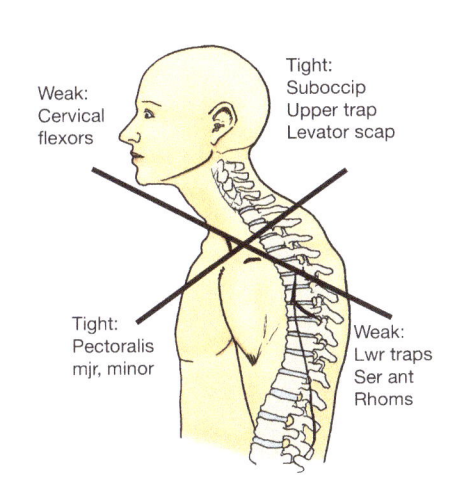

Serratus anterior

The *serratus anterior* arises as finger-like projections from the upper eight or nine ribs and inserts into the underside of the smooth, medial border of the scapula. When the arm swings forward, the serratus anterior muscle assists in bringing the shoulder forward. The serratus anterior has another important, albeit less powerful action that plays a primary role in keeping the lower tips of the shoulders blades on the posterior rib cage.

This action comes from a small section of the lower serratus anterior that specifically draws the bottom tip of the shoulder blade anterior, forward toward the back ribs. As will be discussed in the next chapter, this action is a critical instruction in shoulder integrative alignment.

The serratus anterior muscle receives its nerve supply from the *long thoracic nerve* that originates from the roots of the cervical 5 to 7 spinal nerves. This nerve is subject to injury from stretch or compression trauma. Should the long thoracic nerve become damaged or paralyzed, the effect can be clearly observed in the shoulder blade "winging" off the back.[6] This observation dismisses any controversy that may arise to the importance of the serratus anterior and its principal role in shoulder integration and alignment.

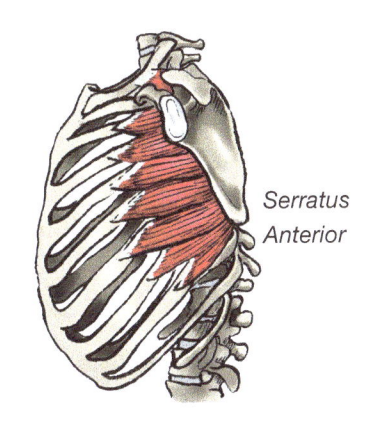

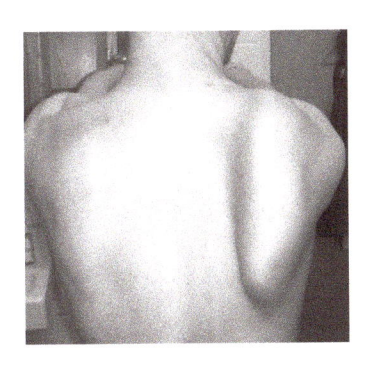

Serratus Anterior

Challenges of the serratus anterior

The good: The *rhomboid major* and the *teres major* form a sling that secures the inferior tip of the scapula to the posterior rib cage. They assist the function of the serratus anterior in drawing the lower tip of the scapula tip forward.

The bad: Many students have a habit of tensing the middle *trapezius* to draw the shoulder blades onto the back. A tense trapezius muscle will limit the actions of the rhomboids and the serratus anterior. It is more effective to *soften* the mid-trapezius and upper back muscles and allow the scapulae to gently slide closer, moving from the musculature that wraps the torso from under the inner armpits. This approach enables the serratus anterior muscles to be more engaged.

The ugly: The *pectoralis minor*, a small but powerful chest muscle, is a direct antagonist with the serratus anterior. It attaches to the coracoid process on the scapula on the front of the shoulder. Its action pulls the shoulder down and toward the middle of the chest, directly in opposition to the serratus anterior. The more anterior the shoulder is positioned, the greater the pectoralis minor overpowers the serratus anterior. Posture that rounds the shoulders increases and perpetuates this imbalance.

The muscles of the front body are approximately 30% stronger than those of the back body in normal circumstances. This ratio is consistent with greater demand by physical activity that front body of humans performs. If the pectoralis minors further overpower the serratus anterior muscles and the shoulders depress, entrapment can occur to the nerve and blood supply that course through the front of the neck and under the collarbones. Rounded shoulders and muscular imbalance between the serratus anterior and the pectoralis minor can often result in pain, weakness, numbness, or high blood pressure.

Latissimus dorsi

An important way to engage back body musculature is to lift the arms primarily from the back muscles and much less from the muscles of the chest and anterior shoulders. When lifting the arms, avoid tightening or shrugging the shoulders. Instead, lift the arms from the muscles below the armpits, drawing the scapulae down the center back and toward the pelvis. This action engages the Latissimus dorsi. It draws down the head of the humerus to the raises the arm. A helpful image is the torso as a building crane. The arm being lifted is the extending boom and jib that suspends in the air. From the middle and lower back and its foundation at the pelvis, the latissimus dorsi supplies the power to pull its tendon down, which in turn, lifts the arm up.

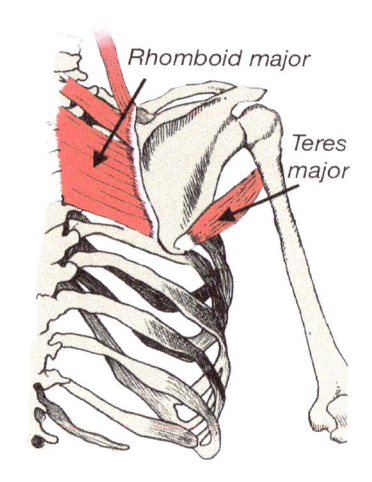

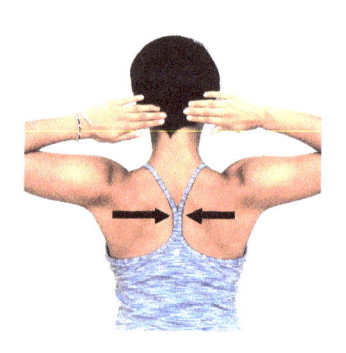

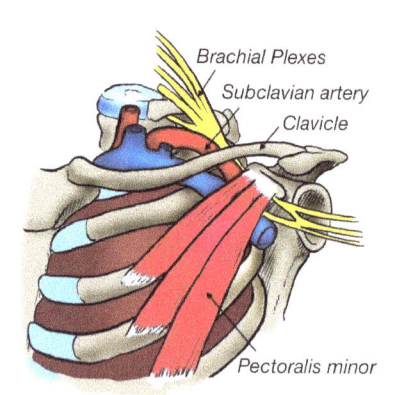

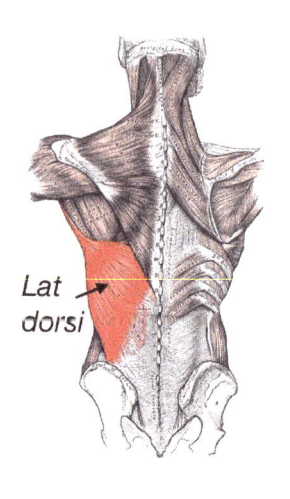

The Rotator Cuff

The *rotator cuff* consists of four short muscles that attach between the scapula and the humerus and not beyond. Their function is to move, yet also stabilize the gleno-humeral joint. The larger muscles of the upper torso assist the rotator cuff muscles. Conversely, the rotator cuff muscles in turn, assist the same large muscles, operating as "keys" that release the large muscle's power. Their inefficiency is highest while at rest and requires assistance to launch their movement.

The first letters of the four rotator cuff muscles spell out the useful mnemonic: "SITS".

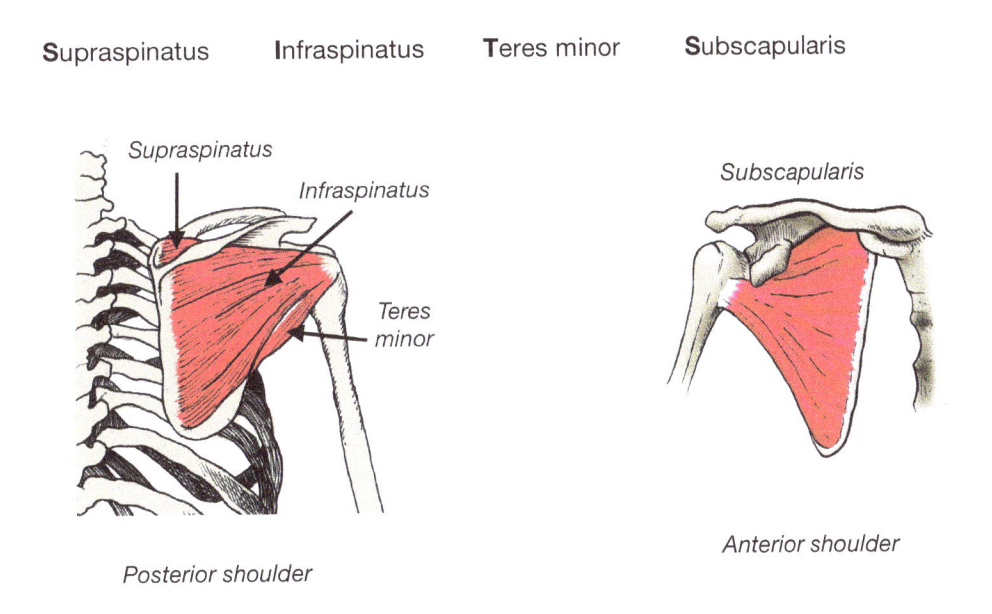

Supraspinatus   **I**nfraspinatus   **T**eres minor   **S**ubscapularis

Posterior shoulder

Anterior shoulder

Actions of the rotator cuff

External rotation: working together, the *infraspinatus* and *teres minor* externally rotate the shoulder. The external rotators receive assistance from the trapezius and rhomboid muscles since the muscles of external rotation is 30% weaker than their opposing internal rotators. When the shoulders shrug, the trapezius muscles lose some of their ability to assist in external rotation. The trapezius and rhomboid also adduct the shoulder, however, they are not considered part of the rotator cuff because they attach to other bones besides the gleno-humeral joint. The infraspinatus and teres minor also stabilize the humerus during abduction.

Internal rotation: the *subscapularis* attaches across the inner surface of the scapula. It stabilizes the shoulder, preventing it from slipping anteriorly. The subscapularis works with the pectoral muscles to internally rotate the humerus, especially when the arm is behind or at the side of the body.

Abduction: the *supraspinatus* spans across the top of the gleno-humeral joint. It initiates abduction when the arm is at rest along the side of the torso. The *deltoid*, the more powerful abductor of the shoulder, is unable to initiate abduction without assistance from the supraspinatus. The supraspinatus' involvement reduces after abduction is initiated but increases again when abduction enters the range of 90-180°.

The supraspinatus also stabilizes the gleno-humeral joint and prevents the head of the humerus from anterior-inferior dislocation, the shoulder's weakest and most unstable direction.

Rotator cuff injury

In 2006, the number of medical patients reporting shoulder and upper arm injuries was approximately 7.5 million. More than 4.1 million of these cases involved the rotator cuff.[7] In 2013, the Journal of Orthopedics reported a study that suggested 22.1% of the general population has had a rotator cuff tear with nearly one-half being asymptomatic.[8]

The presence and the specific location of shoulder pain are diagnostic indicators of rotator cuff injury. Pain from rotator cuff injury is felt at the front of the shoulder. It can also be felt in the neck. Rotator cuff injury will limit shoulder motion, resulting from either an acute muscular tear or degenerative changes to the joint and tendons. Degenerative changes often result from over-use and repetitive injury that cause labral tears, ligament sprain, or loss of blood supply to the rotator cuff's tendons.

The most common rotator cuff injury is a tear of the supraspinatus. The initiating cause is usually poor alignment and insufficient function of the deltoid muscle. If the deltoid muscle does not maintain its alignment centered over the top of the shoulder, the supraspinatus is forced to carry the load normally handled by the deltoid and will become vulnerable to injury. This scenario occurs when the shoulders round and drop forward. Rounded shoulders also compromises and weakens the posterior deltoid muscle, making it unable to stabilize the back of the shoulder. Strengthening the posterior (rear) deltoid can prevent rotator cuff injury and is often a key determinant for successful rotator cuff rehabilitation. *Push ups, Reverse Fly*, and *Bent-over Lateral Raise* are three gym exercises best suited to strengthening the posterior deltoid. In yoga practice, Plank Pose, **Chaturanga** and Reverse Plank can increase posterior deltoid strength.

Chaturanga Dandasana is a valuable asana for rotator cuff rehabilitation, although it is usually not appropriate in the acute phases of injury. To be rehabilitative, Chaturanga requires precise alignment of the shoulders with the elbows not rising beyond to the side body ribs. Without correct alignment, Chaturanga becomes the cause of rotator cuff injury, not its cure!

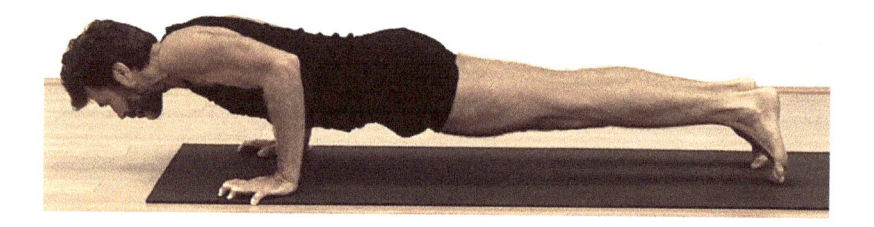

Deep shoulder joint release

An excellent therapy for rotator cuff injuries and tears of the labrum, the cartilage collar that surrounds the gleno-humeral joint. It can help increase shoulder range of motion and resist the development of frozen shoulder (adhesive capsulitis).

1. Place arms in a Headstand Pose position
2. Draw the inner humeral heads posterior
3. Press chest toward the wall; forehead may touch

31 Integrative Alignment of the Shoulders

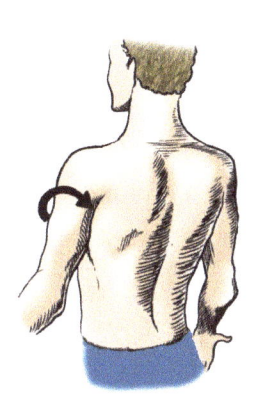

Shoulder alignment follows various step-by-step procedures that fully integrate the shoulders with the chest, upper back, and upper extremities. It establishes a balanced platform to support the head and neck. The shoulders align and its parts integrate in every asana, regardless of the importance the shoulders have in the overall posture.

Shoulder alignment principles would be incomplete if they overlooked the stabilizing role that the upper thoracic spine and rib cage that play as they anchor the shoulder girdle to the rest of the torso. Skillful asana alignment keeps the thoracic cage flexible without sacrificing stability provided for the shoulder girdle.

Presented in this chapter are many instructions for preparation and engagement of shoulder integrative alignment. Some of the principles will be already familiar as they apply to other regions of the body, as well. As will be seen, simply drawing the shoulders back will fall short of effective alignment. Shoulder alignment principles are precise and detailed. It is recommended that you be particularly patient while exploring the numerous groups of instructions that are offered in this chapter, perhaps practicing and integrating only one or two at a time until they all become second nature.

And, not to worry - a simplified summary is provided at the end of the chapter!

Shoulder ligament function

With the extensive ranges of motion afforded the shoulder joints, it is easy to ignore or override their mechanical design. This usually results in strain, compression, or rupturing of its ligaments, tendons, or cartilage. Carefully adhering to the shoulder's design elements reduces the likelihood of injury.

The shoulder ligaments have the same mechanics as do the ligaments of other joints throughout the body. As already discussed, the mechanics of ligament function may seem contradictory to the intended directions of movement, particularly at the initiation of a pose. The movements also engage often at only an isometric or energetic level.

Ligaments Loosen
- Flexion
- Internal rotation
- Adduction

Ligaments Tighten
- Extension
- External rotation
- Abduction

Review: Alignment principles for shoulders/whole body

As fundamental principles presented in earlier chapters for all integrative alignment, the following list reviews the general procedures to be engaged as preparation and during integration and alignment of the shoulders:

1. Establish a stable and aligned foundation
2. Build the foundation of every posture from the ground up, where possible
3. Stabilize the core before moving into asana
4. Draw the bones of the extremities into their joints and toward the midline of the body
5. Correctly engage the ligaments for flexibility - the directions that ligaments pleat and unwrap
6. Stretch muscles from their central bellies, not from the joints where ligaments and tendons attach
7. Once the maximal range of motion is reached, stabilize the posture with strength and stability
8. For stability, unpleat and wrap the joint ligaments by subtly, perhaps isometrically moving in the directions that engage stabilizing actions
9. Contract the muscles from their bellies, drawing to the core of the body
10. Hug the muscles firmly along the shafts of its bones (as if being swaddled with plastic wrap)
11. If the asana is to be taken deeper, repeat procedures above, now from their new, deeper place

Preparation for shoulder alignment

More specific to the shoulders, these instructions prepare the shoulders for alignment:

1. Position the chest on front body and keep the shoulders on back
2. Align the ears over the shoulders
3. Align the center of each shoulder socket and each arm along the *coronal* axis
   If arms are straight, the middle fingers align with the "seam of the pants"
   If the elbows are bent, inner elbows align along with side body ribs
4. Glide the outer edges of the shoulder blades onto upper back and toward spine
5. Hug shoulder blades onto back
6. Draw the lower rib cage back, resisting it from jutting forward

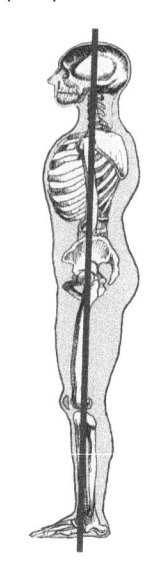

Coronal Axis

Thoracic region preparation for shoulder alignment

Once these preliminary actions are undertaken, the following two steps further prepare the upper torso for integrative alignment:

1. Lengthen both sides of the body equally from the iliac crests to axillae, or *"hips to pits"*.

   - Hollow the armpits (axillae) from underneath and upward while keeping the shoulders square and level; avoid the tendency to shrug
   - Scoop the bottom tip of the breastbone inward to secure the lower rib cage to resist it from jutting forward
   - Side body lengthening flattens the thoracic spine into extension; increasing space in the joints that enables increased mobility

2. Inflate the upper chest (chest expansion)

   - Using the breath, expand the upper quadrants of the lungs at the level just below the collarbones, expanding in all directions, 360° around the chest, below the armpits, and across the upper back
   - Maintain the volume of the expanded upper chest throughout both the inhalation and exhalation phases

Increased chest volume creates an inflated, rigid barrier that prevents the shoulders from collapsing forward.

For students whose upper thoracic curve is flat, breathing across the upper back is a valuable tool to improve their spine and offer stability.

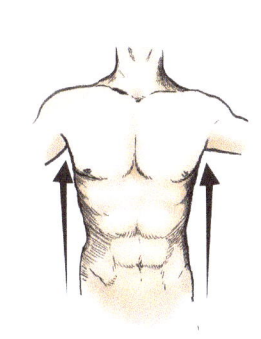

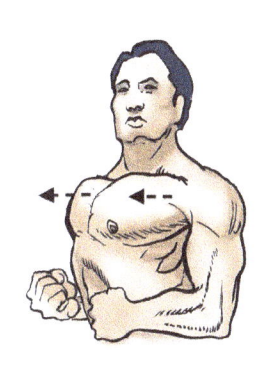

Specific action steps for shoulder alignment

1. The upper chest and shoulder muscles are 30% stronger than the upper back muscles. To reduce the chest muscles' otherwise overpowering effect, visualize engagement of shoulder alignment from the upper back muscles and particularly the *latissimus dorsi*, even as other muscles participate.
2. Slightly lift the shoulders to both initiate movement and release muscular tension (an action of the supraspinatus and deltoid)
3. Draw the inner aspect of the heads of the humerus' posterior into their joints
4. Return the shoulders back to neutral
5. Wrap the shoulder blades onto the back from their outer edges toward the spine. Widen the collarbones to the outer borders of the shoulders. Both of these actions naturally occur together
6. Press the bottom tips of the scapulae forward into the back ribs
7. The chest is propelled forward from action of the scapulae
8. Draw the lower rib cage and lower tip of breastbone (xiphoid process) posterior

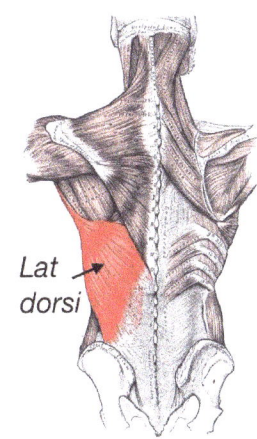

Latissimus dorsi

Details: Action steps for shoulder integrative alignment

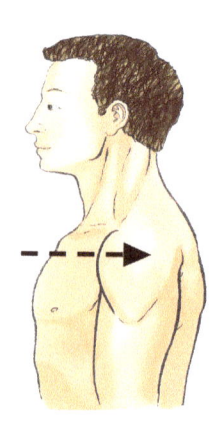

1. The inner head of the humerus draws posterior (a subtle form of flexion) and internally rotates.
   - Both actions cause the ligaments to pleat and unwrap, enabling the shoulder to move freely and safely.
   - The outer aspect of the shoulder moves faster than the inner (axillar) portion. Although it is a small discrepancy, this can initiate external rotation and propel the humeral head forward. To avoid this adverse action, learn to move the head of the humerus specifically and consistently from its inner aspect. Since the inner humeral head forms the armpit, the concept of deepening and drawing the armpit back can accomplish this action.

2. When raising the arm above the shoulder, particularly to the side (abduction), internal rotation should be discontinued before reaching 90° engaging external rotation from that point upward. As explained in Chapter 30, this deviation in its path is necessary to avoid pinching or tearing of the ligaments that attach between the upper greater tuberosity of the humerus and shoulder blade's acromium and coracoid processes.

3. Other than slightly raising the shoulder when initiating arm movement, avoid shrugging, which causes the upper trapezius and levator scapulae muscles to create tension in the neck. Instead, use the musculature of the inner armpit, below the shoulders, to engage the arms. This activates important muscles of the back such as the teres major and the latissimus dorsi.

4. The forearm can rotate the hand 180° independently of the shoulder. However, the position of the hand can orientate the position of the shoulder. Pointing the thumb inward or down causes the palm to face inferior or lateral, which encourages internal rotation and shoulder mobility. Rotating the palm superior or lengthening from the little, 5th finger, influences external rotation.

5. The collarbones lengthen, helping to create space in the joint for mobility. The scapulae move in the opposite direction, gliding medially toward the spine to initiate stabilization of the shoulders. Mechanically, both actions are linked, occurring together at an equal (1:1) rate. This results in a full and broad foundation for the head and neck to rest upon.

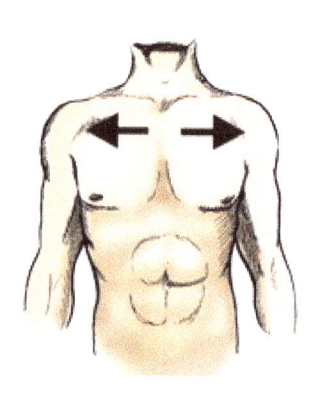

Inner humeral heads draw posterior

Clavicles move lateral to create space for movement

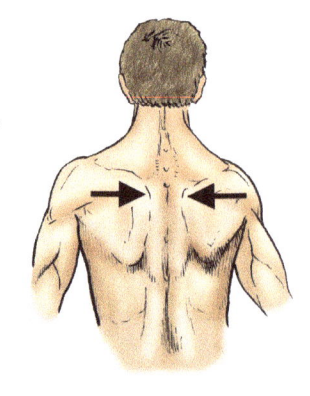

Scapulae move medial to prepare for stability

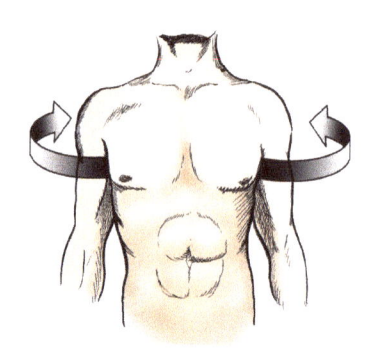

Humeral heads externally rotate once stabilization is required

6. To continue and more substantially stabilize the shoulder, the humeral head externally rotates while remaining deep and aligned within the glen-humeral joint. External rotation is a primary direction of movement to stabilize the shoulder and necessary in the final positions in arms-overhead poses and arm balances. External rotation causes the ligaments to stiffen, which enables joint stability.

   When the humerus externally rotates, the *triceps muscle* (back of upper arm) rotates toward the midline of the body. When the arms are overhead, the triceps move toward the ear. When the arms are at the side of the body, the triceps move toward the side body ribs.

7. Lower tips of the scapulae press forward onto the back. A useful image shared in the Anusara tradition is to press the tips of the shoulder blades forward as if "two hands are cradling the heart from behind". The heart "melts" forward, through the chest, focusing the energy of asana leading from the heart. Mechanically, this action occurs from the lower portion of the serratus anterior, which pulls the lower tips of the shoulder blades forward and onto the posterior rib cage. This process provides the serratus anterior with an important and sufficient mechanical advantage to resist the powerful forward and downward pull of the pectoralis minor muscles that otherwise would round the shoulders forward of the chest.[1]

8. The arms align with central axis, located along the frontal (coronal) plane of the side body, bringing them in the in line between the center of the shoulders and center of the hip sockets. With arms bent: inner elbows align with side body. With arms straight: middle fingers align with hip sockets and seam of pants.

All of these actions combined first mobilize and then stabilize the shoulders in every yoga posture. The actions can be explored and the basic standing pose of **Tadasana,** the seated posture of **Siddhasana**. The qualities of stability and strength can be confirmed in the arm-supported pose of **Chaturanga**.

Cocktail tray arm position: The arm position used to carry a cocktail tray is very effective for correct shoulder integration and alignment. It brings the inner elbow next to the side body, medially rotates the triceps muscles, lengthens the collarbones, and draws the shoulder blades closer to the spine, propels the chest forward - all which wonderfully integrate the shoulders for alignment and stability.

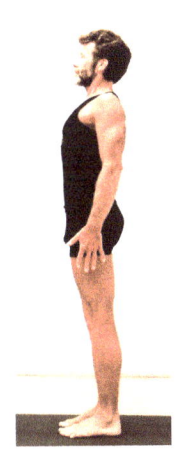

| Tadasana | Siddhasana | Cocktail tray arm |

Examples of asana to demonstrate shoulder alignment

Vasisthasana   Side Plank Pose

In Side Plank Pose, the arms form a "T" position, which support the torso as it lifts up from the hips. The inner heads of the arm bones are drawn back. The clavicles lengthen toward the arms and the shoulder blades glide from their lateral borders toward the spine. Once in the full posture, the humeral heads externally rotate and the tips of the shoulder blades press forward.

It would be more appropriate to not call this pose Side Plank but instead, Side-Facing Downward Dog. From Downward Facing Dog, the transition safely and naturally brings the arms to the "T" position. An exception would be for the yogi who takes a long and deep Downward Dog who should shorten Dog pose before taking Side Plank. Commonly, Side Plank is cued from Plank Pose, which places the lower arm vertical and askew with the upper arm. This compresses the lower shoulder's acromio-clavicular (AC) joint and places substantial force on the shoulder, driving into weakness - anterior and inferior.

Gomukhasana   Cow Face Pose

Students often initiate **Gomukhasana** by rolling their shoulders forward to bring each arm behind the back. To prevent this, inflate the chest to create a barrier that blocks the shoulder from collapsing foreword. Draw the inner heads of arm bones posterior and lengthen the collarbones to create joint space for mobility. The pose's design places the lower arm naturally in internal rotation. Although the upper arm will eventually move into externally rotation, a ligament tightening action, initiate upper arm movement with internal rotation to enable ligament release.

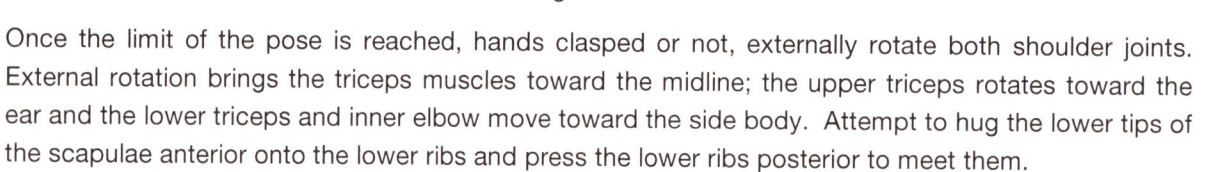

Once the limit of the pose is reached, hands clasped or not, externally rotate both shoulder joints. External rotation brings the triceps muscles toward the midline; the upper triceps rotates toward the ear and the lower triceps and inner elbow move toward the side body. Attempt to hug the lower tips of the scapulae anterior onto the lower ribs and press the lower ribs posterior to meet them.

Tips: When bringing each arm behind the torso, deepen the inner humeral head and lightly rotate the upper chest toward the side being engaged. This prevents the shoulder from rolling forward of the chest. Also, the lower back of hand can press into the opposite back pocket area to re-set the humeral head posterior into its joint before the elbow bends toward the upper spine.

Ardha Matsyendrasana   Seated Spinal Twist

Students often roll their upper back and drop their shoulders forward when attempting to bind the hands or grasp the knee. Instead, instead inflate the upper quadrants of the chest, draw the inner humeral heads back and square and lengthen the collarbones. The elbow over the front bent knee can stabilize the arm when drawing the inner humeral head back. Against this resistance, the chest expands forward.

Bhujangasana assist   Cobra assist

There are many ways to assist the Cobra Pose. A different one that emphasizes lifting into the pose by bringing the lower rib cage posterior is offered in Chapter 15.

1. In this assist, a strap is placed around the student's back, crossing the center of the scapulae. The assistant sits in front of the student and holds the ends of the strap.
2. The assistant uses his feet to firmly support and press back the inner humeral heads while encouraging the student to lengthen wide across her collarbones.
3. The shoulder blades are pressed forward and allow the student to expand her chest forward to complete the pose.
4. The student draws the lower rib cage posterior and engages Uddiyana bandha firmly. To best achieve this action, the heels of the hands isometrically press into the floor, forward and superior.

Chaturanga Dandasana   Four Limbs Staff Pose

Chaturanga Dandasana may seem challenging for yoga students who assume that great arm strength is required to suspend the torso in a straight position. Although strength does play a role, the pose is attainable by increasing the distribution of body weight toward the head, balanced over the fulcrum created by the forearms and hands. Shoulder integration is required to create a strong pose and not to injure the shoulders. The forearms are in a vertical position with the inner elbows hugging snugly to the side body ribs. The humeral heads draw deep into the shoulder sockets. The collarbones lengthen and the shoulder blades hug firmly onto the back body. Additional alignment instructions for the arms will be presented in Chapter 32.

A common Chaturanga misalignment is for the elbows to lose contact with the side body, spread apart or point upward. This forces the shoulders to roll forward, toward the floor. These splayed or posterior elbow positions push the humeral heads in an anterior-inferior direction where the shoulders are most unstable. They also force the student to use their chest muscles instead of their upper back muscles. If Chaturanga is practiced repetitively in this incorrect fashion, the shoulder's tendons and ligaments become overstretched and weakened, predisposing the shoulder joints to dislocation.

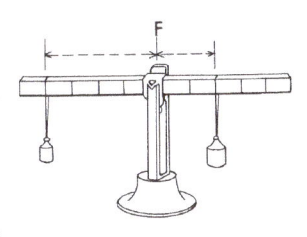

Inner elbows hug onto side body rib cage   *Shown as incorrect – elbows pointing upward*

Avoid shrugging the shoulders

As listed in 'Specific action steps for shoulder alignment", step one instructs for the shoulders to lightly lift to both initiate movement and release muscular tension (an action of the supraspinatus and deltoid muscles). Those are the two muscles that initiate gleno-humeral movement. The challenge is to not engage the levator scapulae and upper trapezius muscles. These muscles will cause the shoulders to shrug and do not contribute to shoulder mobility in any significant way.

Shoulder walking (Bridge Pose variation)

Sometimes referred to as a *tuck and roll*, this action is used to set the foundation for all shoulder balancing asana. It assists the yogi in integrating the shoulders onto the back. It is also an excellent therapeutic for limited upper thoracic mobility and tight musculature. It is a valuable therapy to reverse the habit of forward rounded shoulders.

1. Lie on back with knees softly bent
2. Bend arms at the elbows with the hands pointing toward the ceiling (robot arms)
3. Begin by walking each shoulder under the upper back toward the opposite side
4. Walk the skull toward the shoulders without compressing the neck
5. Repeat the walking sequence- one shoulder, the other shoulder, head. One shoulder, the other shoulder, head
6. Each successive series of movements brings the top of the shoulder, which is capped by the middle deltoid muscle, under the body and closer into direct contact with mat
7. Allow the pelvis to lift and the legs and feet to walk in as far as needed to adjust to the action of the shoulders
8. Once in full position, place feet flat and shins vertical in the Setu Bandha position.

Simplified summary of shoulder integration and alignment

An exhaustive amount of detail has been presented for aligning the shoulders. All the specific steps are presented in this summary, hopefully making it easier to memorize. Learn to engage each instruction in this list and you will very effectively align and integrate the function of the shoulders and upper back.

Step One:
1. Lengthen the sides of the body equally
2. Inflate the upper quadrants of the lungs, 360° around the upper torso

Step Two:
3. Draw the inner head of the arm bones back
4. Lengthen the collarbones outwardly; draw the outer edges of the shoulder blades medially toward the spine
5. Press bottom tips of the shoulder blades forward and onto the back ribs
6. Draw lower rib cage posterior, scooping breastbone inward (Uddiyana bandha)

Step Three:
7. With bent arms, inner elbows align with side body along the coronal axis
8. With straight arms, middle fingers align with hip sockets on coronal axis

32 The Upper Extremities

Out on a limb

With many of our daily activities, such as eating, the extremities and the core move instinctively, such as when the hand brings food to the mouth. And, because these basic actions are instinctive and mostly unconscious, poor mechanical habits can easily develop.

The upper extremities are considered an extension of the shoulders. Whether performing an arm-balancing pose or simply standing in Tadasana, the arms maintain their structural relationship with the shoulders and upper torso. Attempting to lift the arms overhead before integrating the shoulders and upper back will quickly prove challenging. For example, if the shoulders are forward and the arms are bent at the elbows, lifting them overhead will be increasingly difficult the deeper they flex. This principle is apparent and is the limiting factor in poses such as **Urdhva Dhanurasana** or **Chakrāsana**, the Wheel pose. The shoulders align and fully integrate with the upper torso before the arms are engaged. It is common for the cause of chronic strain and injury to the elbows and wrists to find their origins with lack of shoulder integration. The Latissimus dorsi plays a valuable role in reducing elbow strain and should be engaged with the triceps to extend the arm, especially when the arms are overhead.

There are numerous muscles that attach from the shoulders to the arms. The two muscles most associated with the arms are the biceps and the triceps. The *biceps brachii* (from the Latin *biceps,* meaning *two heads,* and *brachii*, meaning *arm*) flexes the elbow. The *triceps brachii* is an elbow extensor. The biceps has two heads (muscle bundles), and the triceps has three heads.

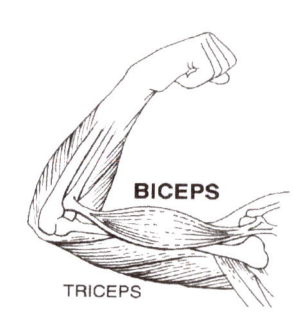

Pre-requisites for upper extremity principles

The following is full review of the general principles of body alignment, some already introduced in Chapter 5, that are engaged before the nuances of upper extremity alignment are undertaken:

1. Move from the core, using the shortest levers

When raising the arm, use muscles that are as close to the shoulder as possible. This action enlists the strong, upper-torso core muscles to raise the arms. When extending the arms from the side body, yogis often unconsciously over use and overstrain the forearm muscles, even though there is no direct mechanical way to lift the arms using these outer extremity muscles.

2. Latissimus dorsi   Use core upper back muscles to lift arms

Because life's activities take place in front of us, the tendency is to lift the arms using the muscles of the chest and anterior shoulders. Unfortunately, this develops and strengthens the front body musculature at the expense of a relative weakening of the back muscles. A primary goal in yoga practice and its effect on body dynamics is to increase upper back strength and flexibility to offset the inherent overpowering imbalance that favors the front.

Lift the arms from upper back muscles, primarily the *latissimus dorsi*, accompanied by the teres major and the posterior deltoid. The latissimus dorsi attaches to the humerus from inside the armpit and expands down the back and into the lumbo-pelvic fascia. Using eccentric contraction, a less obvious action, the arm is pulled down in order to lift up. This method produces a surprising freedom of motion in lifting the arm as opposed to contraction from the chest musculature.

Eccentric contraction of the latissimus dorsi is similar to the action of a building crane. The human arm simulates the crane's extending steel jib. The power to lift up counter-intuitively comes from pulling downward on cables from the base. In the human body, the latissimus muscle belly pulls its tendons downward to lift the arm up.

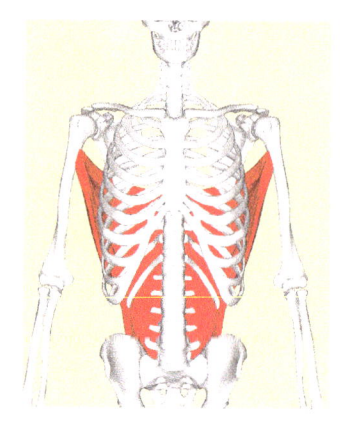

Latissimus dorsi

3. Bones draw to the midline; muscles extend out

This concept ensures movement initiates from the core while it prevents strain on the muscle tendons. When yoga teachers cue students to "reach the arms out," they are incorrectly instructing students to move from their extremities, not the core. Instead, draw the arm bones into the shoulder joint and toward the core; then the muscles extend outward.

When extending the arms in Warrior Two Pose, students will often forcefully contract their forearm muscles rather than engaging the muscles closer to the core. Preferably, use the middle deltoid, supraspinatus, and latissimus dorsi muscles to lift and support the arms. In order to stabilize the extremities, draw the arm bones into the shoulders. Then extend the arm muscles outward.

4. Triceps muscle rotates toward the midline

When the arms are along the side of the torso, as in **Chaturanga Dandasana**, roll the back of each arm (triceps muscle) medially, toward the side body ribs. When the arms are overhead, as in **Urdhva Hastasana**, rotate the triceps muscles medially, which is forward and toward the ears. Both actions employ external rotation of the shoulder and elbow, thereby providing stability to the joints.

5. Use the triceps muscle to extend the elbow

It is common for the elbows to inadvertently bend in poses where the arms are held overhead. Should it be difficult for the arms to straighten, first align and integrate the shoulders onto the back; then engage the triceps muscles. "Draw the outer elbow toward the armpit" is a helpful verbal cue for triceps extension.[1]

Chaturanga Dandasana

Straightening the arms may require more effort than is otherwise expected. This is because the triceps muscle is most efficient and powerful when the shoulder is flexed and the elbow is bent to 20-30°. This position provides the power and readiness that is needed to swing an axe. Even at rest, the triceps retains slight contraction. Bringing the arms overhead lengthens the triceps. Because this must override the elbow's position of efficiency it requires additional effort.

6. Samasthiti Equal tension and balance

Maintain a subtle, energetic balance in length and tension between the entire *radial* (thumb) and *ulnar* (small finger) surfaces of the arm. When the arms are weight bearing, as in Downward Facing Dog, the floor provides a firm foundation from where equal and balanced pressure can be established. When the arms are not weight bearing, as in **Urdhva Hastasana**, Samasthiti must be brought to the pose though muscular action. When the arms find Samasthiti, the entire body integrates and aligns.

Try This: Bring the arms overhead with the palms facing the midline. Note whether the muscular and energetic tension along the thumb-side and small finger-side of each arm or is different.

1. Lengthening through the ulnar (small finger) side of the arms opens and lengthens the front of the body.
2. Lengthening through the radial (thumb) side of the arms rounds the shoulders and upper back and causes the chest to collapse.

Most yoga students will need to lengthen through the ulnar side of their arms more than the radial side. Lengthening through the ulnar side of the arms rotates the elbows and triceps bellies toward the midline. The chest expands, the shoulders hug onto the back, and greater Samasthiti is created in the upper body.

The back of the arm, the triceps belly, always rotates toward the side body, ears or nose. If the back of the palms face forward, it is indicative that the shoulders are rounded forward and Samasthiti is lost.

Ta Da!
Rembering how trapeze artists complete a routine is how the arms complete a pose

The elbow

In its most basic design, the elbow is a simple hinge joint, consisting of three bones: the *humerus*, the upper arm bone, and the bones of the forearm, the *radius* and *ulna*. Because elbow mobility is integrated with the overall movement of the arm and shoulder, the joint may appear to be capable of moving through multiple directions. However, extensive crisscrossed ligaments that bind the radius firmly to the ulna prevent movement of the elbow joint in any other directions than *flexion* and *extension*. Regardless of this limitation, its function is vital to the upper extremity and to the specific action of moving the hand to the mouth.

Epicondylitis is inflammation and/or tearing of the forearm tendon at the elbow. Lateral epicondylitis is called tennis elbow and the medial is golf or baseball elbow. It is important that the latissimus dorsi participates in all arm movement and engaged whenever applying asana as therapy.

Back of palms facing forward rounds shoulders

The ulna alone forms most of the elbow's hinge joint with the humerus, making it the primary forearm bone involved with elbow motion. The radius only minimally participates in elbow mobility and instead focuses more distal, controlling rotation of the forearm and wrist. The radius' control of the forearm and wrist is independent of the arm's position or the elbow's degree of flexion or extension. The forearms can rotate the palms downward, as often performed in sitting, meditation postures, without offsetting shoulder alignment.

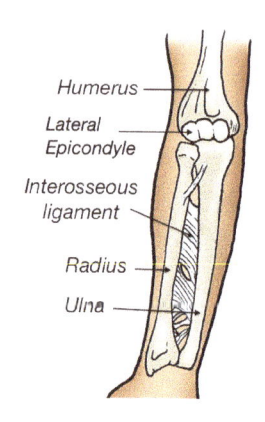

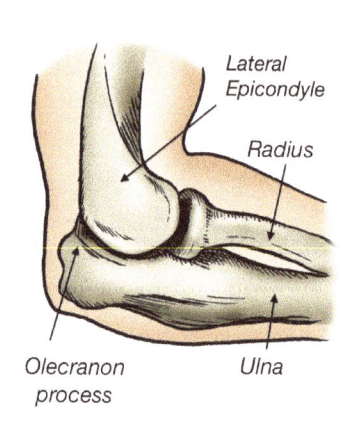

Elbow flexion

Elbow flexion is measured with the arm straight and unbent with that position being the zero degree point. Flexion bends the forearm towards the shoulder and reaches its full range between 140° to 160°.

Elbow extension

Technically, there is no elbow extension. The elbow is capable only of flexion. There is no normal range of movement that extends the elbow beyond the position of the arm being straight at zero degrees. Movement beyond straight is not extension but is *hyperextension*.

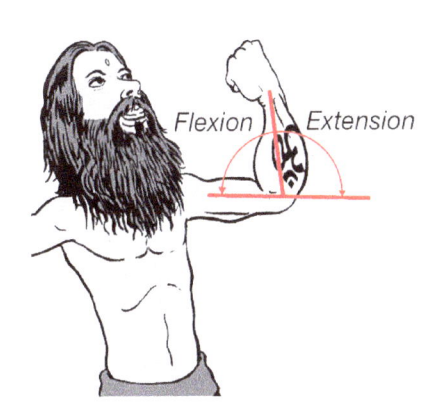

From a flexed position, the elbow can straighten through the direction of extension and technically, is assigned the term "relative" extension. The same considerations are ascribed to the knee joint.

To prevent elbow hyperextension, a large boney hook on the posterior ulna called the *olecranon process* locks into the posterior humerus as the elbow straightens. It acts as a doorstop to limit extension.

Hyperextension of the elbow

Many flexible yogis hyperextend their elbows, habitually moving beyond the straight-arm position with no physical feedback indicator that they have reached neutral. This usually results from overly flexible ligaments that allow the joint to exceed their mechanically desirable range. If hyperextension occurs regularly, the ligaments weaken and the joint becomes unstable. A common result is overstretch and strain with leads to muscle tendonitis. Over many years of habitual hyperextension, the likelihood of degenerative damage to the elbow joint is possible.

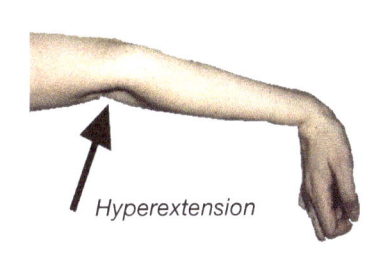

Use the following steps to remedy hyperextension in arm balancing poses:

1. Micro-bend the elbows
2. Squeeze the elbow joints from side-to-side (radius-to-ulna) using their accessory muscles
3. Turn the hands slightly inward
4. Counter-rotate the arms (instructions to follow) to normalize and stabilize the elbow position

The "eyes" of the elbows

The inner elbow creases are sometimes called the *eyes* of the elbows. In Downward Facing Dog and other arm balancing poses, the eyes face inward, pointing toward the clock positions of ten to two o'clock.

"Squeezing a beach ball" between both elbows is a useful visualization that helps to rotate the eyes correctly into position.

As an exception, students who hyperextend their elbows should face the eyes of the elbows more directly toward each other and look straight across - "eye-to-eye". The greater the degree of hyperextension, the more direct the elbow creases should face each other.

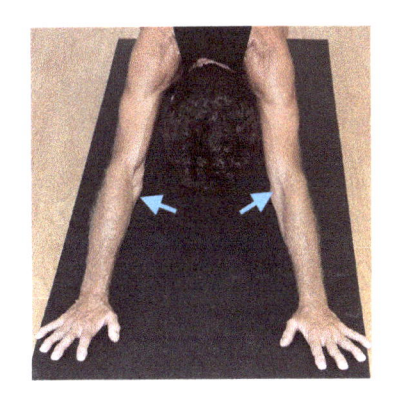

The forearm

The ulna is the primary forearm bone for elbow flexion/extension while the radius contributes only a minor role. The radius, however, is the primary rotator of the forearm and wrist. The ulna does not participate in wrist rotation, but remains stationary as the radius arcs around it. The forearm *pronates* (palm facing down) and *supinates* (palm facing up).

The forearm's full range of rotation is 180°. Although the forearm may appear to have a far greater range, perhaps nearly 360°, the additional range comes from movement of the shoulder and not from the forearm. In binding asana, following this anatomical concept allows the shoulders to remain aligned while the forearms rotate independently to clasp.

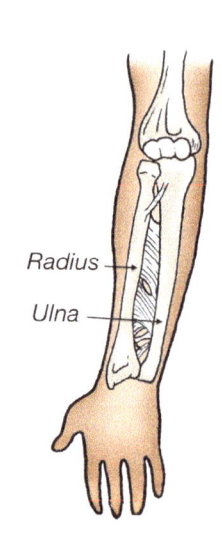

Anatomical terminology

Because the arms adopt various orientations in asana, using anatomically-based terminology helps to eliminate confusion in establishing their aligned position. The palms' facing forward is the universally accepted, anatomically neutral posture for the arms. The *ventral* surface of the body is the anterior or abdominal side; this being the inner palm side of the wrist and forearm. The *dorsal* surface of the body is the posterior of the body (think dorsal fin) and is the term assigned to the back of the hands.

Upper extremity alignment

Similar to the legs, there are also specific alignment principles for the arms. Aligned joints allow them to maximize flexibility and stability and provide a system of therapeutics and rehabilitation for injuries. Individuals may have different ratios of lengths of their upper and lower arm bones, however, these length differences do not affect the rationale for basic upper extremity alignment.

Placing the hands

In Downward Facing Dog, the hands are placed in a specific order to establish the maximum stability. Press the pads of the fingers to the mat, evenly across each nailbed. The metacarpal arches (knuckles) press down while the inter-phalangeal joints remain lifted. The palms remain vaulted as the heels of the palms press down. The wrists lift up from the small arches between the heels of the palms, slightly producing wrist palmar flexion. This lifting action continues up and through the forearms. The energy of the pose runs up through the middle fingers and the underside of the hands, wrists, and forearms, drawing up toward the inner humeral head and into the inner armpit. The energy of the pose runs through the middle fingers and the underside of the hands, wrists, and forearms and draws up toward the inner humeral head and into the inner armpit. Piano students and computer geeks already know the value of these instructions for the prevention of tendonitis and carpal tunnel injury!

The subtleties of wrist and hand positioning are good tools and indicators for developing overall upper extremity and shoulder alignment. Additional details for hand and wrist placement are explored in Chapter 33.

Establishing a stable foundation in arm-balancing postures by engaging precise hand alignment helps prevent forearm tendon strain and wrist injuries. Correct hand and wrist alignment also integrates the shoulders onto the back and helps to shift muscular engagement to the upper back musculature.

Middle fingers in line with the seam of pants

The middle fingers align with the body's lateral central axis. This is the line that courses through the center of the ears and shoulders and vertically over the center of the hip sockets and anklebones. It will usually correspond to the outer seams on a pair of pants. Checking that the middle fingers can tap against this line is a good indicator of alignment of both the shoulders and extremities.

If, in a student's "resting posture", the back (dorsum) of the hands face forward, it typically indicates that the shoulders are incorrectly rolled forward. Although the hands can rotate 180° from the forearms without affecting shoulder alignment, the unconscious, negative habit of moving and orientating from the hands easily take control in the pose.

Tension through the arms, as measured between the thumb and small finger should remain equal. When the thumb side of the arm (radial side) lengthens, the shoulders roll forward and the front body musculature becomes shorten and tight. Lengthening the ulna side of the arm and hand increases muscles energy on the upper back body. The arms overhead in **Urdhva Hastasana** (Upward Hands Pose) is a good pose for observing these differences.

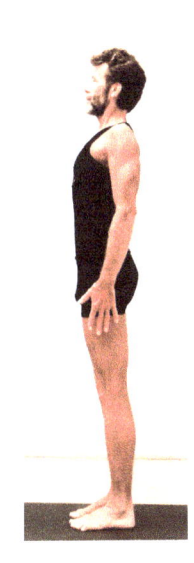

Elbows aligns with the side body rib cage

Leg alignment places the centers of the hips vertically over the ankles, allowing the knees to "find their place". The same holds true for the arms and the elbows. The elbows are not forced into a straight line between the shoulders and hands, but instead, come into their natural position after the center of the wrists align vertically with the shoulders. Usually, the elbows find their position posterior of the central axis.

In postures where the arms are bent, the "end" of the extremity that aligns with the central axis is no longer the hand but becomes the elbow. Whether the hands are resting on the thighs in **Siddhasana** (Seated Pose) or supporting the torso in **Chaturanga Dandasana**, the elbows align on the central axis between the shoulders and hips and directly in contact with the side-body rib cage. As presented in a previous chapter, a finishing step of shoulder *integrative alignment* is for the elbows to align with the side body rib cage. Aligning the elbows directly with the side body ribs is critical to effectively and safely perform Chaturanga Dandasana. This elbow position accesses strength from the upper back's core musculature instead of relying on the less powerful arm muscles. In Chaturanga, the forearms are vertical and form 90° angles at the elbows. If the elbows divert away from the side body, either posterior toward the ceiling or wide of the torso, the integrity of shoulder alignment is lost and strain is placed on the anterior gleno-humeral joint, the weakest and most susceptible point for shoulder injury.

Chaturanga Dandasana

Shown as incorrect

Arm counter-rotation provides stability: Towel Twisting

For increased support and stabilization across the elbows, rotate the forearm radially (in the direction of the thumb) while externally rotating the upper arm (triceps toward the midline). The effect is a "towel-twisting" action that locks the elbow without hyperextension and protects the joints and soft tissues from injury. The musculature of the entire arm is more efficient and powerful when this action is engaged. This is also a good technique to reduce the tendency of chronic hyperextension.

How wide apart are the arms in arm-balancing poses?

In Headstand, Downward Facing Dog Pose, and all other two-handed balancing poses, the center of the wrists align vertically with the outer edge of the shoulders. Also, keep the forearms and wrists energetically lifted away from the mat, dorsally toward the back body.

Carrying Angle

Back in medieval times, the peasants who carried the largest buckets of water or biggest bundles of wood were more highly desired, or so the mythology goes. The further away the forearm deviated from the side of the body, the bigger the bucket that could be carried; hence the name, *carrying angle*. A larger angle is more common with women than men.

Anatomically, angle is determined by the direction of a small impression located at the hinge of the elbow called the *trochlear groove*. A larger angle brings the hands and forearms further away from the body. Larger carrying angles presents challenges in arm balancing poses.

Carrying angle

Carrying on with the carrying angle

When the hands are placed on the mat, an excessive carrying angle creates misalignments throughout the upper extremities. An excessive carrying angle causes the hands to turn inward, the elbows point out, and cause the shoulders roll forward.

- To compensate for a large carrying angle, place the hands on the mat wider than usual. Turn the hands outward, toward the small finger side (ulnar deviation).

- Elbow hyperextension often accompanies a large carrying angle. If this is the case, the position of the hands must compromise. Place the hands wider apart to compensate for the carrying angle. Then, modify for hyperextension by pointing the hands slightly inward and have the elbow creases face directly toward each other.

In the next chapter, the anatomy of the wrist and the hand and their alignment will be explored. Their alignment continues to follow the seamless flow and interdependence of movement that occurs through the rest of the upper extremities and shoulders.

33 The Wrists and Hands

Phalanges
Metacarpal bones
Carpal bones
Ulna
Radius

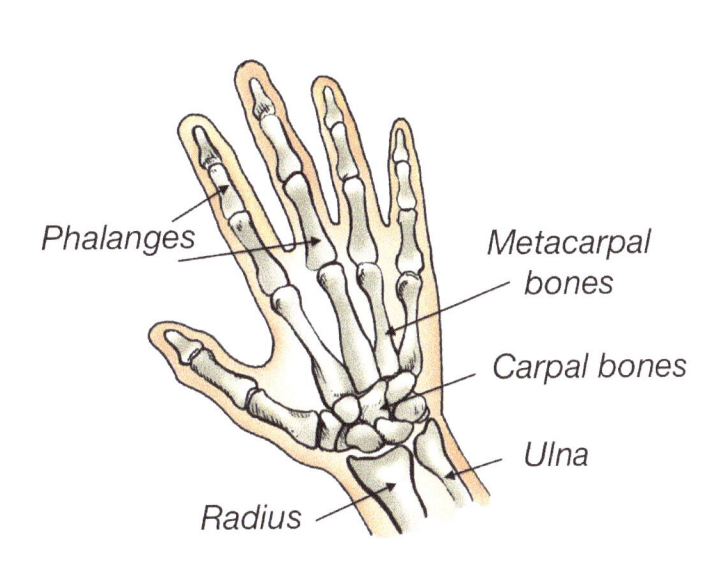

The hands are the extraordinary end users of a complex coordination of movements within the upper extremities, all needed to perform our most basic daily tasks. The ability of the hands to hold, seize, and grasp is called *prehension*. The hands demonstrate a level of sophistication that single-handedly, has elevated the human experience far above our closest animal brethren. Not only is the hand a masterful, mechanical apparatus but also, through touch and feel, the hand provides vast amounts of sensory information to the brain and nervous system.[1]

The anatomy of the wrists and hands is comparable to that of their counterparts of the lower extremities, the ankles and feet. Although there are clear differences between the two regions, alignment strategies are similar. This chapter reviews the wrists and hands together because their functions are interconnected and unified.

The bones that form the wrist are called *carpal bones*. The bones that comprise the region of the palm are called *metacarpals*. They correspond in the feet to tarsal bones and metatarsals, respectively. The bones of both the fingers and toes are called *phalanges*. In the hand, the first phalange, the thumb, is positioned on the heel of the hand instead of in line with the other phalanges, as is found on the feet.

The carpal bones of the wrist are a tightly-fitted group of eight, irregularly-shaped bones that arrange in two distinct sequences: a *distal row*, consisting of bones close to the hand that articulate with the metacarpals; and a *proximal row*, bones that adjoin the radius and ulna.

Mechanically, the two rows display subtly different actions. The distal row is more engaged in wrist extension while the proximal row's role is more pronounced in flexion.

- Distal row: Extension
- Proximal row: Flexion

This information may seem somewhat obscure and unnecessary but it can become a useful tool when trying to pinpoint the cause of wrist pain and devise a yoga therapy plan to resolve it.

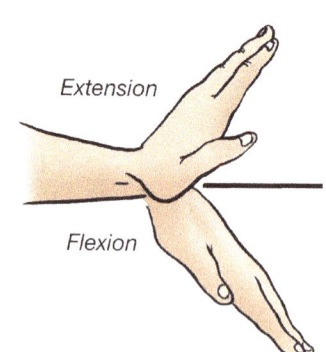

Wrist and hand relationship

The central line of the hand passes through the phalanges of the third finger, continuing through the third metacarpal. The central line of the hand meets the wrist crease perpendicularly and forms an inverted "T". Because injuries of the fingers are common and can cause them to be bent, it is best to observe the central line from the phalangeal-metatarsal joint, or more simply, the third knuckle. This subtle alignment between the hand and wrist helps to form the arches of the center palm and heels of the hands.

In arm-balancing asana, such as Downward Facing Dog, wrist and hand alignment bolsters weight bearing and avoids muscle strain, tendonitis, or other injuries to the wrist and rest of the upper extremities.

When setting the foundation in arm-balancing poses, the first priority is to align both wrist creases parallel to the front of the yoga mat. The central lines of the hands may not remain parallel and they should not be forced to conform. As discussed in the last chapter, hand position may need modification in cases of elbow hyperextension or increased carrying angle.

The muscles responsible for the fine motor movements of the wrists and hands originate in the forearms and pass over the wrists. The tendons are held in place by the flexor retinaculum, a thick, fibrous ligamentous band that crosses over the carpal bones. If the upper arms are not first aligned and integrated with the shoulders, the forearm muscles are often forced to provide stabilization duty that they are not designed to handle. This can pull the tendons from the wrists and cause dysfunction.

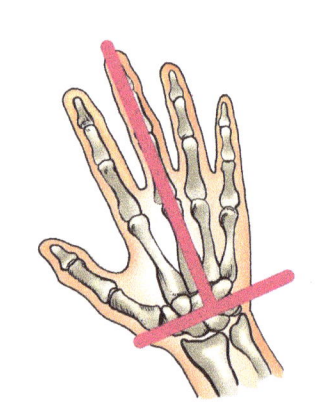

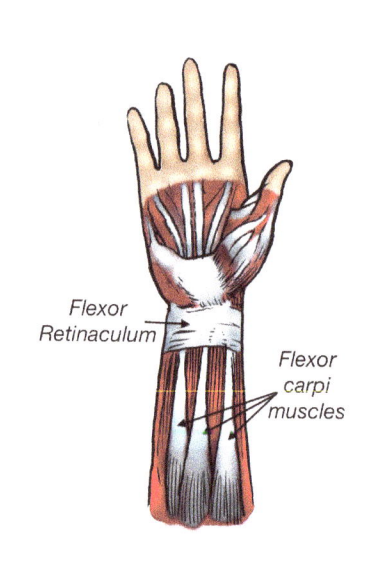

THIRTY-THREE: THE WRISTS AND HANDS | 243

> **Anatomical terminology for the hands**
>
> - Ventral or palmar: refers to the palm of the hand
> - Dorsal: refers to the back of the hand
> - Supination: palm turning upward
> - Pronation: palm turning downward

Arch of the heel of the palm

Comparing the musculature and flesh of the inner and outer heels of the hand, it is easily observed that the heel of the thumb (*thenar*) are more pronounced than the heel on the small finger side (*hypothenar*).

The space formed between the two heels is referred to as the *arch of the heel of the palm*. This arch can best be viewed by placing the hands in an upside down Namaste position, the *Anjali Mudra*. When all four heels come together, the space created at the center of the heels is approximately the size of a kidney bean.

The arches of the heels of the palms are maintained and press evenly into the floor with arm weight bearing postures. Since the hypothenar heel is smaller and less developed than the thenar heel, increasing the pressure into the outer heel may be necessary to create balance across the palm. Pressure into the outer heel of the palm will also assist in stabilizing the upper extremities and shoulders by activating external rotation and stabilizing the ligaments.

The knuckles form the *transverse*, or *metacarpo-phalangeal* arch. It supports the hand, toning and vaulting the palmar fascia and muscles in a fashion similar to the foot's transverse arch. To align, spread the arch wide (frontal plane) and press evenly into the ventral aspect of each knuckle.

A handprint exerting correct pressure resembles an upside down horseshoe through the palm and heels of the hand.

Four-step hand placement for weight bearing

As with the feet, there is an ideal sequence for pressing the hands into the floor in order to best engage the arches:

1. Index finger knuckle (metacarpophalangeal joint)
2. Heel of thumb (thenar heel)
3. Outer knuckle (metacarpophalangeal joint)
4. Outer heel of the hand (hypothenar heel)

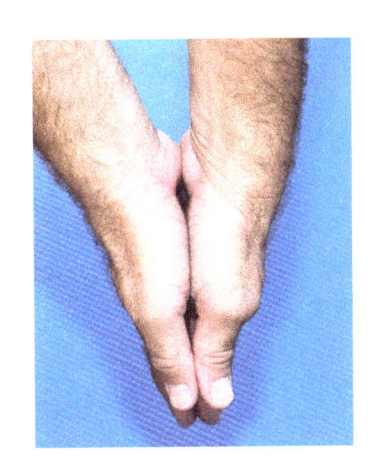

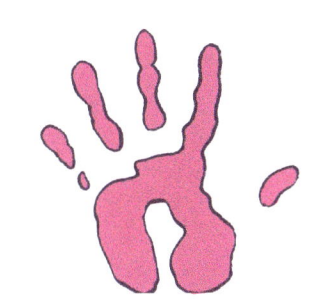

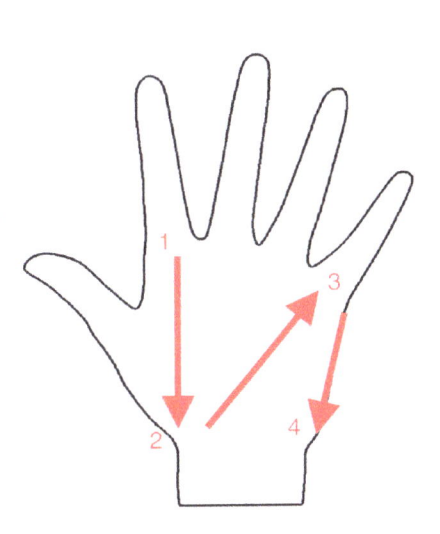

Additional refinements in hand placement

Precise hand placement is important for postural stability. These specific steps become important tools when applying yoga asana for rehabilitation or prevention of hand and wrist injuries.

- Fingers do not spread apart, but instead, extend straight from each respective knuckle
- Each outer finger pad (thumb excluded) presses evenly, making the nail bed of each finger flat and level from side-to-side
- Metacarpal arches (knuckles) press down
- The joints within each finger, the *inter-phalangeal* joints, lift to make fingers appear claw-like
- Palms remain vaulted as the heels of the palms press down
- Follow the four-part order of hand placement for the previous three steps
- Wrists lift up from the heel of the palm arches, slightly producing wrist palmar flexion
- Lifting action and energy continues up and through the forearms and humeri, reaching the inner armpit to stabilize with the shoulder blade

Keep the foundation stable

Once the foundation for an arm-balancing asana is established in the hands, they remain stable, well grounded, and unaltered. The palms remain vaulted and muscularly engaged. The hands supply their weight-bearing support primarily in the heels and transverse arches. The support that the fingers offer is as "outriggers" that apply subtle pressure through the finger pads to re-establish a stable foundation each time that the body shifts its position slightly.

Spider-fingers

The wrist position that produces maximum strength and efficiency is 40° extension and 15° ulna deviation (laterally flexed toward the ulna). This configuration is often referred to as *spider-fingers*. With the wrist in this position, the palmar arches are well formed, the fingers claw-like, and body weight is distributed evenly across each finger pad. For most types of injuries to the wrist and hands, *spider-fingers* is an excellent therapeutic position for the wrist.

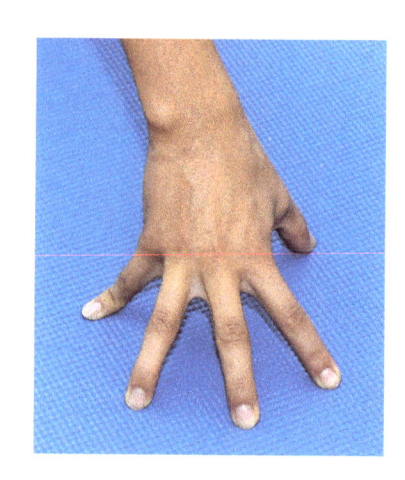

Flexion of the wrist

Palmar flexion, the "limp wrist" position, cannot support significant weight bearing but is necessary in gripping and clasping actions. Hand strength will diminish as the degree of flexion increases.

Pianists and computer keyboard workers recognize this vaulting of the wrist as necessary for performing well and avoiding injury in their crafts. Deep palmar flexion is a beneficial therapy to offset the strain that can develop from overuse of wrist extension.

Wrist extension – less than you think

Wrist extension, or *dorsi-flexion*, is the bending back of the hand. In arm-balancing postures such as Handstand or **Urdhva Dhanurasana** (Upward Bow Pose), it may appear that the wrists are able to extend 90° or greater. Anatomically, the wrist's actual range of motion is closer to 85°, both for flexion and extension. The range is further reduced when the wrist is either in pronation or supination.[2] Despite the apparent ease with which some yoga students will hyperextend their wrists beyond anatomical limits, it is not safe, especially if repetitive or habitual.

Wrist hyperextension causes overstretching and weakening of the wrist ligaments. A potential result of ligament weakness is carpal bone instability and a condition known as *Carpal Tunnel Syndrome*. Additionally, the tendons of the forearm muscles, which are also held firmly to the wrist by these same ligaments, can lift away from the wrist, causing increased strain and tendonitis. Wrist hyperextension can cause the arches of the palms to collapse and strength of the hand's intrinsic muscles to reduce.

Wrist hyperextension is very easy to exploit during yoga practice. Students unaware of the anatomical design of the wrists may purposely force the wrists to extend a full right angle (90°) or beyond throughout their practice. For flexible students, hyperextension is easy and therefore may feel natural, making these students unaware of the insidious damage they are doing to their wrists.

Asana practice should avoid exploiting the loose structural design of the wrists, which is tantamount to Asteya (non-stealing). If one carefully observes even the most advanced arm-balancing poses, it is clear that it is not necessary for the wrists to reach a full right angle (90°). Aligning the shoulders onto the back and extending the upper thoracic spine is what provides the extra degrees needed for the wrists to establish the foundation of the pose without exploiting wrist extension.

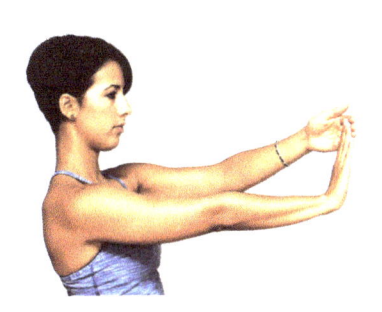

85° dorsi-flexion

Urdhva Dhanurasana (Upward Bow/Wheel Pose) can place substantial pressure on the wrists. It especially challenges yoga students who have highly mobile wrists to resist the tendency to extend them beyond the normal limits. Often, students complain of wrist pain immediately after performing this pose. Although Wheel Pose requires the hands to rest firmly on the floor, the wrists can still maintain an angle of extension that is closer to 85° if the shoulder integrative alignment principles are fully engaged before full weight bearing is placed on the hands and wrist.

Thoracic extension, anterior chest expansion, and shoulder integrative alignment, especially pressing the lower tips of the scapulae anterior, are essential actions that will reduce wrist pressure.

246 | YOGA ALIGNMENT PRINCIPLES AND PRACTICE

Urdhva Dhanurasana

Eka Pada Urdhva Dhanurasana

Wrist wrap

A fibrous band of connective tissue wraps around the wrist. It is called the *transverse carpal ligament*, also referred to as the *flexor retinaculum*. On the palmar side, the band overlaps a major nerve supply to the hand called the *median nerve*. The band also anchors the flexor tendons of the ventral forearm compactly to the forearm bones and wrist. The transverse carpal ligament tightly wraps around the carpal bones, helping to prevent them from spreading apart.

The muscles that provide gross hand movements originate in the forearm. They are either long muscles originating near the elbow or short muscles that arise mid-forearm and closer to the wrist. The muscles between the wrist and finger bones and between the fingers are called interosseous muscles. They originate distal to the forearm and provide the more finely tuned, detailed movements of the hands.

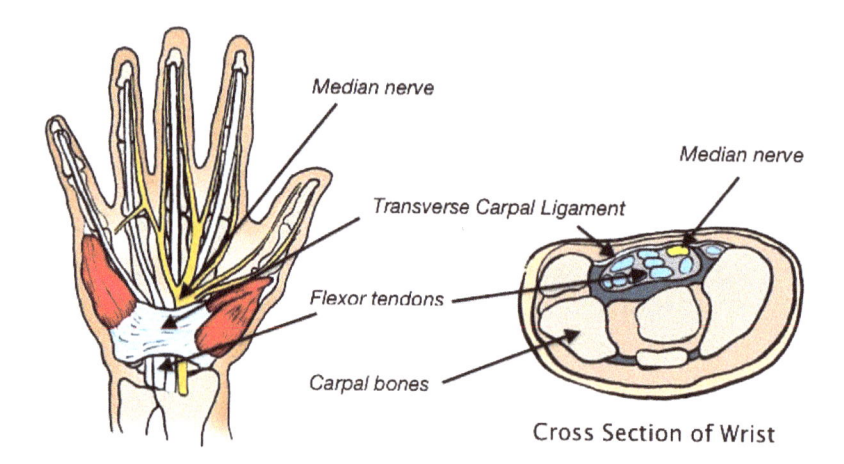

Cross Section of Wrist

Carpal Tunnel Syndrome

A frequent injury to the wrist is *Carpal Tunnel Syndrome* (CTS). Reported occurrences range between 5% in the general population to nearly 50% in occupations that require repetitive movements, firm grasping, or exposure to constant vibration. [3,4] Yoga practice can aggravate an existing CTS injury. When the upper extremities or shoulders are poorly aligned, yoga asana can also become the cause.

Carpal Tunnel Syndrome often occurs when the transverse carpal ligament becomes traumatized and/or overstretched. The weakened ligament allows the carpal bones to shift from their tightly aligned configuration and spread apart. This collapses the tunnel between the inner wrist bones and entraps the median nerve. Inflammation often engorges the limited space and further compresses the nerve.

If carpal tunnel syndrome is suspected, there are some general diagnostic signs of the condition. Common complaints of CTS are pain, numbness, or weakness in the hand and fingers. Symptoms increase if the wrist is held firmly in flexion for more than a minute of time. Tapping on the inner wrist can produce sharp or tingling pain. If symptoms are present, it is important to neurologically test the function of the median nerve to accurately determine the presence of CTS. Even with medical testing, however, CTS is easily misdiagnosed. Carpal tunnel syndrome can be confused with other painful conditions of wrist and hand or *tendonitis* from the forearm muscles.

The transverse carpal ligament not only secures the carpal bones and maintains a patent tunnel for the median nerve; it also anchors the forearm muscle tendons to the radius and ulna before they pass through the wrist. As mentioned previously, incorrect shoulder and upper extremity alignment and integration can cause the tendons to strain and become inflamed, develop adhesions, or be torn. In these instances, the diagnosis of carpal tendon syndrome is mistaken when the condition is tendonitis.

Other localized conditions confused with CTS are bone fractures of the wrist and arthritis of the thumb. With these other conditions, there is usually no significant nerve compression or positive findings in neurological testing. Additional causes of CTS-like symptoms can originate in the central nervous system and brain. These conditions are rare, but possible. It is best to begin any inquiry by considering issues localized in the wrist before anticipating a more serious condition.

If carpal tunnel syndrome is present, yoga students often find postures such as Downward Facing Dog and the Handstand painful to perform. The alignment principles presented earlier in this chapter - the sequencing of hand placement, preventing wrist hyperextension, maintaining the palmar arches - are useful for limiting trauma or rehabilitating the wrists if damage has already occurred. In poses where it is possible, engaging *spider-fingers* is especially therapeutic. As in all situations, the overall asana must be correctly aligned, especially shoulder integrative alignment. If it is not possible to modify a position or find one that significantly reduces or eliminates the symptoms, performing that asana should cease.

Quick tips for reducing wrist strain

- Hug the transverse carpal ligament firmly to the ventral side of the wrist. This action increases vaulting of the palm's arches and strengthens the muscles of the fingers.
- Avoid wrist flexion beyond 85°
- Before fully weight bearing on the hands, align and integrate the shoulders: bring the inner heads of the humerus' back; draw the outer edges of the shoulder blades toward the spine on the back; press the bottom tips of the shoulder blades forward; lower ribs press back.
- Most people have a small fat pad that overlays the transverse carpal ligament that is usually obvious when the wrist is dorsi-flexed (extension). To protect the wrist, always draw the fat pad inwards until it is appears less protruding. This is quite valuable in many poses and physical activities, in general.

Besides carefully preparing the hand and wrist positions in every asana, the following two postures are therapeutic for carpal tunnel syndrome, tendonitis, and overall wrist and hand health:

"Popeye™ arms" therapy

This procedure can help relieve strain and discomfort in the wrists and the flexor tendons. It provides great relief after poses that fully dorsiflex the wrists such as **Urdhva Dhanurasana** (Upward Bow/Wheel Pose).

To engage:

1. Bring the arms straight out to a "T"
2. Bend the elbows to 90°
3. Fully flex the wrists
4. Fingers form either a closed fist or the fingers can be straight with all the pads pressing together
5. Maintain wrist flexion and the chosen finger position while slowly straightening the elbows

Eccentric contraction is activated by this therapy in the muscles of the forearms, which is effective for tendon rehabilitation

Sphinx Pose

Although relatively easy to perform, Sphinx Pose is a useful rehabilitative posture for the wrists. If the severity of an injury or pain makes pressure on the hands in hand weight-bearing postures not possible, Sphinx Pose is a valuable way to engage all of the wrist and hand alignment principles. It effectively draws the transverse carpal ligament inward while also bringing the shoulders into integrative alignment. This is an excellent position for strengthening injured ligaments and tendons.

To engage Sphinx pose:

1. Draw the shoulders onto the back, extend the upper spine, and press the chest forward
2. Press the hands onto the mat using the four-step hand placement
3. Isometrically twist the Hands inward to bring the scapulae fully onto the back and glided toward the spine
4. Once the shoulders are firmly on the back, slightly turn the hands isometrically outward. This lifts the wrists, arches the heel of the palms, and draws the carpal ligament and fat pad inward

> Pain and injury are our two greatest teachers
> If we listen carefully to the advice they give, yoga will help heal and maintain our bodies

34 The Head and Neck

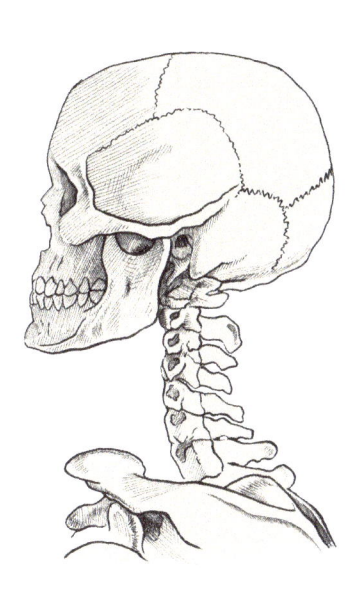

For most of us, the reflection that we see looking back from the bathroom mirror represents our most personal identity. What we see – our face, head, and neck - is the greatest expressive and interactive region of our body. The body below is sometimes more disassociated, being perhaps only a framework for displaying clothes. Along with the hands, our head and neck are the way we interact with the world.

Yoga practice is an opportunity for the rest of the body to become animated, conscious, and articulate. Keeping the head and neck from dominating the rest of our physical experience requires awareness of the body below and applying the integrative alignment principles already explored in this book. With their own set of alignment principles, the head and neck can function most safely and effectively while integrating seamlessly with the body as a whole.

A balancing act

The average head of an adult weighs between 9-11 pounds. Propped up on a stack of seven cervical vertebrae with each weighing between 2-4 ounces, the head must constantly negotiate gravitational compressive forces with heel strike forces that are rising up from below.

The seven vertebrae form an anterior convex curve called a *lordosis*. The curve provides stability and shock absorption yet does not deepen to a degree that would compress the spinal discs and nerves.

The shoulders provide the foundation for the head and neck. When the shoulders are aligned, the weight of the head can be safely supported.

The cervical spine

The cervical spine is comprised of seven small vertebrae. Its facet joints are positioned at an oblique angle, which best enables multi-directional movements in the neck. The lower five cervical vertebrae have essentially the same structural design as the rest of the spinal vertebrae except for the added openings for the vertebra arteries. The top two vertebrae - *atlas* and *axis* - are unique in both their design and function.

Atlas, the first cervical vertebra (C1)

Analogous to the mythological god Atlas that braces the earth upon his upper back and shoulders, the *atlas* vertebra supports the base of the skull at the top of the spine. The atlas is the only vertebra that does not have an anterior body. Instead, it is forms a ring of bone with two lateral masses affixed at the opposing junctures of the anterior and posterior sections of the ring.

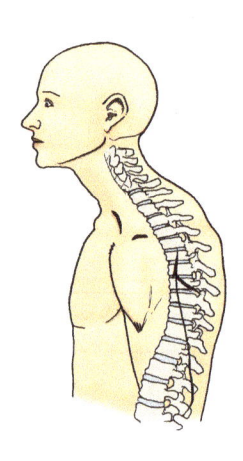

Incorrect, forward-positioned head

Cervical Ranges of motion:

- Flexion 45°
- Extension 65°
- Left Lateral flexion 45°
- Right Lateral flexion 45°
- Left Rotation 80°
- Right Rotation 80°

The Axis (C2)

The second cervical vertebra is the *axis*. As its name implies, it is the neck's swivel point and where most rotation occurs. A large, anterior, tooth-like process called the *odontoid process*, or *dens* (Latin for *tooth*) acts as a pivot point on which the atlas rotates. The brain stem, the oldest and most vital portion of the nervous system, descends from the skull and reaches the level of the C2 vertebra. Improper movement, misalignment, or trauma to the C1 or C2 vertebrae can interfere with brain stem signals and impair the basic body functions that they control, including those of the heart and lungs.

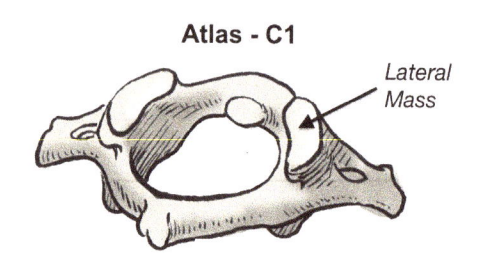

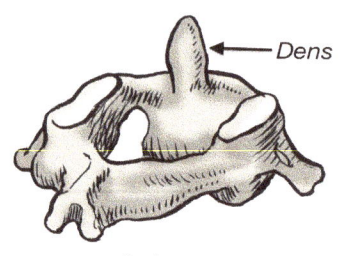

Atlas - C1
Lateral Mass

Axis - C2
Dens

The anterior compartment of the neck

Many of the major nerves and blood vessels that serve both the head and upper extremities travel through the lower anterior region of the neck, through a region referred to as the *anterior compartment*. Shoulder integrative alignment principles are essential for keeping the anterior compartment open, avoiding compression of these delicate structures. Broadening the chest and lengthening the collarbones out to their lateral edges keep the anterior compartment open.

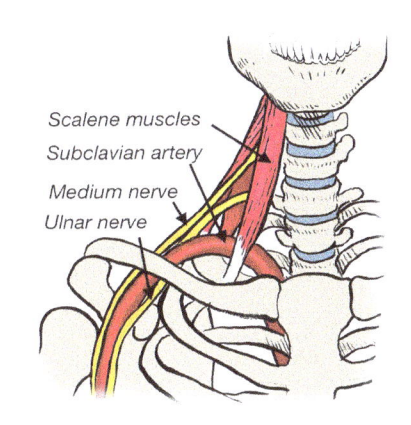

Can you see the collarbones?

As presented in Chapter 30, a deep, visible recess above protruded collarbones is a sign that the shoulders have rounded forward and lost their integrative alignment. The anterior compartment of the neck is likely to be compromised when the collarbones are markedly visible. Nerve and blood supply compression becomes greater when the arms and shoulders are at the far end of their ranges.[1] Collarbone visibility can be used to assess proper shoulder alignment.

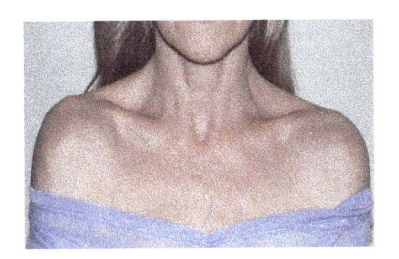

Foundation for the head and neck

The shoulder girdle and upper thoracic spine create the foundation for the head and neck. Many of the muscles that move the head and neck also anchor to the shoulders and upper back. All strategies for aligning the head and neck must begin with fully integrated and aligned shoulders.

Head and neck posture

The following steps align the head and neck after shoulder foundation has been set:

- Sagittal plane: align head and neck along the body's central axis with "third eye" over breastbone
- Coronal plane: align ear canals over the middle deltoid muscle (gleno-humeral joint)
- The tempromandibular joint, or TMJ abuts the ear canal and aligns over the shoulder. This can be felt by placing a finger in the canal and opening and closing the mouth
- Align the posterior fontanelle of the skull vertically over ear the canals. Lift vertically from here to complete alignment through the entire body's central axis
- Roof of mouth aligns with the ear canal and base of occiput - Jaralandhara bandha
- Center of throat draws back and lifts to behind ear to base of occiput. This action completes Jaralandhara bandha. Refer to the whiplash section for a possible exception to this instruction

Head games

Teachers often lead students into postures with instructions such as, "turn your head", or "look left or right". These directives encourage movement to initiate from the head. More precise instruction would be to move from the neck instead of the head and let the head follow. An even more precise instruction would be to initiate head movement from the "eyes" of the upper chest that are located just below the collarbones and between ribs 2 and 3.

Managing the cervical curve

At every stage in every asana, the yogi must determine whether mobility or stability is required. The depth of the cervical curve determines how mobile verses how stable the spine will be. A deep lordotic curve provides stability but mobility is reduced. A lengthened, flatter cervical spine creates mobility but at the loss of stability. Too straight a curve causes weakness and instability while too deep a curve can compress the spinal discs and nerves. For example, a curved spine is necessary in Headstand Pose for stability, however too deep a curve can cause compression.

> For most yoga students, the ideal cervical curve is one that, when lying in **Savasana,** would comfortably round over a *small-sized lemon*. The same-sized curve is ideal in **Setu Bandha Sarvangasana**

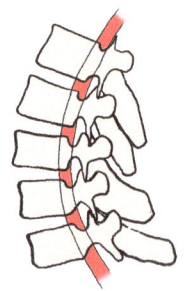

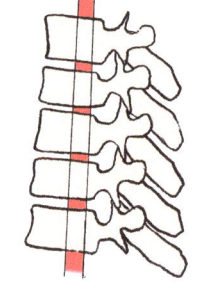

Normal Lordotic curve *Flat Hypolordotic curve*

Excessive extension of the neck can compress the cervical spine and damage the vertebral discs. In **Matsyasana** (Fish Pose), "tuck and roll" the shoulder blades onto the back, walk the occiput inferior toward the shoulders, and expand the chest. Draw the throat posterior to resist hyperextension.

Matsyasana variation

Setu Bandha Sarvangasana – Bridge Pose

Shoulder Stand – Plow Pose – Bridge Pose

In these asana, none of the neck vertebrae or any portion of the spine should rest on the floor. These poses rest on the back of the skull, the shoulder blades, and the back of the arms. A small, lemon-sized cervical curve is maintained and does not flatten. The throat draws posterior. To avoid spinal compression when coming out of these poses, lengthen the neck but do not flatten it to the floor. Allow the head to touch and draw the shoulders away from the skull.

Pencil test

A pencil should be able to slide under the neck and be retrieved from between the shoulder blades if the spine is properly lifted from the floor. The spinous process of the 7th cervical vertebra, being oversized and prominent, serves as a convenient marker. If space can be maintained below C7, the neck is properly lifted. Bruises or calluses are sometimes visible on the skin that covers C7 for yoga students who, for many years, had not kept the spine properly lifted.

Whiplash and cervical spine instability

A whiplash injury can occur when the head thrashes forward and backward with considerable speed and force. A whiplash injury usually alters the shape of the cervical curve, causing it to become straight or reversed and referred to as *hypolordosis*. A trauma-induced, straight or reversed curvature cannot support the weight of the head nor deliver the proper spinal mechanics. In the initial stages of whiplash recovery, the muscles are weak and inflamed and often unable to support the head. An immobilizing, orthopedic collar may be necessary.

Whiplash is not the only cause of a flattened cervical curve. Working positions where the head and neck are bent forward and the back rounded or poor postural habits, particularly those acquired while the spine was developing, can result in the formation of a flattened curve.

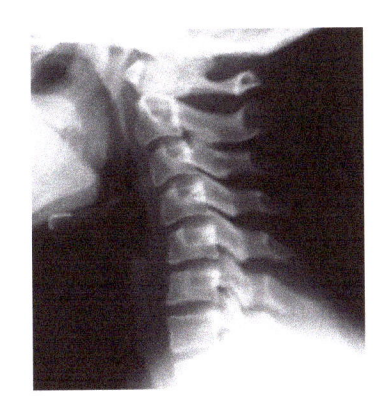

Whiplash - Hypolordosis

When a flattened curve first develops, the spine is often overly flexible due to it being hypermobile and unstable. If a "lemon-sized" curve is not established or restored within a number of years, degenerative changes in the spine develop. This causes the spine's character to change from being excessively mobile to range-restricted as degenerative adaptations attempt to remedy the chronic instability.

Hypermobility

Besides arising from poor postural habits or injury, hypermobility can also result as a compensatory response to regions where mobility has been compromised. The normally moving cervical spine can be recruited to do additional work for thoracic spine immobility, poor shoulder alignment, or even for other cervical segments that have lost their full mobility, often due to trauma or degeneration.

Limited thoracic spine mobility is a common stress for the neck. Not fully engaging the upper thoracic spine, which may be due to habit or some degree of thoracic joint fixation, forces the cervical spine to become hypermobile. Thoracic spine immobility can also cause a compensatory counter-response to occur in the lumbar spine.

Is Headstand Pose safe?

In the ongoing debate concerning the general safety of yoga, detractors who believe asana are inherently dangerous - a snake pit of trauma - usually make **Sirsasana** (Headstand) the central point of their argument. To them, the Headstand Pose is simply unsafe; end of story. Far on the other side of the argument are those who believe that yoga is divinely inspired and that every asana unconditionally creates health and eternal wisdom. The question of safety in Headstand Pose is important to yoga teachers, students, and health professionals alike. To be clear, there is no answer that is incontrovertibly correct but the question of Headstand safety can be explored in a way that will help individuals make a wise personal choice. In general, inverted postures will provide many health benefits. For yoga students whose cervical spine and circulation are healthy, a Headstand is beneficial and a valuable part of one's practice.

Blood circulation in the legs is under constant gravitational demand. Inversions temporarily reduce pressure on the veins, offering the valves and smooth muscles a healthy respite from the pressure. Although specific scientific studies may not provide confirmation, inversions may help strengthen the heart and major blood vessels, since modulating demand will increase muscle resiliency.

Inversions improve lymphatic drainage. The *lymphatic system* is a secondary circulatory system to the cardio-vascular system. Lymph is a watery fluid created from blood that is located within body tissue. Lymph is important for circulating the white blood cells of the immune system and for carrying away metabolic waste products from the cells. Lymph re-enters the blood system at two points located under the clavicles in the upper chest. Since there is no built in lymphatic pump, inversions can assist in re-circulating the lymph, especially if edema (tissue swelling) is present.

Inverted postures improve balance. While inverting, muscles take a different orientation and work from a new center of gravity. For example, when standing, the feet naturally pronate (flatten) and the inner heels drop to the floor. When inverted, the feet are non-weight bearing and the inner heels supinate (sickle). In inversion postures, yoga students can learn to press through the inner heels, an important refinement in asana practice for efficient muscle stretching as well as safer knee function.

Headstands increase bone density. Loading the bones of the skull and spine with weight stimulates bone density. In many indigenous cultures, people routinely carry over half their body weight on the top of their head. Incidences of bone density loss (osteoporosis) and spinal fracture are very low in these populations, often where nutrition is poor. Back pain is also rarely reported. Compared to epidemic levels of bone loss and back pain in Western societies, these findings strongly support the benefits of axially loading the spine and skull. In first-world cultures where loading weight onto the head is not common practice, a Headstand practice may be the best option.

Alignment essential for weight bearing

Women walking with heavy bundles on their head in indigenous cultures have nearly flawless posture regardless of age. No doubt they learn early in life that correct alignment is necessary to safely carry their loads. Maintaining the full spinal "S" curve is critical for safely loading the head, neck, and spine. Headstand provides a similar benefit since it also requires precise spinal alignment. With balanced curves and precise shoulder alignment, the spine can best balance the weight of the body. Headstand is essentially a reversed **Tadasana**, the Mountain Pose. As with all other asana, Headstand uses Tadasana's alignment principles to equally distribute the weight of the body.

Where is the head placed for Sirsasana One?

Correctly aligned, the spine's natural curves allow it to support ten times more weight than does a straight spine. If the curves are too deep, spinal compression occurs. Too shallow and the spine is unstable. Finding the spine's *sweet spot* for correct head placement is essential in keeping Headstand safe and therapeutic.

The most precise method to determine the sweet spot is best performed with the help of an assistant. Bend into Half Forward Fold and press the top of the head against a wall. Keep the torso in **Tadasana,** maintaining the cervical and lumbar curves. An assistant offers feedback on correct curve formation and identifies the exact point of contact of the head with the wall. This spot is the point of contact with the floor in the full Headstand posture. As a reference, count the number of finger-pad widths that the point of contact is distanced from the tip of the nose.

Without assistance, a reasonable approximation for head placement is approximately twelve finger-pad widths from the tip of the nose. If the student knows that their cervical curve is flatter than normal, as often results from a whiplash neck injury, then the point of contact is one or two finger-pad widths closer to the forehead. Moving the contact point forward causes the cervical curve to deepen by shifting the center of gravity forward. This adjusted position makes Headstand rehabilitative for a flattened cervical curve, helping to develop a deeper arch.

Conversely, if the cervical curve is already too deep -more than lemon-sized - the contact point on the head moves one or two finger pad-widths posterior toward the center of the skull. This position uses gravity to flatten the curve.

In Summary: to deepen the cervical curve, the contact point is closer to the forehead. To reduce the curve, the contact point moves toward back of head. However, only a small degree – one to two finger pad distances - is necessary and appropriate.

The rest of the foundation for **Sirsasana One** is the arm placement from elbows to wrists. The skull and two forearms construct three points of contact that form the corners of an equilateral triangle. The ulnar edge of each forearm presses firmly to the mat. The wrists are perpendicular or slightly turned inward in relation to the floor. The wrists slightly cock up and press down through the pisiform wrist bones, located proximal to the heel of the pinky side of the palm. The palms firmly cup together as if holding a tennis ball. The thumb heels press firmly into the back of the skull.

*Although some yoga styles employ an open, flat-hands base, cupping the hands resists the forearms and hands from externally rotating, which causes the foundation to collapse and lose its stability.

Considerations for the Headstand

There are a few conditions may make Headstand and other inversions contra-indicated:

- Glaucoma, retinal detachment, or any condition where intraocular pressure is increased
- Uncontrolled high blood pressure
- Low blood pressure, which can cause the student to faint and fall out of the pose
- Any condition where vertebral or cerebral artery blood flow is compromised, such as advanced atherosclerosis
- Spinal disc prolapse or advanced degenerative disc disease
- Spinal arthritis that has produced significant degeneration or severe bone loss
- Acute inner ear or sinus infections
- Menstruation. Some traditions of yoga routinely caution against any inversion practice during menstruation or pregnancy. Medicals research has yet to determine the validity of this prohibition

Musculature of the neck

The head's center of gravity drops slightly forward of the body's central axis. This can be experienced if falling asleep while sitting and "nodding off". As the neck muscles release, the head unwittingly collapses to the chest. To compensate for the natural anterior position of the head, the muscles that attach to the back of the head and neck evolved to be stronger than their anterior counterparts. This muscular imbalance can predispose the muscles of the neck and shoulders to residual tension. Adding to this tension, muscles, such as the trapezius appear to be the common location where emotional stress imbeds.

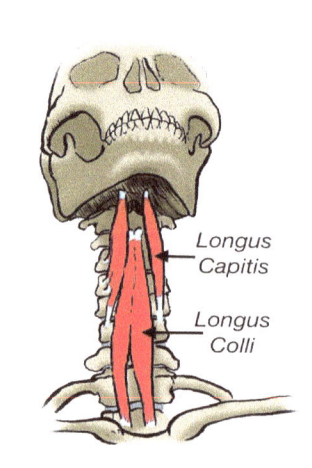

At least five deep cervical muscles stabilize the anterior cervical spine: the sterno-cleido-mastoid (SCM), scalenes, rectus capitis, longus colli, and longus capitis.

The *longus colli* straightens the neck from a curved position. It maintains the neck's neutral position and prevents the cervical curve from becoming too deep. The longus colli is often injured and weakened in a whiplash accident. This results in the characteristic reversal of the curve. Engaging the longus colli protects the neck from hyperextension and compression. To engage and tone the longus colli, draw the throat back. The longus colli and other neck muscles follow the alignment principles that engage the horseshoe-shaped bone suspended within the muscles of the anterior neck - the *hyoid bone*.

The hyoid bone

Many people do not know that the *hyoid bone* exists. It is located in the center of the neck and the only structural bone in the human body that is non-articular: meaning it does not attach directly to any other bones.

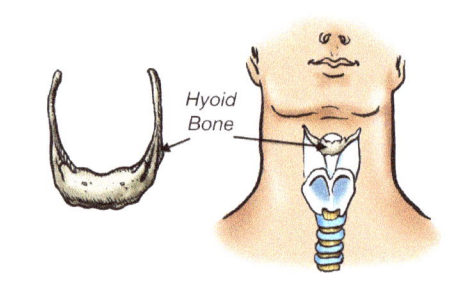

The hyoid bone anchors muscular attachments from the tongue to the larynx (voice box). It is involved with the intricate tongue movements and regulates air through the larynx to create speech. The position of the hyoid bone influences the swallowing mechanism and function of the tempromandibular joint, or TMJ. Anecdotally noted: either jutting the hyoid bone anteriorly or not properly engaging its alignment may adversely affect digestive activity.

Hyoid alignment   The smiling throat of Buddha

Instructions to move the head and neck are often, incorrectly given in reference to moving from the chin. This can inadvertently compress or tighten the tempromandibular (TMJ) joints. Instead, to move the head, it is best to move from the neck; and more specifically, the hyoid bone.

The hyoid bone is aligned by a similar muscular action that creates a soft, subtle smile - one that also mimics the smile we see depicted on images of Buddha or on the Mona Lisa. Hyoid bone alignment has been referred to as the *smiling throat*.

To engage the hyoid bone:

1. Gently lift the outer corners of the hyoid bone, drawing back to the occiput behind the lower ears. This prompts a small, soft smile to form on the lips. Hyoid bone engagement creates the lower portion of Jalandhara bandha. It joins the upper portion of Jalandhara bandha at the lower occiput, which forms by a straight-line alignment with the roof of the mouth (soft palate), ear canal, and base of the occiput.
2. Big smiles of the mouth tend to engage muscles of the face and jaw and less from those of the throat. Sufferers of TMJ syndrome will benefit from practicing and maintaining a small Buddha smile while the head and neck are in motion.
3. Drawing the throat back from the hyoid bone has a two-fold effect: it increases cervical spine mobility and reduces the likelihood of damaging compression to the discs and nerves from too deep a curve.
4. Hyoid bone engagement may not be beneficial in cases of whiplash or hypolordosis. Instead, the focus should be on the upper portion of Jalandhara bandha alone.
5. Another way to engage the hyoid bone is to touch the tip of the tongue softly to the roof of the mouth.

Subtle actions for moving the neck

- Initiate movements of the neck from the hyoid bone, drawing it back and up with a *smiling throat*. This action flattens the cervical curve and increases neck flexibility
- Rotation: Once the hyoid bone is engaged, rotate the neck from below the ear canals, as if moving from where Frankenstein's electrodes had been located. This engages the Axis (C-2) vertebra. As its name suggests, the Axis is cervical vertebra that is most responsible for rotation
- Flexion and extension: tilt the head from above the ear canals. This action glides the occipital condyles on the lateral masses of the C1 vertebra (atlas). Although this movement is minimal as compared to flexion/extension from the center of the cervical spine, it is an integratve action that is otherwise overpowered and lost if not first engaged

To rotate the head forward in **Warrior Two**, first draw the hyoid bone back; then rotate from below the ear, engaging the C2 vertebra.

In **Triangle Pose**, draw the hyoid bone back. To open the facet joints on the upper the side of the neck the side being turned toward, tilt the lower ear directly down to the lower shoulder. This is the action of lateral flexion toward the lower side of the neck. Maintain the position of lateral flexion. Rotate the neck to look toward the hand. Lateral flexion will softly release back to neutral as the rotation is completed.

Warrior Two

Triangle Pose

How to align the head

- With minimal effort, the head *floats* on the neck, centered between the shoulders
- The head lifts along the body's central axis, drawing up from the center of the ear canal through the posterior fontanelle, located at the back of the top of the skull
- Horizontally, align the roof of mouth (soft palate) with the ear canals and base of occiput
- Keep the eyes horizontal, deep, and soft in their sockets. Eyelids remain in line with each other
- With a subtle "nod" that does not disturb the alignment of the roof of the mouth, lift the posterior ridge of the occiput (back of skull), as if a hand were gently lifting the hair at the nape of the neck[2]

Therapeutic movements of the neck

The cervical spine has nearly fifty individual vertebral movements at its facets. Methods of assessment used by chiropractors and other professionals evaluate the movement of each vertebra and determine if it is normal, immobile, or hypermobile. Students can learn to evaluate their own cervical spines and modify their neck movements and restore balance to the cervical spine.

A procedure that can be easily used utilized by yoga teachers and students is one that moves the head in a stair-step or "turtle-neck" fashion. It can be performed by oneself or with assistance.

"Turtle Neck" procedure

1. The student can be either in a supine position or seated upright
2. The head and neck start in a flat position
3. The head does not nod or tilt throughout the procedure
4. Slowly glide the head anterior, segment-by-segment, keeping the face flat at all times
5. The lowest vertebra of the neck, C-7, will be first to move
6. Avoid any extraneous or nodding motions. "Stair-stepping" engages movement of specific pairs of spinal facets that are located between each pair of vertebrae
7. At each stair-step, move the flat head slowly and meticulously, from side-to-side in a figure-eight motion without adding tipping, rotation or nodding movements
8. The figure-eight movements shear across each vertebral facet surface to engage all six major directions of motion (excluding axial extension)
9. This action provides an analysis of the cervical spine's mobility and loss of motion can be address with the same procedure
10. If the joints are moving normally, the steps will be smooth and fluid from one to the next
11. Steps between vertebrae that feel "stuck" or bumpy, or where the spine cannot smoothly glide indicates limited facet mobility in that region
12. If a stuck section is found, repetitively glide over that specific section with forward-and-back and figure-eight movementhgjios until fluidly is restored

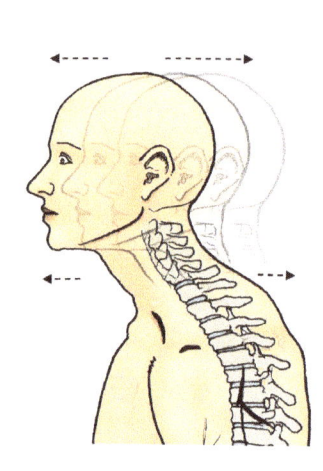

Alterative scenario: stair-stepping "jumps" over a vertebral region without engaging. In this situation, remain at that spot and use slow, figure-eight rocking until some isolated movement is restored. The sensation is best described as "catching an edge". With patience and precision, this procedure can be very effective. When the procedure jumps or skips over vertebral segments, this represents a section when there is a greater loss of movement between two vertebrae, what is referred to as a *fixation*.

The Turtle Neck procedure is best performed with an assistant and the student lying supine and completely passive. The assistant gently stair-steps the head anterior, keeping the face flat to the ceiling. The assistant supports the head with hands placed below the occiput. The gliding pressure is gentle. The direction of movement for stair-stepping is anterior and inferior, up towards the ceiling and down toward the feet. The assistant performs the figure-eight, side-to-side action and administers the reparative repetitions in the same manor as described with the self-administered procedure.

Easy on the eyes

Some students habitually move their eyes before engaging the other, less mobile regions of the upper body. This behavior inhibits movement in the upper thoracic spine, preventing it from fully participating in a pose. A helpful instruction is for the student to imagine that their eyes are located on the upper, medial chest, just below the collarbones. The principle is to "look" first from the "eyes of the chest" and allow the head and neck to behave more passively and simply "go along for the ride".

In some yogic traditions, students are taught to keep their eyes deep in the sockets, maintaining a soft passive gaze. This encourages students to metaphorically be receptive; taking in the world, instead of forcing themselves onto it. The point of view of receptivity is not only a valuable for how to approach life but as a mechanical application, it prevents the eyes from dominating the movements of asana.

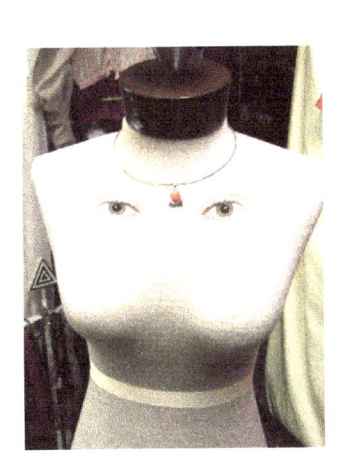

Eye pillow for Savasana

The eyes are delicate organs. They are filled with fluid and changes of pressure within eyeballs can increase internal stress on the retinas and lenses. Gentle pressure on the eyeballs can activate the 5th and 10th cranial nerves; both play a role in lowering the cardiac pulse rate. This phenomenon, called the *oculo-cardiac reflex*, is calming to the entire body. An eye pillow can stimulate this response effectively.

In the presence of any eye disease that increases ocular fluid pressure, such as glaucoma, it is important to be cautious when placing direct pressure on the eyes. An eye pillow may not be helpful or appropriate in some of these cases.

That's it! You made it! Congratulations!

If you started from the beginning of the book and ended up here, you have made it through a vast amount of material that was condensed into this one text. There were many details presented and you probably noticed important concepts were repeated. This was to help reinforce the learning process. I hope you have made this a compassionate journey and remember that most of our learning comes by doing. If you bring the principles to your regular practice, in a surprisingly short amount of time, most of this material will seem second nature. A successful way to integrate the material is to take one idea at a time and working with it for a week or more; then moving on to the next one.

My sincerest hope is that you will continue your studies and the application of alignment, integration, and postural mechanics with ceaseless and rigorous dedication as your practice. If there is any doubt regarding a particular detail concerning alignment, come back to the book as a reference and make it clear and refined. Better yet, go back to page one and begin a second reading!

My intention in writing this book has been to share as much information as I could to assist you in your yoga practice. I hope you see asana practice as a life-long experience that will continually provide you with greater health and increased wellbeing. The information in this book will support you on that journey. I encourage you to apply your own unique skills and knowledge to all you have read and add greater insight to any topic that you can shed more light. Then, share with everyone all ideas freely!

Namasté

Steven Weiss, MS, DC, C-IAYT

Footnotes and References

Chapter 1
1. "Yoga in America" Market Study." *Yoga Journal*, Feb. 26, 2008.
2. "Yoga is Fastest Growing Sport in America", *Bloomberg TV*, 8/23/2010.
3. vox.com/2018/11/8/18073422/yoga-meditation-apps-health-anxiety-cdc
4. Singleton, Mark, Yoga Body, Origins of Modern Posture Practice, Oxford, Oxford University Press, 2010.
5. Public lecture at Omega Institute, Rhinebeck, NY. Ram Dass explains the Hindu concept that the most we can ever know of God is no more than the broad direction that the finger points. Beyond what we can see, the rest is mystery. July, 1993.
6. *Pranayama*: breath control practice that moves life force through the tissues of the body
7. Paraphrased excerpt of B.K.S. Iyengar from film, "Enlighten Up", Kate Churchill, Balcony Releasing, 2008.
8. Paraphrased excerpt of B.K.S. Iyengar from film, "Enlighten Up", Kate Churchill, Balcony Releasing, 2008.
9. "Common yoga injuries…How to Avoid Them", Herndon, James, MD, Revolution Health, 2012.

Chapter 2
1. "How Yoga Can Wreck Your Body!", Broad, William, NY Times Sunday magazine, Sheila Glaser, 2012.
2. Candace Pert, PhD, Workshop at Omega Institute, Rhinebeck, NY,1995.
3. Patel, North, "Randomised controlled trial of yoga and bio-feedback in management of hypertension", Lancet, 1975 Jul 19;2(7925):93-5. 1975.
4. Rajain, Archana, India, Rajesh, "Beneficial Effects of Yogasanas and Pranayama in limiting the Cognitive decline in Type 2 Diabetes", Natl J Physiol Pharm Pharmacol. 2017;7(3):232-235. Epub 2016 Sep 24.

Chapter 3
1. Freeman, Richard, The Mirror of Yoga, Shambhala Publications, 2010, pg. 3.

Chapter 4
1. "Highlander", Film director Russell Mulcahy, 1986.

Chapter 5
1. The concept of a universal alignment blueprint is attributed to John Friend, presented during his international national workshops from 2009-2012. More details on the blueprint are presented in Chapter 10.
2. John Friend, founder of Anusara Yoga, Universal Principles of Alignment, live at Covens Center, Miami, FL, 2010.
3. "Goldilocks and the Three Bears" was first published in 1837 by British author, Robert Southey.
4. Suzie Hurley, "Divine Play of Anusara", Garden of the Heart Yoga Center, Sarasota, FL, July, 2011.
5. The periphery of the body moves faster than at the core. This concept is frequently presented in Anusara yoga classes and workshops.
6. Concept presented by B.K.S. Iyengar. John Friend formulated a similar set of principles using the terms "Muscular energy and Organic energy".

Chapter 6
1. "Form Follows Function", Attributed to American architect Louis Sullivan,1896- Wikipedia online.
2. Robin, Mel, A Physiological Handbook for Teachers of Yogasana, Fenestra Books, 2002.

Chapter 7
1. Robin, Mel, A Physiological Handbook for Teachers of Yogasana, Fenestra Books, 2002, pg. 273.
2. Eighty to ninety percent of the body's connective tissue consists of these four main types. There are numerous variations in these classifications and the percentages vary.
3. Robin, Mel, A Physiological Handbook for Teachers of Yogasana, Fenestra Books, 2002, pg. 275.
4. Robin, Mel, A Physiological Handbook for Teachers of Yogasana, Fenestra Books, 2002, pg. 173-274.
5. MedicineNet, Medical defination of elastin, medicineNet.com/script/main/art.asp?articlekey=24541
6. Robin, Mel, A Physiological Handbook for Teachers of Yogasana, Fenestra Books, 2002, pg. 27.

Chapter 9
1. Teach PE.com, Anatomy, structure, skeletal muscle, #17.
2. Synerstretch: For Total Body Flexibility, Health for Life, 1984.
3. Robin, Mel, A Physiological Handbook for Teachers of Yogasana, Fenestra Books, 2002.
4. Robin, Mel, A Physiological Handbook for Teachers of Yogasana, Fenestra Books, 2002.
5. Norkin & Levangie, Joint Structure and Function: A Comprehensive Analysis,1992.
6. Carvalho et al, Journal of Strength and Conditioning Research, "Acute Effects Of A Warm-Up Including Active, Passive, And Dynamic Stretching On Vertical Jump Performance", 11/5/2011.
7. webmd.com/fitness-exercise/news/20110217
8. Morton et al, Journal of Strength and Conditioning Research, "Resistance training vs. static stretching: effects on flexibility and strength", 12/25/2011.
9. Cole, Roger, Ph.D, "Why Won't My Head Reach My Legs in Standing Forward Bend?", Yoga Journal Q&A, Aug 28, 2007.
10. exrx.net/WeightTraining/Tidbits (internet reference).
11. Crago, P. E., Houk, J. C., and Rymer, W. Z., (1975). Influence of motor unit recruitment on tendon organ discharge. Neuroscience Abstracts, 1: 280.
12. Signorile, Joseph, PhD, Bending the Aging Curve, Human Kinetics, 2011.
13. Bass, Clarence, Carol, "Ripped", Ripped Enterprises, 2011.

Chapter 10
1. When cartilage receives blood supply, hydroxyapatite crystals form and turn it into bone. This is the mechanism for infant-to-adult development of the long bones. The jaw, clavicle, and flat bones (skull, sternum, ribs, scapula and pelvis) form directly from periosteum and do not use this system.
2. Heinegård, Dick, "Molecular Events in Cartilage Formation and Remodeling", Arthritis Research, Supplement A , 2001.
3. NIH Osteoporosis and Related Bone Diseases National Resource Center, "Bed Rest and Immobilization: Risk Factors for Bone Loss", 1/2012.
4. Levy et al, "Associations of Fluoride Intake with Children's Bone Measures at Age 11", Community Dentistry and Oral Epidemiology, 2009. Vol. 37, pg. 416-426.
5. Becker, Robert, Selden, Gary, The Body Electric: Electromagnetism and the Foundation of Life, William Morrow,1985.

Chapter 11

1. John Friend, Yoga Therapeutics Workshop, Miami, FL, March, 2009.
2. Christopher Baxter, "Mula and Meditation" workshop, Mandala Yoga, Sarasota, FL, 12/2011.
3. Anusara Yoga uses a system of "loops and spirals" in its depiction of the Universal Principles of Alignment. Much of the inspiration for the Alignment Grid is drawn from that model. B.K.S. Iyengar is purported to have formulated similar concepts to loops and spirals in the 1970's (Joan White). Iyengar's work, however, was considerably expanded and codified by John Friend.
4. Friend, John, Anusara Teaching Training Manual, Anusara Press, 2008.

Chapter 12

1. Education. Yahoo .com, "The Femur", Gray's Anatomy of the Human Body, 2009.
2. Sesej, Nahhas et al, "The Influence of Age at menarche on Cross-sectional Geometry of Bone in Young Adulthood", Science Direct.com, Bone, Vol 51, Issue 1, Pg.s 38-45, 2012.
3. Baxter, Christopher, "The ADC's of Core", Sarasota, FL, 1/2012.
4. Schafer, DC, R.C., Clinical Biomechanics- Musculoskeletal Actions and Reactions, Baltimore: Williams & Wilkins, 1983.
5. Woodley, Kennedy, "Anatomy in Practice: the Sacrotuberous Ligament", New Zealand Journal of Physiotherapy, Vol. 33,3, 11/2005.
6. Cole, Roger, Ph.D, "Protect the Sacroiliac Joints in Forward Bends, Twists, and Wide-Legged Poses", Yoga Journal (On-line teachers/1027).
7. Cole, Roger, Ph.D, "Protect the Sacroiliac Joints in Forward Bends, Twists, and Wide-Legged Poses", Yoga Journal (On-line teachers/1027).
8. Sacro Occipital Reasearch Society International, (Internet- our technique defined).
9. DeJarnette, Major,DC , Sacro Occipital Technique, Nebraska City, NE, 1984, pg.s Preface, 66, 95.
10. John Friend, Yoga Therapeutics Workshop, Miami, FL, March, 2009.

Chapter 13

1. This aspect of hip release is fully inspired by Anusara's *Inner Spiral*.
2. Forward tailbone scoop is fully inspired by Anusara's *Scoop the Tailbone*.
3. Schafer, DC, R.C., Clinical Biomechanics- Musculoskeletal Actions and Reactions, Baltimore: Williams & Wilkins, 1983. There are more sophisticated variations of this test used in chiropractic and orthopedic settings where eight separate points of reference are observed.
4. DeJarnette, Major,DC , Sacro Occipital Technique, Nebraska City, NE, 1984, pg.s Preface, 66, 95.
5. DeJarnette, Major,DC , Sacro Occipital Technique, Nebraska City, NE, 1984, pg.s Preface, 66, 95.

Chapter 14

1. Yoga teacher and researcher Doug Keller orientates this action from the front hips points, what are called the *anterior superior iliac spines* (ASIS). In inward hip release, the ASIS move closer together.
2. The specific location in the upper inner thighs from which the *roll in* and *spread apart* actions initiate is the lesser trochanter, a small boney prominence that juts out from the upper, medial surface of the femur bones. It is the attachment point for the iliopsoas and psoas major muscles.
3. The psoas major, along with its companion hip flexor muscles, attaches to the lesser trochanter located on the upper medial thigh. More about the psoas major muscle and its importance to structural alignment is presented in chapter 26.
4. Offering another approach, yoga teacher and author Richard Freeman describes the action of *forward tailbone scoop* as lifting the second sacral segment towards the navel. Doug Keller follows his own methodology for creating an action similar to *forward tailbone scoop* by advising students to *spread* the two ASIS away from the midline.

Chapter 15
1. A detailed review of the anatomy and structural mechanics of the rib cage and the thoracic spine is presented in Chapter 28.
2. The bottom tip of the breastbone forms a tail-like structure called the xiphoid process. In CPR classes, the xiphoid process is identified as the place to avoid when performing chest percussion as it can break off and puncture the liver underneath.

Chapter 16
1. A simple reflection of a phrase offered by Anusara yoga teacher Sianna Sherman in workshop, Feb 2007 Sarasota, FL.

Chapter 17
1. The major hip ligament is the "Y" ligament, which consists of three parts: iliofemoral, pubofermoral, and ischiofermoral. The ligamentum teres, another femoral ligament, attaches to the femur head, offers minor additional support, and, from the ligament's hollow core, brings arterial supply to femur head. A minor ligament, the iliotrochanteric, responds with the opposite action.
2. Ray Long, MD. Yoga Anatomy Workshop discussion, St. Petersburg, FL, September, 2106.

Chapter 18
1. Calais-Germain, Blandine. Anatomy of Movement, Eastland Press, 1993.
2. This stretch resembles Gaenslen's test, an orthopedic examination procedure that evaluates the stability of the sacroiliac joints. Forward tailbone scoop protects the sacroiliac joints from sprain while stretching the iliopsoas and extending the hip.
3. Morton et al, Journal of Strength and Conditioning Research, "Resistance training vs. static stretching: effects on flexibility and strength", 12/25/2011.

Chapter 19
1. *Shins in-Thighs apart* is a term commonly used in Anusara Yoga.
2. Numerous muscles are involved in *Thighs apart*: the gluteus maximus and medius, tensa fascia lata, and the adductor group.
3. Kapangji, I.A., The Physiology of the Joints, Volumes 1-3. NY: Church Livingston, 1982.
4. Draganich-LF; Jaeger-RJ; Kralj-AR Department of Surgery, University of Chicago
5. Baratta et al, "Muscular coactivation. The role of the antagonist musculature in maintaining knee stability", American Journal of Sports Medicine, 1988, pgs. 113-122.
6. Cole, Roger, Phd. Iyengar Workshop, Prana Yoga, Sarasota, FL. 9/10/15.

Chapter 20
1. en.wikipedia.org/wiki/Ham
2. Dictionary.com
3. Woodley, Kennedy, "Anatomy in Practice: the Sacrotuberous Ligament", New Zealand Journal of Physiotherapy, Vol. 33,3, 11/2005.
4. John Friend, Anusara therapeutics workshop, Miami, 2007.

Chapter 21
1. The TFL "locks" the knee in full extension and laterally rotates it when in flexion.
2. Gage, McIlvain, et al. Epidemiology of 6.6 million knee injuries presenting to United States emergency departments from 1999 through 2008. Acad Emerg Med. 2012 Apr;19(4):378-85.
3. American Academy of orthopedic surgeons, (internet- OrthoInfo), 2007.
4. Additional support and alignment is provided by the patella via the menisco-patellar ligament
5. Numerous smaller ligaments attach to the knee, its menisci and the patella and can be a common source of pain. In cases where knee pain cannot be isolated to the cruciates or the collaterals, these smaller tissues merit evaluation.

6. Lateral muscles: iliotibial band, biceps femoris (primarily the long head). Medial muscles: gracilis, sartorius, semimembranosus, semitendinosus. bi-lateral muscles: quadriceps (these fibers attach to both collateral ligaments).
7. The angle used to measure tibial torsion is that created by a line drawn from the knee to the back outside ankle with the leg in neutral position, neither flexed nor bent, and the femur centered over the knee.
8. Bursas are small pads between tendons that serve as spacers and prevent tendons from rubbing against each other and causing irritation.
9. Tuckerman et al, "Outcomes of meniscal repair: minimum of 2-year follow-up", Bull Hospital of Joint Diseases, NYU-Hospital for Joint Diseases Department of Orthopaedic Surgery.

Chapter 22
1. Kapangji, I.A., The Physiology of the Joints, Volume Two, NY: Church Livingston, 1982, pgs. 136-150.
2. Anusara Yoga instructors refer to this action as the *calf loop*.
3. Kapangji, I.A., The Physiology of the Joints, Volume Two, NY: Church Livingston, 1982, pgs. 136-150.
4. Deltoid refers to a type of quadrilateral shape, similar in shape to a *leaf*. In the body, there exists the deltoid muscle of the shoulder, the deltoid of the hip (the joined portion of the gluteus maximus and the TFL) and the deltoid ligament of the ankle.
5. American Academy of Orthopaedic Surgeons; American Orthopaedic Foot and Ankle Society, ©1995-2012.

Chapter 23
1. D'Costa, Krystal, Scientific American, What makes the human foot unique?, October, 2018.

Chapter 24
1. Kapangji, I.A., The Physiology of the Joints, Volume 3. NY: Church Livingston, 1982, pg. 20. According to the Dlema Index, resistance to axial compression on a column is directly proportional to the square of the number of curvatures plus one.
2. Shamji, Mohammed, Phd, "Surprising Find May Yield New Avenue of Treatment for Painful Herniated Discs", Duke Medicine News and Communications, DukeHealth.org, 6/2010.
3. Kapangji, I.A., The Physiology of the Joints, Volume Three, NY: Church Livingston, 1982, pg.. 108.
4. Sato et al, "In Vivo Intradiscal Pressure Measurement in Healthy Individuals and in Patients With Ongoing Back Problems", Spine, Volume 24, number 23, 1999, pg.s 2468-2474. Breath retention may help initiate the Valsalva effect but should not continue while postures are being performed. Breath retention is best utilized while in sitting or lying postures, the usual positions for pranayama.
5. Kapangji, I.A., The Physiology of the Joints, Volume Two, NY: Church Livingston, 1982, pg. 170-185.
6. Being able to evaluate all 336 subtle, segmental movements may seem like an overwhelming task, however it is a skill, readily learned, and the day-to-day practice of a chiropractor.
7. Doug Keller, DoYoga.com. Workshop presentation on the anatomy of back muscles in yoga practice, Sarasota, FL, 2012.

Chapter 25
1. "Chartbook on Trends in the Health of Americans 2006, Special Feature: Pain", National Centers for Health Statistics.
2. Vallfors B. "Acute, Subacute and Chronic Low Back Pain: Clinical Symptoms, Absenteeism and Working Environment". Scan J Rehab Med Suppl 1985,11: 1-98.
3. Kapangji, I.A., The Physiology of the Joints, Vol. Three, NY: Church Livingston, 1982, pg. 126.

Chapter 27
1. Koch, Liz "The psoas is NOT a hip flexor", Pilates Digest (internet source), 9/8/2009.

Chapter 28
1. Kapangji, I.A., The Physiology of the Joints, Volume 3, NY: Church Livingston, 1982, pg 132.
2. Magee, David, Orthopedic Phyical Assessment, 5th Edition, Saunders, 2008, pg. 483-490.

Chapter 29
1. Kapangji, I.A., The Physiology of the Joints, Vol 3, NY: Church Livingston, 1982, pg. 146-161.
2. Kapangji, I.A., The Physiology of the Joints, Volume 3, NY: Church Livingston, 1982, pg. 148.
3. Kapangji, I.A., The Physiology of the Joints, Volume 3, NY: Church Livingston, 1982, pg. 150.
4. Kolar et al, "Stabilizing function of the diaphragm: dynamic MRI and synchronized spirometric assessment", Journal of Applied Physiology, 109:1064-1071, 2010.
5. Kolar et al, "Stabilizing function of the diaphragm: dynamic MRI and synchronized spirometric assessment", Journal of Applied Physiology, 109:1064-1071, 2010.
6. Entnet.org/content/your-nose-guardian-your-lungs, 2018.
7. Douillard, John, DC, PhD, "The Invincible Athlete", Workshop, Westport CT, 1991.
8. Douillard, John, DC, PhD, "The Invincible Athlete", Workshop, Westport CT, 1991.
9. Raman, MD, Krishna, A Matter of Health, Integration of Yoga and Western Medicine for Prevention and Cure. Madras, India: Eastwest Books, 1998.
10. Ornish, Dean, MD, Program for Reversing Heart Disease, Mass Paperback Books, 1996.
11. Freeman, Richard, Omega Institute Workshop - Astanga Flow, 1998. Richard Freeman describes hatha yoga as the general term for the physical forms of yoga. "Ha" means sun and "tha" means moon. Hatha joins together and interpenetrates these two opposite patterns, awakening the kundalini (energy serpent) of the body.
12. Freeman, Richard, Yoga Breathing, Shambala Press, 2002.
13. A pelvic floor exercise named after Arnold Kegel, MD thought to improve prolapse and muscular weakness in various genital, urinary and uterine conditions. Engaging Mula bandha may help in cases of prostatitis or incontinence.
14. The floor of the pelvis (perinium) contains the vestigial musculature of tail wagging. Engaging Mula bandha and may trigger a phantom energy, perhaps stirring up archetypal body "memories".

Chapter 30
1. Kapangji, I.A., The Physiology of the Joints, Vol 1, NY: Church Livingston, 1982, pg. 60-65
2. Kapangji, I.A., The Physiology of the Joints, Vol 1, NY: Church Livingston, 1982, pg. 44-52.
3. Kapangji, I.A., The Physiology of the Joints, Vol 1, NY: Church Livingston, 1982, pg. 38-42
4. Kapangji, I.A., The Physiology of the Joints, Vol 1, NY: Church Livingston, 1982, pg. 38-42
5. Kapangji, I.A., The Physiology of the Joints, Vol 1, NY: Church Livingston, 1982, pg. 44-52.
6. Duncan et al, "Incidence, Recovery, and Management of Serratus Anterior Muscle Palsy after Axillary Node Dissection", Physical Therapy, Vol 63 /Number 8, August 1983, pgs. 1243-7.
7. American Academy of Orthopaedic Surgeons , 2006 Common Shoulder Injuries, topic A00327
8. Minagawa, Yamamoto, Abe, et al., Prevalence of symptomatic and asymptomatic rotator cuff tears in the general population: From mass-screening in one village. Journal of Orthopedics, March 10(1): 8-12.

Chapter 32
1. White, Joan, Iyengar Yoga workshop, Omega Institute for Holistic Studies, Rhinebeck, NY, July 2009.

Chapter 33
1. Kapangji, I.A., The Physiology of the Joints, Volume 1, NY: Church Livingston, 1982, pg. 164.
2. Kapangji, I.A., The Physiology of the Joints, Volume 2, NY: Church Livingston, 1982, pg. 134.
3. deKrom MC, Kester AD, Knipschild PG, "Risk factors for carpal tunnel syndrome". American Journal of Epidemiology. Dec 1990; 132(6) 1102-10. Medscape Research.
4. Miller, BK, "Carpal tunnel syndrome: a frequently misdiagnosed common hand problem", Nurse Practitioner. ASU, Tempe College of Nursing, 1993 Dec:18(12):52-6.

Chapter 34
1. John Friend, Anusara Therapeutics workshop, Tuscon, AZ 2007.
2. Anusara yoga teachers refer to this action as *skull loop*.

Additional source material:

- Martin, Jaye, 2002-2019, Weekly instructions in yoga classes that continually bring insight into refinement of yoga asana.
- Draganich-LF; Jaeger-RJ; Kralj-AR Department of Surgery, University of Chicago
- Dimon, Jr., Theodore, Anatomy of the Moving Body. Berkeley, CA: North Atlantic Books, 2001.
- Earle, Roger, Baechle, Thomas, NSCA's Essentials of Personal training, National Strength and Conditioning Commission, Human Kinetics, 2004.
- Earle, Roger, Baechle, Thomas, Essentials of Strength Training and Conditioning
- Human Kinetics, 2000.
- ExRx.net. "Weight Training Glossary". (Online) http://www.exrx.net/WeightTraining/Glossary.html#anchor1279833.
- Freeman, Richard, The Mirror of Yoga, Shambhala Publications, 2010.
- Friend, John, Anusara Yoga-Therapy Training Manual, Tucson, AZ, 1007.
- Friend, John, Anusara Therapeutic and Alignment trainings, personal notes, 2004-2008.
- Iyengar, BKS, Tree of Yoga Shambhala Pubilcations, Boston 1988.
- Iyengar, B.K.S., Light on Yoga- British Edition. London: Thorsons, 2001.
- Iyengar, B.K.S., Light on Life. USA: Rodale Press, 2005.
- Keller, Doug, Anusara Yoga, Hatha Yoga in the Anusara Style. VA: DoYoga Productions, 2001.
- Keller, Doug, Yoga as Therapy. VA: DoYoga Productions, 2004.
- Schafer, DC, R.C., Clinical Biomechanics- Musculoskeletal Actions and Reactions, Baltimore: Williams & Wilkins, 1983.
- The Body Worker. "Anatomy and Kinesiology". (Online) http://www.thebodyworker.com/muscleslegchart.htm.
- University of Washington. "Musculoskeletal Atlas". (Online) http://depts.washington.edu/ventures/UW_Technology.
- Wells, Katharine, Luttgens, Kathryn, Kinesiology- Scientific Basis of Human Motion. Philadelphia: W.B. Saunders Company, 1976.
- http://education.yahoo.com/reference/gray/subjects/subject/57
- http://education.yahoo.com/reference/gray/subjects/subject/59
- Singleton, Mark, Yoga Body, Origins of Modern Posture Practice, Oxford, Oxford University Press, 2010.

Photographic Acknowledgements

Cover	Shutterstock_287266130
Preface	Cart at Mae Taeng, Thomas Kriese, 2010, Wikimedia Commons
Introduction	Hindu Om Symbol, 2008, Wikimedia Commons
Chapter 1	B.K.S. Iyengar, Mutt Lunker, 2008, Wikimedia Commons
Chapter 2	Vesalius, Andreas, Skeleton Contemplating a Skull from De Humani Corporis, pinterest.es/aliganaya/public-domain/.
Chapter 2	Siddhasana, Mirzolot2, Yoga, art and science, Wikimedia Commons
Chapter 3	Fire- P3200034, Jgisbert, MorgueFile
Chapter 4	Larus Canus, Nyman, Bengt, Stockholm, 2010, Wikimedia Commons
Chapter 5	Iceberg Baffin Island, Ansgar Walk, 2000, Wikimedia Commons
Chapter 5	Interior door hinge, Infrogmation, New Orleans, 2011, Wikimedia Common
Chapter 5	Leyland Station with train, Ben Brooksbank, 1963, Wikimedia Commons
Chapter 6	Sports fishing, US National Oceanic and Atmospheric Administration
Chapter 7	Swamp Landscape, 16071, PublicDomainPictures.net
Chapter 7	Bodybuilder girl, Gamer1606, 2007, Wikimedia Commons

Chapter 9	Overspagat, 2006, Wikimedia Commons
Chapter 11	Lock Nr.9 - Moskva River, A. Savin, 2011, Wikimedia Commons
Chapter 12	Gears DCS_3537, K. Connors, morgueFile
Chapter 12	Commons.wikimedia.org/wiki/Equator#/media/File:World_map_with_equator.jpg
Chapter 14	Hello Kitty PEZ dispenser, Deborah Austin, 2009, Wikimedia Commons
Chapter 18	Body builder-Kevin Sperling, MCS Seaman, Eric Cutright, USN, 2007
Chapter 19	Gold Diggers of 1933, Warner Brothers "42 Street", Wikimedia Common
Chapter 21	Two-part iron hinge, Audrius Meskauskas, 2006, Wikimedia Commons
Chapter 21	Joni Mitchell, Paul C Babin, Whoknoze, 1974, Wikimedia Commons
Chapter 22	Left ankle sprain, Hildgrim, 2006, Wikimedia Commons
Chapter 24	Scoliosis patient in Cheneau brace, Weiss, HR, 2007, Wikimedia Common
Chapter 24	Scoliosis Cobb Skoliose Info Forum, Germany, 2005, Wikimedia Common
Chapter 24	Scoliosis, Weiss, Goodall, BioMed Central, Ltd, 2008, Wikimedia Common
Chapter 25	Baby playing with feet, Anita Peppers, morgueFile
Chapter 26	Big Butts, Alias 0591- Netherlands, 2009, Wikimedia Commons
Chapter 26	3-D spine. Created by Kjpargeter - Freepik.com
Chapter 26	IMG_7278 springs, morgueFile
Chapter 29	Exhausted runner, Mary K Baird, 2007, morgueFile
Chapter 29	Throat, Creative Commons Attribution-Share, Wikimedia Commons
Chapter 29	Heart, Idhayam, 2005, Wikimedia Commons
Chapter 29	Parrot cartoon-1872, Lear, Edward, Wikimedia Commons
Chapter 29	Hindi-Yoga darshan parmarthik trust, 2011, Wikimedia Commons
Chapter 29	Siddhasana, Mirzolot2, Yoga, art and science, Wikimedia Commons
Chapter 30	Right Scapula, Body Pts 3D Anatomography, 2012, Wikimedia Commons
Chapter 30	Aarohuttunen.com/ kolmipainen-olkalihas-triceps- brachii/
Chapter 30	Tabattoo II, Lauren Liston, 2012, Wikimedia Commons
Chapter 30	Latissimus dorsi, Wikimedia Commons
Chapter 31	Fulcrum, P., S. Foresman, #2010061110041093 Wikipedia Foundation
Chapter 31	Public domain pictures.net/view-image.php?image=212326&picture=young-waitress
Chapter 32	Biceps Pearson S. Foresman, #2010061110041093, Wikipedia Foundation
Chapter 32	Latissimus dorsi muscle frontal2. Nove,12,2012, Anatomagraphy, Wikimedia.com
Chapter 32	Ta-da! Dombrowski, Quinn, Russavia, 2011, Wikimedia Commons
Chapter 32	Panorama Neptunwerft crane/hall,ifeKirsche/Template:cc-by-nc-3.0-de, May, 2014
Chapter 32	i-l-fitness-jp.com/comment/sa/jo-trochlea-of-humerus.html
Chapter 33	Sarah Beth Briggs (piano), Clive Barda, OTRS #2011070810011038, 2010
Chapter 34	Buddha, Kittyela_P1000528_h, morgueFile
Chapter 34	Turtle IMG_2142, morgueFile
Chapter 34	mf118, Jeotocski, 2007, morgueFile
Chapter 34	Auge (Eye)- 1983, Leviathan, Wikimedia Commons- 2009.

Acknowledgements

I am grateful to the many people whose expertise and support have been invaluable to the creation of this text, Yoga Alignment Principles and Practice.

I want first to acknowledge and honor the teachers that I have studied with directly and have guided my practice and professional development over forty years. Some inspired my personal growth and yoga practice while others shared information that influenced the direction of this book. The most notable are Joan White, Kofi Busia, Glenn Black, Betsey Downing, Jaye Martin, John Friend, Doug Keller, and Christopher Baxter. I am particularly grateful to Jaye Martin, who has been my weekly instructor for nearly twenty years. I treasure his exceptional ability to impart knowledge of alignment in a clear, accessible fashion. Jaye also appears as a model for many of the photographs in this book.

Special appreciation is given to the Iyengar and Anusara yoga traditions. Their inspiring research and teachings have provided groundbreaking approaches to alignment as the basis of asana.

A special thank you to the Omega Institute for Holistic Studies in Rhinebeck, New York. Being staff and faculty for nearly twenty-five years as a wellness practitioner and yoga teaching faculty introduced me to many great teachers and provided a unique opportunity to explore and develop my craft.

Thank you to Christin Neisler, Jaye Martin, and Catherine Barefoot who gracefully modeled for many of the photographs in the book.

Thank you to Walter Fritz for his photography and photo enhancements.

I appreciate having Ben Schikowitz's share his skillful and creative illustrations, which are featured in the first edition of the book. Ben's abilities and willingness to work under a tight deadline was much appreciated. I am honored to have his work centrally featured.

I have deep appreciation and admiration for Esther Veltheim, co-founder of BodyTalk, who had been a constant support and relentless motivator through every stage of the first edition. Esther provided editing, proofreading, photography and equipment, referrals and a steady stream of creative suggestions. I have so much gratitude for Esther for helping make this project a reality. Thanks also to the International BodyTalk Association for their support and their graphic designer, Cliff Berry for his cover design in the first edition.

To Debra Gitterman, the primary editor and all around wordsmith, for her patience and professionalism and for being a voice of reason in leading us through the arduous editing process.

To Ronni Geist for the style and formatting set up, her editing skills and advice. To Carol Weiss, for graphic design and photographic assistance in both editions. To my son, Joshua Nodiff, for his photo studio work and helping to make sense out of Word for Mac!

Thank you, all!

About the author:

Steven Weiss has nearly 40 years of experience as a holistic chiropractor and nutritionist, yoga teacher and therapist. Dr. Weiss presents programs in anatomy, alignment principles, postural mechanics, and yoga therapeutics for yoga teacher trainings internationally. Dr. Weiss has spent twenty-five years as a resident faculty member at the Omega Institute for Holistic Studies in Rhinebeck, New York. He teaches regularly with the Sivananda and Integral yoga organizations. As a certified yoga therapist, Steven's brings a unique, hands-on approach to his teaching that incorporates skills from his diverse background. More information can be found at AlignYoga108.com on the web and on FaceBook.

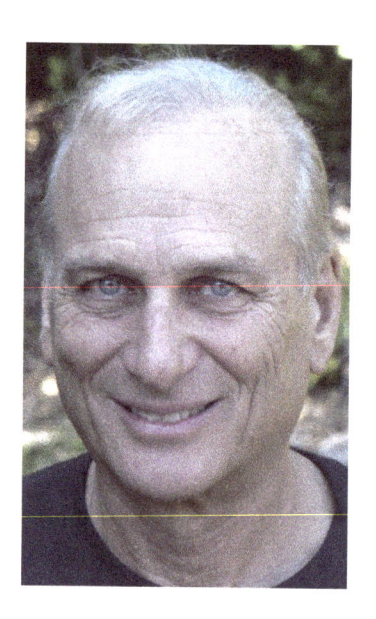

About the Illustrator:

Ben Schikowitz is an artist and massage therapist who lives with his wife and three kids near Chapel Hill, North Carolina.

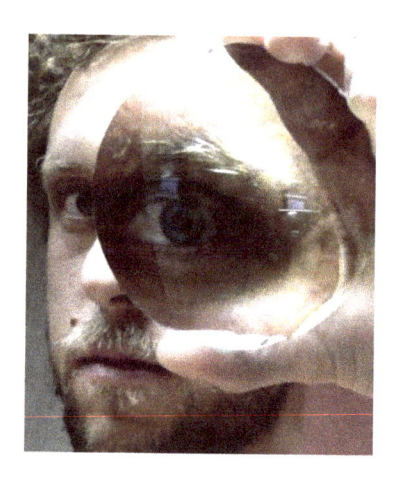